AF710987

MOSFET MODELING & BSIM3 USER'S GUIDE

MOSFET MODELING & BSIM3 USER'S GUIDE

by

Yuhua Cheng
Conexant Systems, Inc.

and

Chenming Hu
University of California, Berkeley

KLUWER ACADEMIC PUBLISHERS
Boston / Dordrecht / London

Distributors for North, Central and South America:
Kluwer Academic Publishers
101 Philip Drive, Assinippi Park
Norwell, Massachusetts 02061 USA
Telephone (781) 871-6600
Fax (781) 871-6528
E-Mail <kluwer@wkap.com>

Distributors for all other countries:
Kluwer Academic Publishers Group
Distribution Centre
Post Office Box 322
3300 AH Dordrecht, THE NETHERLANDS
Telephone 31 78 6392 392
Fax 31 78 6546 474
E-Mail <orderdept@wkap.nl>

Electronic Services <http://www.wkap.nl>

Library of Congress Cataloging-in-Publication Data

Cheng, Yuhua, 1958-
MOSFET modeling & BSIM3 user's guide / by Yuhua Cheng and Chenming Hu.
p. cm.
Includes bibliographical references.
ISBN 0-7923-8575-6 (alk. paper)
1. Metal oxide semiconductor field-effect transistors--Computer simulation. 2. Electronic circuit design--Data processing. I. Hu, Chenming. II. Title. III. Title: MOSFET modeling and BSIM3 user's guide.
TK7871.95.C57 1999
621.3815'284--dc21 99-37157
CIP

Copyright © 1999 by Kluwer Academic Publishers

All rights reserved. No part of this publication may be reproduced, stored in a retrieval system or transmitted in any form or by any means, mechanical, photo-copying, recording, or otherwise, without the prior written permission of the publisher, Kluwer Academic Publishers, 101 Philip Drive, Assinippi Park, Norwell, Massachusetts 02061

Printed on acid-free paper.

Printed in the United States of America

Contents

Preface

At the dawn of its fifth decade, the semiconductor industry continues to grow at an amazing pace. High-speed and low-power integrated circuits (IC) are used in an ever expanding plethora of applications, permeating every aspect of human life. A critical part of this technology is high-quality circuit design.

Circuit simulation is an essential tool in designing integrated circuits. The accuracy of circuit simulation depends on the accuracy of the model of the transistors. Reduction in transistor size continually complicates the device physics and makes device modeling more challenging and sophisticated. Recently, BSIM3v3 (BSIM for Berkeley Short-channel IGFET Model) was selected as the first MOSFET model for standardization by the Compact Model Council, consisting of many leading companies in the semiconductor industry such as Advanced Micro Devices, Analog Devices, Avant!, BTA Technology, Cadence design Systems, Compaq, Conexant Systems (formerly Rockwell Semicondutor Systems), Hewlett Packard, Hitachi, IBM, Intel, Lucent Technologies, Mentor Graphics, Motorola, NEC, Philips, Siemens, Texas Instruments, and TSMC. This is a historic milestone in device modeling for circuit design.

As two of the principal developers of BSIM3v3, the authors have received hundreds of comments and questions from device engineers and circuit designers. They revealed to us the areas and the points that require explanations and clarifications. We realized the need for a reference book on BSIM3 that takes the readers from device physics through model equations to applications in circuit design.

This book explains the important physical effects in MOSFETs, and presents the derivations of the model expressions. The purpose is to help the model users understand the concepts and physical meanings of the model equations and parameters. The book emphasizes the BSIM3 compact model for use in digital, analog and RF circuit design. It covers the complete set of models, i.e., I-V model, capacitance model, noise model, parasitic diode model, substrate current model, temperature effect model and non-quasi-static model. The book also addresses model implementation and new applications such as technology prediction using BSIM3. As a special feature of this book, many helpful hints based on our personal knowledge and experience are presented at the end of chapters 3 through 12 to help readers understand and use the models correctly and effectively.

This book is a summary of the contributions from many former and current colleagues and students. One of us (CH) had the distinct pleasure of collaborating with Prof. Ping K. Ko, Hong Kong University of Science and Technology (formerly with University of California, Berkeley) on the development of BSIM1, BSIM2, and BSIM3, and MOSFET physics research over a period of 15 years. His contributions to BSIM3 are countless. Dr. Jianhui Huang is one of the principal developers of the first version of BSIM3. It is a pleasure to acknowledge the following contributors to the development of BSIM3v3: Mansun Chan, Zhihong Liu, Minchie Jeng, Kelvin Hui, Weidong Liu, Xiaodong Jin, Jeff Ou, Kai Chen, James Chen, Ya-chin King, and Michael Orshansky.

We would like to acknowledge many colleagues in the Compact Model Council, Britt Brooks, Bhaskar Gadepally, Keith Green, Tom Vrotsos, Colin McAndrew, David Newmark, Marc McSwain, Pratheep Balasingam, Bob Daniel, Mishel Matloubian, Sally Liu, Shiuh-Wuu Lee, Chris Lyons, Joseph Watts and many more for their valuable inputs and comments that have been incorporated in BSIM3 and, indirectly, into this book.

We thank Prof. Michael Shur of Rensselaer Polytechnic Institute, Prof. Tor A. Fjeldly of Norwegian University of Science and Technology, and Dr. Mishel Matloubian of Conexant Systems, for reviewing the manuscript and giving helpful comments. We also thank Dennis Sylvester for technical editing, Jeff

Ou, Weidong Liu, Xiaodong Jin, Sandeep D'souza, Michael Orshansky, Mark Cao, and Pin Su for reading and commenting on individual chapters. Finally we would like to give our thanks to our families for their patience, support and help that made this book possible.

Yuhua Cheng
Conexant Systems, Inc.
Newport Beach, CA

Chenming Hu
University of California
Berkeley, CA

CHAPTER 1 Introduction

This chapter presents a brief review of the history and recent developments of MOSFET compact modeling for circuit simulation. We first survey the scene of open MOSFET compact models. Then we discuss the trends of compact model development.

1.1 Compact MOSFET Modeling for Circuit Simulation

During the 1970's, MOS technology emerged as the major driving force for VLSI [1.1]. At the same time, circuit simulators, principally SPICE, appeared as tools for circuit design. The development of SPICE started at the University of California at Berkeley in the late 1960s, and continued into the 90's [1.2].Today, commercial SPICE simulators such as HSPICE (Avant!), SPECTRE (Cadence), ELDO (Mentor Graphics), SSPICE (Silvaco), PSPICE (MicroSim) are widely used for circuit simulation. In addition, many semiconductor companies use their proprietary circuit simulators. There is also a class of fast circuit simulators such as Starsim (Avant!) and Timemill (Synopsis) that can achieve much faster simulation speeds than SPICE. In order to use these simulators, device models are needed to describe the device behaviors in the circuits. In other words, device models are the link between the physical world (technology, manufacturing, ...) and the design world (device level simulation, timing simulator model, macro model, synthesis, ...) of the semiconductor industry. In the development of modern MOS technology, many new processing techniques have been introduced into IC fabrication, and the channel length of MOSFETs has been scaled down to the 0.1μm range. The increasing level of complexity of the device structure and the appearance of new physical mechanisms that govern the characteristics of advanced devices have increased the difficulty of device modeling.

Today, circuit simulators are employed to optimize circuit performance, and verify timing and functionality of circuits.The accuracy of SPICE simulation is mainly determined by the accuracy of the device models since the simulation algorithms and convergence techniques in circuit simulators have become mature [1.3, 1.4]. As a result, there is a strong need for accurate device models to predict circuit performance. In addition to accuracy, it is desirable for a compact model to have some predictive capability. For example, a device model should ideally predict the effect of device size fluctuations and technology modifications so that it can be used by the circuit designers to study the statistical behavior of the circuits, and to explore circuit design for a modified or more advanced technology.

There are three categories of device models, (1) numerical models, (2) table lookup models, and (3) analytical (or compact) models.

Two or three dimensional numerical models use device geometry, doping profiles, and carrier transport equations, which are solved numerically, to get the device electrical characteristics [1.5, 1.6, 1.7]. Since these numerical models are computationally intensive they are not used for simulation of large circuits. Instead, numerical models may be used, for example, to explore the effects of a new transistor structure on the speed of a small ring oscillator. Table look-up models present the measured device current and capacitance as functions of bias voltages and device sizes in a tabular form for access by the circuit simulators [1.8]. There are many advantages to generate the data tables from analytical models rather than directly from the measured data. Table look-up models are used in the fast circuit simulators. Numerical models and pure table look-up models are far less popular for circuit simulation than the compact model and are not the subject of this book.

Analytical or compact models are based on device physics. However, some analytical models have such a poor grounding in device physics that they are not much more than empirical curve fitting equations. This type of empirical model [1.9] is no longer popular. The compact model equations are necessarily long and complex in order to describe the device characteristics accurately in all the operation regimes [1.10-1.25]. Fitting parameters are introduced to improve the accuracy of the model. Most of the models used in today's simulators are examples of such physical compact models.

Very sophisticated models such as the Pao-Sah model have been available to explain the device characteristics for a long time [1.26, 1.27]. These models were too complicated, i.e. too time consuming for circuit simulation and they require iteration or integration and are only semi-analytical models. These models played an important role in describing the physics of MOSFETs but were not intended nor suitable for use in circuit simulators. Simpler models have also been developed to clarify the device physics [1.28]. The explicit development of MOSFET compact models for circuit simulation started with the appearance of circuit simulators in the 1970's. Since then, more than 100 MOSFET models, including, MOS 1 [1.10], MOS 2 [1.11], MOS 3 [1.12], MOS9 [1.20], PCIM [1.18], EKV [1.21], Level 28 [1.29], ISIM [1.19], BSIM1 [1.13], BSIM2 [1.14], and BSIM3 [1.15, 1.16, 1.17] have been reported. Many of these models have been implemented in various circuit simulators but only a small number are used widely [1.29, 1.30, 1.31]. Some of these are closed (meaning the model equations are known only to the owner of the model) and proprietary (meaning the access to the model is controlled) models, such as Level 28 in HSPICE [1.29]. Implemented in nearly all circuit simulators, MOS 1, MOS 2, MOS 3, BSIM1, BSIM2, and BSIM3 are examples of open MOSFET models in the sense that the model equations are public knowledge and free licenses of the source code are provided to all users [1.10-1.17]. Some of these models can be downloaded from world-wide-web sites freely, such as BSIM3v3 [1.32]. We now give a brief review of some of these public compact models.

MOS 1 is a very simple MOSFET model based on device physics appropriate for long-channel and uniform-doping devices used two decades ago [1.10]. Because the model equations are simple and easy for circuit designers to understand, MOS 1 is still used occasionally for hand calculation and preliminary circuit simulation.

MOS 2 includes more device physics than MOS 1 [1.11]. However, it is still not accurate for devices with submicron geometries.

MOS 3 introduced many empirical parameters to model short channel effects [1.12]. However, the accuracy and scalability (the ability to model devices over a wide range of channel length and width using one set of model parameters) of the model is not entirely satisfactory to the circuit designers. The short channel and narrow width effects are not modeled accurately in the MOS 1, 2, and 3 models and high field effects are not considered properly because of the limited understanding of the physics of short channel devices at the time these models were developed.

BSIM1 (Berkeley Short Channel IGFET Model 1) was developed for 1μm MOSFET technology [1.13]. It incorporated some improved understanding of the short channel effects, and worked well for devices with channel length of 1 μm and above. However, it also introduced several fitting parameters for each model parameter just to enhance the model scalability. Even then, the model scaleability was not totally satisfactory. Also, circuit designers did not like the use of the many fitting parameters which did not have any physical meaning.

BSIM2 improved upon the BSIM1 model in several aspects such as model continuity, output conductance, and subthreshold current [1.14]. However, the model still cannot use one set of parameters for a wide range of device sizes. Users typically need to generate a few or many sets of model parameters (process files), each covering a limited range of device geometries in order to obtain good accuracy over the full range of device sizes. This makes the parameter extraction difficult. Also it is difficult to use these parameters to perform statistical modeling or extrapolation of the model parameters from the present technology to a future one.

To address these issues, BSIM3 was developed from a coherent quasi-two-dimensional analysis of the MOSFET. The device theory has been developed over a number of years [1.15, 1.17]. The model explicitly takes into account the effects of many device size and process variables for good model scalability and predictability. The short channel and narrow width effects as well as high-field effects are well modeled.

The first version of BSIM3 was released in 1994 [1.15]. BSIM3v2 (BSIM3 version 2) has been implemented in many circuit simulators [1.29, 1.30, 1.31]. BSIM3v2 has better model accuracy and scalability than the previous BSIM models but still suffers from discontinuity problems such as negative conductance and glitches in the g_m/I_d vs. V_g plot at the boundary between weak inversion and strong inversion. In the mean time, the need for a good open MOSFET model had been widely recognized by the semiconductor companies.

To eliminate all the kinks and glitches in BSIM3v2, BSIM3v3 (BSIM3 version 3) uses a single-equation approach with enhanced modeling of small size and other physical effects [1.16, 1.33]. The first version, BSIM3v3.0, was released in Oct. 1995 [1.32]. The model is scalable and may even be considered predictive, and can be used for statistical analysis [1.34, 1.35, 1.36]. It has been verified extensively by both model developers and model users from many different companies [1.37, 1.38], and has been selected as the first compact MOSFET model for industry standardization [1.39]. The next version, BSIM3v3.1, was released in Dec. 1996 with improvements in the robustness of model implementation, modification of source/drain diode models, parameter checking etc. [1.40]. The convergence performance of BSIM3v3.1 was enhanced in comparison with BSIM3v3.0, according to tests on many benchmark circuits. It is now used widely in the semiconductor industry. The latest version, BSIM3v3.2, was released in June 1998. It introduces a new charge/capacitance model that accounts for the quantization effect, and improves the threshold voltage model, the substrate current model, the Non-quasi-static (NQS) model, etc. [1.41]. The research of compact MOSFET modeling is continuing with efforts from both academia and industry. Table 1.1 gives a performance comparison of the models discussed above.

Table 1.1 Performance Comparison of Models

Model	Minimum L (um)	Minimum Tox (nm)	Model Continuity	Id Accuracy in Strong Inversion	Id Accuracy in Subthreshold	Small signal parameter	Scalability
MOS1	5	50	POOR	POOR	NOT MODELED	POOR	POOR
MOS2	2	25	POOR	POOR	POOR	POOR	FAIR
MOS3	1	20	POOR	FAIR	POOR	POOR	POOR
BSIM1	0.8	15	FAIR	GOOD	FAIR	POOR	FAIR
BSIM2	0.35	7.5	FAIR	GOOD	GOOD	FAIR	FAIR
BSIM3v2	0.25	5	FAIR	GOOD	GOOD	GOOD	GOOD
BSIM3v3	0.15	4	GOOD	GOOD	GOOD	GOOD	GOOD

1.2 The Trends of Compact MOSFET Modeling

1.2.1 Modeling new physical effects

For more than 20 years, continuous scaling of CMOS devices to smaller dimensions has resulted in higher device density, faster circuit speed, and lower power dissipation. Currently, 0.25μm CMOS technologies are widely used in the manufacturing of ICs [1.42, 1.43]. Meanwhile, 0.1μm CMOS devices have been developed and the technology will be transferred to production in the first decade of the next millennium [1.44].

Many new physical effects become significant as the device size shrinks. Examples are the normal and reverse short-channel and narrow-width effects [1.45, 1.46, 1.47, 1.48], channel length modulation (*CLM*) [1.49], drain induced barrier lowering (*DIBL*) [1.50], velocity saturation [1.51], mobility degradation due to the vertical electric field [1.52], impact ionization [1.53], band-to-band tunneling [1.54], velocity overshoot [1.55], self-heating [1.56], channel quantization [1.57], polysilicon depletion [1.58], and so on. Some of these have been modeled well in compact models, such as the short channel effect, and velocity saturation. Some of these have been studied extensively but have not yet been widely implemented in compact models for circuit simulation, such as velocity overshoot and band-to-band tunneling. The study of new physical effects in MOS devices will continue as devices become smaller. The quantization effect and radio frequency (RF) behaviors are the new frontiers at the present. Established models for some well-known physical effects such as *DIBL* and *CLM* may need to be reinvestigated for devices with channel length of 0.1μm or less. We will discuss the important physical effects in modern MOS devices in Chapter 2. It is likely that future compact MOSFET models will include more and more physical effects.

1.2.2 High frequency (HF) analog compact models

With the fast growth of the RF wireless communications market, the demand for high performance and low cost RF solutions is high. Because small MOSFETs fabricated on silicon offer ultra-large-scale integration capability and a high cut-off frequency, RF designers have already done a lot of work to explore the use of CMOS in RF circuits [1.59]. To design and optimize circuits operating at radio frequency, accurate high frequency MOSFET models are required. Compared with the modeling of MOSFETs for both digital and analog application at lower frequencies, compact RF models are more difficult to develop and do not presently exist in commercial circuit simulators. A common modeling approach for RF applications is to build sub-circuits based on "low-frequency" MOSFET models. The accuracy of such a model depends on having the right topology for the sub-circuits, and having a methodology for extracting the parameters for the elements of the sub-circuit. Recently, work has been reported for modeling the RF performance of submicron MOS devices [1.60-1.64]. However, further study is still needed to bring HF compact modeling to the level of maturity comparable to the modeling for digital and analog circuit design at the low and intermediate frequency ranges.

Besides the well known requirements for a compact MOSFET model in lower frequency applications, such as accuracy and scalability of the DC model [1.65, 1.66], there are additional requirements unique to the RF model. They include [1.67]:

(1) the model should accurately predict the bias dependence of small signal parameters at high frequency.

(2) the model should correctly describe the nonlinear behavior of the devices in order to permit accurate simulation of intermodulation distortion and high-speed large-signal operation.

(3) the model should accurately predict HF noise.

(4) the components in the sub-circuit, if the subcircuit approach is adopted, should be physics-based and scalable.

1.2.3 Simulation robustness and efficiency

It is a common understanding of circuit designers and model developers that models should have good mathematical continuity for the circuit simulators to obtain robust simulation results with fast convergence [1.66, 1.68]. It is known that the discontinuity of model equations can result in non-convergence in circuit simulation [1.68]. Many model developers have been working on the improvement of the model equation's continuity [1.16-1.25]. In the past two decades, most of the MOSFET models implemented in circuit simulators used piece-wise equations based on the regional approach, in which different equations are used for different operation regions, such as subthreshold and strong inversion regions as well as the linear and saturation regions. Such models can describe the device characteristics quite accurately in each region, but cannot guarantee the higher order continuities of the model at the transition point from one region to another. Even if the equations and their first derivatives are continuous at the boundary of the two regions, the model can still introduce non-physical peaks and valleys or small negative values in transconductance (g_m) and conductance (g_{ds}) [1.69]. Some notorious examples are g_m/I_d glitches at the boundary of the subthreshold and the strong inversion regions and the sharp bend in g_{ds} at the boundary between the linear and saturation regions [1.70]. It is suspected that such behavior not only influences the simulation accuracy, but also results in numerical convergence problems during the iteration process in circuit simulation [1.66,1.68]. The robustness of model implementation also needs to be investigated to ensure efficient simulation performance of the model. Good model implementation must avoid numerical problems such as divide-by-zero and overflow/underflow as well as any inadvertent discontinuities in the model equations caused by implementation [1.70].

1.2.4 Model standardization

As mentioned above, over a hundred MOSFET models have been developed [1.12-1.31]. Realizing the difficulty and waste in supporting a large number of compact models, model developers and users have recently made a joint effort to establish a standard compact MOSFET model with good robustness, accuracy, scalability and computational efficiency to meet the needs of digital, analog and mixed analog/digital designs [1.69-1.73]. It is clear that a standard model common to all or most semiconductor manufacturers and circuit simulators is desirable to facilitate inter-company collaborations [1.39].

Since 1996, an independent Compact Model Council, consisting of many leading companies in the semiconductor industry, has spearheaded the model standardization process [1.39, 1.69-1.73]. The Compact Model Council is affiliated with the Semiconductor Industry Alliance [1.74], and consists of Advanced Micro Devices, Analog Devices, Avant!, BTA Technology, Cadence Design Systems, Conexant Systems (formerly Rockwell), Hewlett Packard, Hitachi, IBM, Intel, Lucent Technologies, Motorola, NEC, Philips, Siemens, Texas Instruments, and TSMC. The standard model will simplify the interactions between foundry/client, technology partners, and even groups within large companies.

References

[1.1] J. Y. Chen, "CMOS - The Emerging Technology," *IEEE Circuit and Devices Magazine*, March 1986, p. 16.

[1.2] A. Vladimirescu, *The SPICE Book*, John Wiley & Sons, Inc., New York, 1994.

[1.3] K. Lee, M. Shur, T. A. Fjeldly, and T. Ytterdal, *Semiconductor Device Modeling for VLSI*, Prentice Hall, Englewood Cliffs, New York, 1993.

[1.4] N. Arora, *MOSFET Models for VLSI Circuit Simulation*, Springer-Verlag, Wien New York, 1994.

[1.5] S. Selberherr, *Analysis and Simulation of Semiconductor Devices*, Spring-Verlag, Wien, New York, 1984.

[1.6] R. Dutton, "Modeling silicon integrated circuits", *IEEE Trans. Electron Devices,* ED-30, p.935 1983.

[1.7] T. Toyabe et al., "Three dimensional device simulator CADDETH with highly convergent matrix solution algorithm," *IEEE Trans. Computer-Aided Design*, CAD-4, pp. 482-488, 1985.

[1.8] T. Shima, H. Yamada, and R L. M. Dang, "Table look-up MOSFET modeling system using 2-D device simulator and monotonic piecewise cubic interpolation," *IEEE Trans. Computer-Aided Design*, CAD-2 pp. 121-126, 1983.

[1.9] R. F Vogel, "Analytical MOSFET model with easily extraction parameters," *IEEE Trans. Computer-Aided Design*, CAD-4, pp. 127-134, 1985.

[1.10] H. Shichman and D. A. Hodges, "Modeling and simulation of insulated-gate field-effect transistor switching circuits," *IEEE Journal of Solid-state Circuits*, vol. SC-3, pp. 285-289, 1968.

[1.11] A. Vladimirescu, and S. Liu, *The simulation of MOS Integrated Circuits Using SPICE2*, ERL Memorandum No. UCB/ERL M80/7, University of California, Berkeley, Feb. 1980 (Rev. Oct. 1980).

[1.12] S. Liu, *A unified CAD model for MOSFETs*, ERL Memorandum No. UCB/ ERL M81/31, University of California, Berkeley, May 1981.

[1.13] B. J. Sheu, D. L. Scharfetter, P. K. Ko, and M. C. Jeng, "BSIM: Berkeley short -channel IGFET model for MOS transistors," *IEEE J. solid-state Circuits*, vol. SC-22, pp.558-565, 1987.

[1.14] M. C. Jeng, *Design and modeling of deep-submicrometer MOSFETs*, ERL memorandum ERL M90/90, University of California, Berkeley, 1990.

[1.15] J. H. Huang et al., *BSIM3 Manual (Version 2.0)*, University of California, Berkeley, March 1994.

[1.16] Y. Cheng et al., *BSIM3 version 3.0 User's Manual*, University of California, Berkeley, 1995.

[1.17] Y. Cheng et al., *BSIM3 version 3.1 User's Manual*, University of California, Berkeley, Memorandum No. UCB/ERL M97/2, 1997.

[1.18] N. D. Arora, R. Rios, C. L. Huang and K. Raol, "PCIM: a physically based continuous short-channel IGFET model for circuit simulation," *IEEE Trans. Electron Devices*, vol.41, pp. 988-997, 1994.

[1.19] D. H. Cho and S. M. Kang, "A new deep submicrometer compact physical model for analog circuits," in *IEEE Custom Integrated Circuits Conf.*, pp.41-44, 1994.

[1.20] R. M. D. A. Velghe, D. B. M. Klassen, and F. M. Klassen, "Compact MOS modeling for analog circuit simulation", In *IEEE IEDM 93, Tech. Dig.*, pp.485-488, Dec. 1993.

[1.21] C. C. Enz, F. Krummenacher and E. A. Vittoz, "An analytical MOS transistor model valid in all regions of operation and dedicated to low voltage and low-current applications", *J. Analog Integrated Circuit and Signal Processing*, Vol. 8, pp.83-114, 1995.

[1.22] J. A. Power and W. A. Lane, "An enhanced SPICE MOSFET model suitable for analog applications", *IEEE Trans. Computer-Aided Design,* vol.CAD-11 pp.1418-1425, 1992.

[1.23] A. R. Boothroyd, S. W. Tarasewicz and C. Slaby, "MISNAN-A physically based continuous MOSFET model for CAD applications", *IEEE Transactions on CAD*, vol. 10, pp.1512-1529, 1991.

[1.24] M. Miura-Mattausch, "Analytical MOSFET model for quarter micron technologies", *IEEE Trans. Computer-aided Design of Integrated Circuits and Systems*, Vol. 13, (5), p. 564, 1994.

[1.25] M. Shur, T. A. Fjeldly, T. Ytterdal, and K. Lee, "A unified MOSFET model," *Solid-State Electronics*, 35, pp.1795-1802, 1992.

[1.26] H. C. Pao and C. T. Sah, "Effects of diffusion current on characteristics of metal-oxide(insulator)-semiconductor transistors," *Solid-State Electron.* Vol. 9, p927, 1966.

[1.27] J. R. Brews, "A charge-sheet model of the MOSFET", *Solid-state Electronics*, Vol.21, pp345-355, 1978.

[1.28] G. Baccarani et al., "Analytical i.g.f.e.t. model including drift and diffusion currents," *IEE Journal on Solid-State and Electron Devices*, vol.2, p.62, 1978.

[1.29] *Star-Hspice user's manual*, Avanti Corporation, 1997.

[1.30] *Cadence Spectre User's manual*, Cadence Design Systems, 1996.

[1.31] *Eldo user's manual,* Mentor Graphics, 1996.

[1.32] http://www-device.eecs.berkeley.edu/~bsim3.

[1.33] Y. Cheng, T. Sugii, K. Chen, and C. Hu, "Modeling of small size mosfets with reverse short channel and narrow width effects for circuit simulation", *Solid State Electronics*, vol. 41, (9), pp. 1227-1231, 1997.

[1.34] Y. Cheng et al., "An investigation on the robustness, accuracy and simulation performance of a physics-based deep-submicrometer BSIM model for analog/digital circuit simulation", *CICC'96*, pp. 321-324, May 1996.

[1.35] J. Chen, C. Hu, C. Wan, P. Bendix, and A. Kapoor, "E-T based statistical modeling and compact statistical circuit simulation methodology," *IEDM Tech. Dig*., pp. 635-639, 1996.

[1.36] Y. Cheng et al., "A Study of deep-submicon MOSFET technology prediction and scaling with BSIM3", *Techcon'97*, 1996.

[1.37] A. Dognis and C. Lyons, *Compact Models Workshop*, Washington D. C, Dec., 1995.

[1.38] M. Jeng and Z. Liu, *Compact Models Workshop*, Washington D. C., Dec., 1995.

[1.39] *Compact Model Workshop*, Washington D. C., Dec, 1995.

[1.40] C. Hu et al., *BSIM3v3.1 release note*, http://www-device.eecs.berkeley.edu/~bsim3.

[1.41] C. Hu et al., *BSIM3v3.2 release note*, http://www-device.eecs.berkeley.edu/~bsim3.

[1.42] T. Yuan et al., "CMOS scaling into the manometer regime", *Proceedings of the IEEE*, Vol.85, pp. 486-504, 1997.

[1.43] M. Bohr et al., "A high performance 0.25um logic technology optimized for 1.8V operation," *IEDM Tech. Dig.*, pp. 847-850, 1996.

[1.44] A. Shijiro, and W. Yasuo, "Technology challenges for integration near and below 0.1um", *Proceedings of the IEEE*, Vol.85, pp. 505-520, 1997.

[1.45] C. Duvvury, "A guide to short channel effects in MOSFETS," *IEEE Circuit and Systems Magazine*, p.6, 1986.

[1.46] C. Y. Lu and J. M. Sung, "Reverse short channel effects on threshold voltage in submicron salicide devices", *IEEE Electron Device letters*, EDL-10, p. 446, 1989.

[1.47] E. H. Li et al., "The narrow channel effect in MOSFET with semi-recessed oxide structures," *IEEE Trans. Electron Devices*, ED-37, p. 692, March 1990.

[1.48] L. A. Akers, "The inverse narrow width effect," *IEEE Electron Device Letters*, EDL-7 (7), p. 419, July 1986.

[1.49] W. Fichtner and H. W. Potzl, "MOS modeling by analytical approximations -subthreshold current and subthreshold voltage," Int. J. Electronics, Vol. 46, p33. 1979.

[1.50] R. R. Troutman, "VLSI limitations from drain-induced barrier lowering," *IEEE Trans. Electron Devices*, Vol ED-26, p.461, 1979.

[1.51] C. G. Sodini, P. K. Ko, and J. L. Moll, "The effects of high fields on MOS device and circuit performance," *IEEE Trans. Electron Devices*, ED-31, p1386, 1984.

[1.52] M. S. Liang et al., "Inversion layer capacitance and mobility of very thin gate oxide MOSFETs," *IEEE Trans. Electron Devices*, ED-33, p409, 1986.

[1.53] C. Hu, "Hot Carrier Effects," Chpt.3 in *Advanced MOS Device and Physics*, N. G. Einspruch and G. Gildenblat, EDS., Vol. 18, *VLSI Electronics Microstructure Science*, Academic Press, San Diego, CA, p. 119-160, 1989.

[1.54] I. C. Chen et al., "Interface-trap enhanced gate-induced leakage current in MOSFET," *IEEE Electron Letters*, EDL-10, p216, 1989.

[1.55] F. Assderaghi et al., "Observation of velocity overshoot in silicon inversion layers", *IEEE Electron Device letters*, Vol. 14, p. 484, 1993.

[1.56] Y. Cheng and T. A. Fjeldly, "Unified physical I-V model including self-heating effect for fully depleted SOI/MOSFET's," *IEEE Trans. Electron Devices*, vol. 43, pp. 1291-1296, 1996.

[1.57] Y. King, H. Fujioka, S. Kamohara, W. C. Lee, and C. Hu, "AC charge centroid model for quantization of inversion layer in NMOSFET," *Int. Symp. VLSI Technology, Systems and Applications, Proc. of Tech. Papers*, Taipei, Taiwan, pp. 245-249, June 1997.

[1.58] K. Chen et al., "Polysilicon gate depletion effect on IC performance", *Solid-State Electronics*, pp. 1975-1977, Vol. 38, No. 11, November 1995.

[1.59] A. A. Abidi, "Low power RF-ICs in wireless transceivers", *1994 IEEE Symposium on low power electronics*, pp. 18-21, 1994.

[1.60] W. Liu et al., “RF MOSFET modeling accounting for distributed substrate and channel resistances with emphasis on the BSIM3v3 SPICE model”, *IEDM Tech. Dig.*, pp. 309-312, 1997.

[1.61] D. Pehlke et al., “High frequency application of MOS compact models and their development for scalable RF model libraries,” *CICC’98*, pp. 219-222, 1998.

[1.62] J. Ou et al., "CMOS RF modeling for GHz communication IC's," Digest of Technical Papers, *1998 Symposium on VLSI Technology*, June 1998.

[1.63] C. Enz and Y. Cheng, “MOS transistor modeling issues for RF IC design”, Workshop of Advances in Analog Circuit Design, France, March 1999.

[1.64] S. H. Jen et al., “Accurate modeling and parameter extraction for MOS transistor valid up to 10GHz”, *ESDERC’98*, Sept. 1998.

[1.65] C. McAndrew and M. McSwain, *Compact Model Workshop*, Sunnyvale, CA, Aug., 1995.

[1.66] Y. Tsividis and G. Masetti, "Problems in precision modeling of the MOS transistor in analogue applications", *IEEE Trans. CAD,* vol.-3, pp. 72-79, 1984.

[1.67] Y. Cheng et al., “RF modeling issues of deep-submicron MOSFETs for circuit design,” *1998 International Conference of Solid-state and Integrated Circuit Technology*, pp.416-419, 1998.

[1.68] A. Vladimirescu, and J. J. Charlot, "MOS analogue circuit simulation with SPICE", *IEE Proc. Circuits Device Systems*, Vol 141, No.4 pp. 265-274, 1994.

[1.69] *Compact Model Workshop*, Dallas, TX, March, 1995.

[1.70] *Compact Model Workshop*, Austin, TX, June, 1995.

[1.71] *Compact Model Workshop*, Sunnyvale, CA, Aug., 1995.

[1.72] *Compact Model Workshop*, Austin, Taxis, Mar, 1996.

[1.73] *Compact Model Workshop*, Burlington, Vermont, Aug, 1996.

[1.74] http://www.eia.org/eig/CMC.

CHAPTER 2

Significant Physical Effects In Modern MOSFETs

This chapter will describe the important physical phenomena that ought to be accounted for in a compact model of a modern MOSFET. They are: (1) Non-uniform doping effect; (2) Charge sharing and *DIBL*; (3) Reverse short channel effect; (4) Normal narrow width effect; (5) Reverse narrow width effect; (6) Body effect; (7) Subthreshold conduction; (8) Field dependent mobility; (9) Velocity saturation; (10) Channel length modulation; (11) Substrate current due to impact ionization; (12) Gate-induced drain leakage; (13) Polysilicon gate depletion; (14) Inversion layer quantization effect; (15) Velocity overshoot; (16) Self-heating effect.

2.1 MOSFET Classification and Operation

A Metal-Oxide-Semiconductor transistor is shown in Fig. 2.1.1, where an n-channel device is illustrated as an example. The MOS transistor is a four terminal device, and the four terminals are called drain (D), gate (G), source (S), and body (B). The source and drain junctions are connected to the inversion layer in the channel region. The length of the channel between the source and drain is called the channel length (L). The width of the channel, in the direction normal to the channel length, is called the channel width (W). Usually, the MOS transistor is symmetric, meaning there is no difference between the source and drain in the design and fabrication, and they are assigned as the drain or source according to their functions in the circuit and the applied bias conditions. MOSFETs with asymmetric source and drain have also been reported in some special applications.

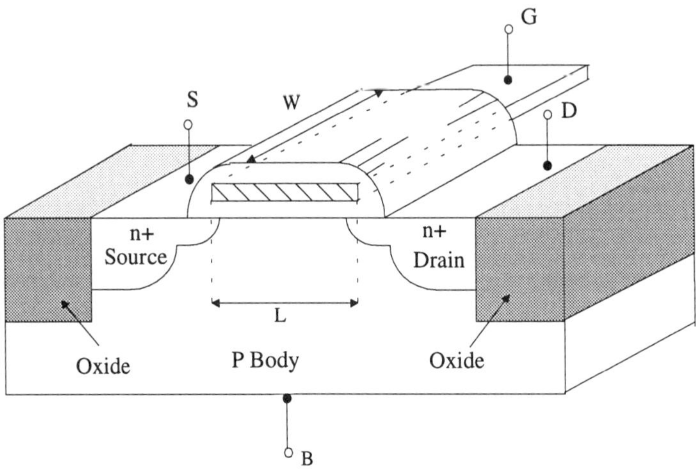

Fig. 2.1.1 A MOSFET structure.

When an appropriate voltage V_g is applied to the gate terminal, an inversion layer is formed between the drain and source to conduct current. The current depends on the applied gate and drain voltages when the source is biased at a fixed voltage (e.g. grounded for an nMOSFET). In fact, the gate and drain voltages and the electric fields (both vertical and lateral directions) produced by these voltages in the channel of the device control the device operation in all bias regimes which we will discuss next. That is the reason why it is called a MOS Field-Effect-Transistor (MOSFET). Because the gate is electrically isolated from the other parts of the device a MOSFET is also known as an Insulated-Gate Field-Effect Transistor (IGFET) [2.1]. We may note that BSIM stands for Berkeley Short-channel IGFET Model.

The critical gate voltage, at which an inversion layer is formed, is called the threshold voltage (V_{th}). When the voltage between the gate and source, V_{gs}, is larger than V_{th} by several times the thermal voltage v_t (K_BT/q), the device is said to be in the strong inversion regime. When V_{gs}=V_{dd} (the power supply voltage), the device is in the "on" state. When V_{gs} is less than V_{th}, the device is in the subthreshold (or weak inversion) regime. When V_{gs}= 0, the device is in the "off" state. When V_{gs} is biased near V_{th}, the device operates in the moderate inversion regime, which is an important operation region in low power analog applications.

Depending on the type of charge carriers conducting the current in the channel, MOSFETs can be either n-channel or p-channel devices. For an n-channel MOSFET, electrons are the charge carriers and the source and drain regions, which are heavily doped with n-type impurities, are formed on a p-type substrate. In a p-channel MOSFET, the source and drain regions are doped heavily with p-type impurities in an n-type substrate. In circuit applications, a combination of n-channel and p-channel MOSFETs are used. This technology is called complementary MOS (CMOS) technology. Compared with NMOS or PMOS technologies, CMOS technology has a huge advantage in its low power consumption, and has been the workhorse of VLSI since the early 1980's [2.2].

MOSFETs can be either surface channel devices or buried channel devices [2.3]. It is commonly believed that the names refer to the location of the conduction channel in the devices. A MOSFET is a surface channel device if its conduction channel is at the SiO_2-Si interface. In a buried channel device the current flows along the path in the bulk of the device. In reality, the names only refer to the net doping type of the surface layer of the substrate. For example, the surface channel PMOSFET has an n-type doped substrate, while the buried channel PMOSFET has a p-type doped but depleted surface layer in the n-type substrate. However, the current in a buried channel device is still conducted by a surface charge layer when the device is in the on state.

A MOSFET is called an enhancement-mode device if it does not conduct significant current when the gate voltage V_g is zero [2.4]. It is a normally-off device, and a minimum gate voltage, the threshold voltage, is needed to induce an inversion layer in the channel to conduct the current. If a MOSFET is normally-on, i.e. a conducting channel is present in the device even when V_g=0V, it is called a depletion-mode device because a gate bias is needed to deplete the channel to turn the device off [2.5]. Depletion mode devices are rarely used in CMOS circuits.

Next, we will give a qualitative description of the operation of a long channel enhancement n-channel MOSFET to illustrate the device characteristics in the various operating regimes, and to introduce some important terms used in compact modeling.

Depending on the bias voltages applied to the drain for fixed gate, source and body biases, a MOSFET may operate in the linear, saturation, or breakdown regions as shown in Fig. 2.1.2. Likewise, depending on the bias voltages applied to the gate for fixed drain, source, and body biases, a MOSFET may operate in the weak inversion, moderate inversion, or strong inversion regions as shown in Fig. 2.1.3. Next, we will discuss the device characteristics in the strong inversion region. Then we will discuss the device characteristics in the weak inversion and moderate inversion regions.

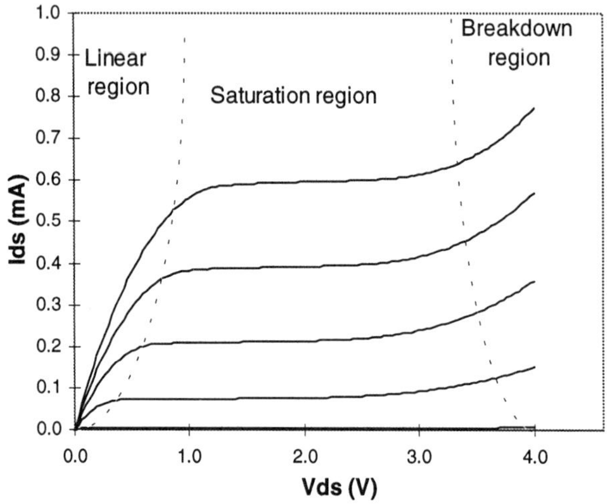

Fig. 2.1.2 I_{ds}-V_{ds} characteristics of an n-channel MOSFET.

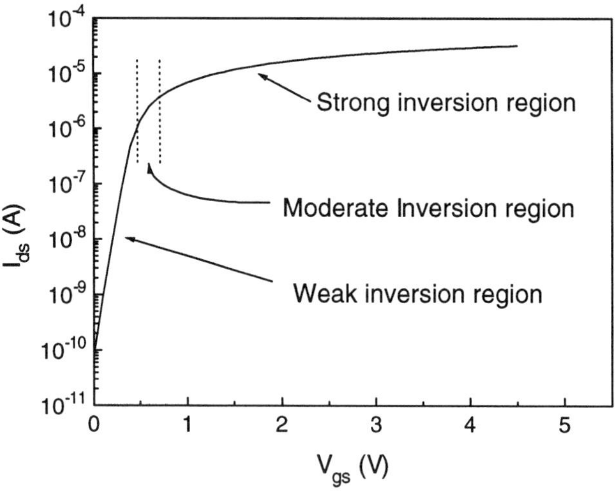

Fig. 2.1.3 Log(I_{ds})-V_{gs} characteristics of an n-channel MOSFET.

2.1.1 Strong inversion region ($V_{gs}>V_{th}$)

When the bias applied to the gate terminal of an n-channel MOSFET is larger than V_{th}, an inversion layer is induced at the Si/SiO_2 interface. Current can flow through the channel from the source to the drain if a positive drain voltage is applied (the source is normally used as the voltage reference). Depending on the magnitude of the drain voltage, the device may be in the linear, saturation, or breakdown regions of operation.

a. Linear region ($0<V_{ds}<V_{dsat}$)

As shown in Fig. 2.1.2, when V_{ds} is small (i.e., $V_{ds}<<V_{gs}-V_{th}$), the inversion channel behaves like a simple resistor. The drain current I_{ds} increases linearly as the drain voltage V_{ds} increases. However, when V_{ds} is larger it will cause an increase of the voltage in the inversion layer at all points along the channel (except for the singular point at the source edge). This reduces the voltage across the gate capacitor and the inversion charge density is reduced. The smaller amount of mobile inversion charges results in a decrease in channel conductance, which leads to a smaller slope in the I_{ds}-V_{ds} characteristics as V_{ds} increases. Eventually, V_{ds} reaches the saturation voltage V_{dsat}, at which point the mobile carriers at the drain side disappear in this first order model, and the channel is "pinched off" at the drain side [2.4, 2.6]. The condition of no mobile carriers at the pinch-off point has traditionally been used to obtain the analytical saturation voltage expressions for long channel compact models [2.7].

b. Saturation region ($V_{dsat}\leq V_{ds}<V_{bk}$)

When $V_{ds}>V_{dsat}$, the pinched-off region of the channel increases and extends towards the source. The excess drain voltage beyond V_{dsat} will drop across this pinched-off region and the drain current remains approximately constant as shown in Fig. 2.1.2. However, we need to point out that the constant saturation current behavior is only an approximation. The small but non-zero slope of the I_{ds}-V_{ds} characteristics in the saturation region is very important to analog circuit performance and must be accurately modeled by a compact model. In addition to the finite length of the pinch-off region (channel length modulation), drain induced barrier lowering and substrate current induced body effect must be accounted for in modeling the current in the saturation region [2.8]. They will be described in sections 2.6 and 2.7.

c. Breakdown region ($V_{ds}\geq V_{bk}$)

When V_{ds} is much larger than V_{dsat}, the device may enter the breakdown region [2.9]. When V_{ds} is larger than the "breakdown voltage", V_{bk}, as shown in Fig. 2.1.2, the current increases dramatically as V_{ds} increases. The definition of V_{bk} in compact modeling is often vague and arbitrary for practical reasons. It should be noted that the breakdown mechanism may be different between long channel devices and short channel devices. Breakdown in a long channel device is caused by the breakdown of the drain-body p-n junction [2.10]. Breakdown of a short channel device may be caused by the breakdown of the parasitic bipolar transistor. The latter is triggered by the substrate current produced by impact ionization [2.11]. We will give further analysis of this in section 2.7.

2.1.2 Weak and moderate inversion or the subthreshold region

The characteristics discussed above are all related to the strong inversion operation, where the gate bias is much larger than V_{th} and surface potential is larger than $2\phi_B$ (here ϕ_B is a doping parameter relating the potential of an electron at the Fermi level to the doping concentration) [2.4, 2.6]. When the gate bias is less than V_{th}, the mobile charge density is very small but nonzero. In this case, a depletion layer exists and the surface potential is smaller than $2\phi_B$. This device operation region is called the weak inversion or the subthreshold regime. Unlike the strong inversion region where drift current dominates, subthreshold conduction is dominated by diffusion current.

The current increases exponentially as the gate bias increases in the weak inversion region ($V_{gs} << V_{th}$) [2.12], as shown in 2.1.3. However, this exponential relationship between the source-drain current and the gate bias breaks down as the gate bias approaches V_{th}. Another operation region called "moderate inversion" exists between the weak inversion and strong inversion regions [2.13, 2.14]. Usually, this region is defined as several v_t around V_{th} [2.13]. In moderate inversion, both the drift and diffusion currents are significant, making compact modeling very difficult.

2.2 Effects Impacting the Threshold Voltage

In this section we discuss the physical effects impacting the threshold voltage of a MOSFET. Unless we point out otherwise, we will base our discussion on an enhancement mode n-channel MOSFET.

For a device with a long and wide channel geometry and uniformly doped substrate, V_{th} can be derived easily by solving the one dimensional Poisson equation in the vertical direction [2.4, 2.14]:

$$V_{th} = V_{th0} + \gamma(\sqrt{\phi_s - V_{bs}} - \sqrt{\phi_s}) \tag{2.2.1}$$

$$V_{th0} = V_{FB} + \phi_s + \gamma\sqrt{\phi_s} \tag{2.2.2}$$

where ϕ_s is the surface potential at threshold, V_{FB} is the flat band voltage, V_{th0} is the threshold voltage of a long channel device at zero body bias, and γ is the coefficient of the body bias effect which is given by

$$\gamma = \frac{\sqrt{2\varepsilon_{si} q N_a}}{C_{ox}} \tag{2.2.3}$$

where N_a is the substrate doping concentration. C_{ox} is the gate oxide capacitance. The surface potential ϕ_s at V_{gs}=V_{th} is given by [2.4, 2.14]

$$\phi_s = 2v_t \ln(\frac{N_a}{n_i}) \equiv 2\phi_B \tag{2.2.4}$$

where v_t is the thermal voltage, and n_i is the intrinsic carrier density. This model is valid only when the substrate doping concentration is constant and the channel length is long. Under these conditions, the potential is uniform along the channel. But in reality, these two conditions are not always satisfied in today's MOSFET because of the small device size and the complex doping profile employed to improve device performance.

2.2.1 Non-uniform doping effects

In making a VLSI MOSFET, many ion implantation doping process steps are used to adjust the threshold voltage value and suppress the punch-through and hot carrier effects [2.15, 2.16, 2.17]. A non-uniform doping density in both the vertical and lateral directions is typical. In that case, the threshold voltage characteristics cannot be described by the classic V_{th} model given above. The influence of the non-uniform doping effect on V_{th} should be carefully considered.

1. Vertical non-uniform doping effects

To adjust the threshold voltage and suppress punchthrough, multiple implantation steps are often used in VLSI fabrication. A shallow implantation of channel dopants of the same type as that of the substrate is designed to achieve a suitable V_{th} value, and another deep implantation of the same type of dopant is used to suppress punchthrough and *DIBL* [2.18]. Two doping profiles produced by implantation are given in Fig. 2.2.1. The doping concentration in one case is higher near the silicon and silicon dioxide interface than deep in the substrate. In the other case, the doping concentration is lower at the surface and higher in the bulk. This is called the retrograde doping profile [2.19]. One extreme example is the delta doped device [2.20].

It is well known that the threshold voltage of a MOSFET with a uniformly doped substrate is proportional to $\sqrt{\phi_s - V_{bs}}$, with the proportionality constant equal to γ as given in Eq. (2.2.1). However, the experimental data shown in Fig. 2.2.2 displays a non-linear dependence. The slope γ becomes smaller as the body bias V_{bs} becomes more negative (for NMOS). This non-linearity comes from non-uniform substrate doping in the vertical direction. This non-uniformity makes γ in Eq. (2.2.3) a function of the substrate bias.

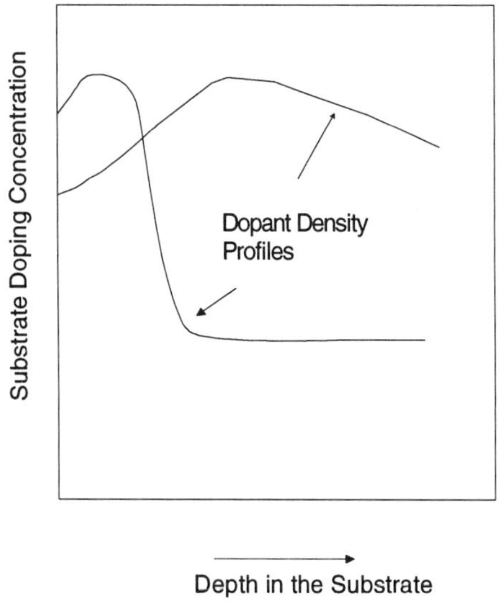

Fig. 2.2.1 Two examples of the vertical doping profiles in MOSFETs.

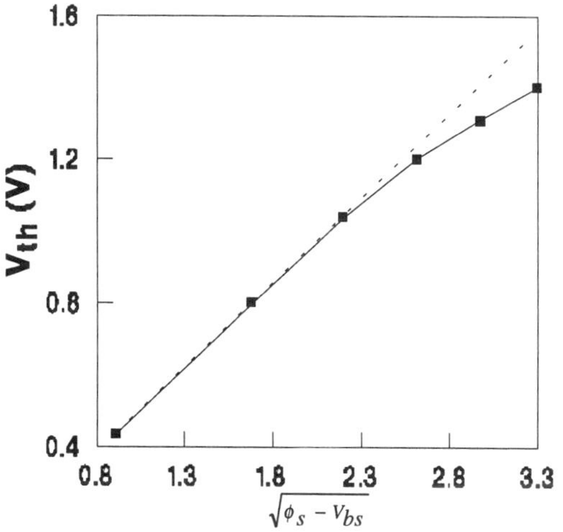

Fig. 2.2.2 The threshold voltage versus $\sqrt{\phi_s - V_{bs}}$. After Huang et al. [2.8].

Detailed considerations are needed to derive an expression of the threshold voltage for a device with a non-uniformly doped substrate. Strictly speaking, an integration of bulk charge density distribution along the vertical direction is needed to obtain the total bulk charge as given in the following:

$$V_{th} = V_{FB} + \phi_s + \frac{q}{C_{ox}} \int_0^{x_{di}} n(x)dx \tag{2.2.5}$$

where n(x) is the doping profile along the vertical direction, and X_{di} is the width of the depletion layer at V_{gs}=V_{th}.

Eq. (2.2.5) states that the threshold voltage in a non-uniformly doped device is dependent on the integral of the doping profile rather than the doping profile function itself. In other words, the threshold voltage is determined by the total amount of bulk charge, which may be obtained with some approximation for the charge density function, without using the real doping profile in the device! This approach has been used in compact modeling to get simple analytic expressions for V_{th} in the device with non-uniformly doped substrate, and will be discussed in Chapter 3.

In addition to the third term in Eq. (2.2.5), the parameters V_{FB} and ϕ_s, which have clear definitions for uniformly doped substrates, also need to be examined for devices with vertical non-uniform doping.

a. Flat band voltage V_{FB}: The definition of the flat band voltage is simple for a uniformly doped substrate. It is the gate voltage at which the surface potential is zero, the surface electric field is zero, and the entire substrate is charge neutral. For a non-uniformly doped device, no gate voltage can cause the substrate to be charge neutral at all depths. V_{FB} in a non-uniformly doped MOSFET may be taken as the gate voltage at which the sum of all the charge in the substrate is zero. At V_{FB}, the surface potential may be nonzero, but the surface electric field is zero. The V_{FB} parameter is important in compact modeling. It is often considered a model parameter that is to be extracted from the device characteristics of a given technology to reflect the real process and device details.

b. Strong inversion condition: For a uniformly doped device, the strong inversion condition is the well known $\phi_s=2\phi_B$ criteria [2.4, 2.13]. For non-uniformly doped MOSFET it is appropriate to use

$$\phi_s = 2v_t \ln(\frac{N_{ch}}{n_i}) \tag{2.2.6}$$

where N_{ch} is an average doping concentration in the channel.

2. Lateral non-uniform doping effects

To reduce the short channel effects, local high doping concentration regions near the source/drain junction edges have been employed recently. This may be called lateral channel engineering, or more specifically halo [2.21] or pocket implantation [2.22]. Significant suppression of the short channel effects has been demonstrated in 0.1μm n-channel and buried p-channel MOSFETs with Large-Angle-Tilt-Implanted (*LATI*) pocket technology [2.22, 2.23]. The pocket implant technology is a promising new option to tailor the short channel performance of deep submicron MOSFETs.

With the introduction of lateral doping engineering, the doping concentration in the channel along the channel length becomes non-uniform as shown in Fig. 2.2.3. The non-uniform doping with higher doping concentration near the source/drain regions will result in an increase in the average doping concentration in the channel and hence cause an increase in the threshold voltage. As the device channel length becomes shorter, this non-uniform lateral doping profile may cause a significant increase in the threshold voltage. This is know as the Reverse Short Channel Effect (*RSCE*) as will be discussed later in this section [2.24].

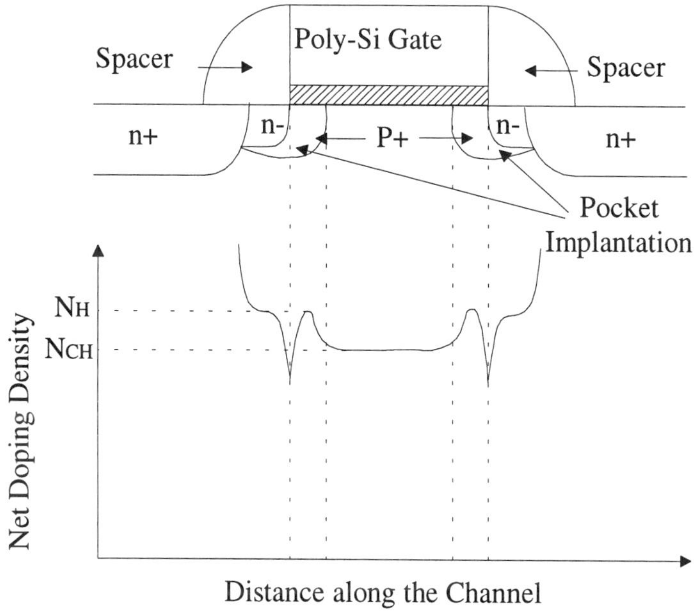

Fig. 2.2.3. Doping profile along the channel of a MOSFET with pocket implantation.

2.2.2 Normal short channel effects

The threshold voltage of a long channel device is independent of the channel length L and the drain voltage V_d. However, experimentally it is observed that V_{th} decreases as L decreases or V_d increases, as shown in Fig. 2.2.4. This effect is called the short channel effect [2.25] (in this section, we will discuss only the so called normal short channel effect, that is, V_{th} decreases monotonically as the channel length L decreases [2.26, 2.29-2.40].) It is also known as V_{th} roll-off. The dependence of V_{th} on L and V_{ds} in short-channel devices cannot be ignored. MOSFETs are normally designed with a V_{th} around 0.5V. If the value of V_{th} drops greatly as the channel length and V_{ds} vary, the device may exhibit excessive drain leakage current even when V_{gs} =0V [2.27, 2.28]. The modeling of V_{th} roll-off will be discussed in Chapter 3.

2.2.3 Reverse short channel effects

In the pervious section we discussed the so-called normal short channel effect. That is, the V_{th} of a MOSFET decreases monotonically as device channel length decreases. However, in devices using halo or pocket implantation, it has been found that, as shown in Fig 2.2.5, V_{th} initially increases with decreasing channel length. This is called the reverse short channel effect (*RSCE*), or V_{th} roll-up [2.41]-[2.44]. V_{th} reaches a maximum value at a certain channel length, and as L decreases further, V_{th} starts to decrease. The latter is the V_{th} roll-off [2.45]. The combined *RSCE* and V_{th} roll-off effects result in a 'hump' in the characteristics of V_{th} vs. L, as shown in Fig. 2.2.5 [2.46]. The cause of *RSCE* is the non-uniform lateral doping [2.47, 2.48, 2.49]. As discussed in section 2.2.2, for some technologies such as pocket implantation, the channel doping concentration near the source and drain is higher than in the middle of the channel. The increased doping concentration in these regions can result in an increase in V_{th} if the channel length becomes small.

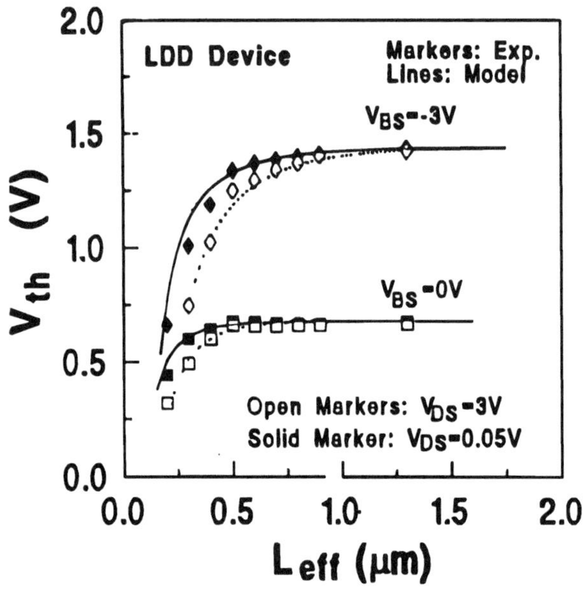

Fig. 2.2.4 Decrease of V_{th} caused by the short channel effect as device channel length decreases. After Liu et al. [2.36].

Even without an intentional pocket implant, channel doping concentration may still be higher near S and D, leading to *RSCE* [2.50, 2.51]. The cause is the transient-enhanced-diffusion (*TED*) of dopants near the source and drain from the deep buried peak to the surface due to the implant-induced interstitial defects produced by S and D implantation [2.43, 2.51]. Compact modeling of *RSCE* will be discussed in Chapter 3.

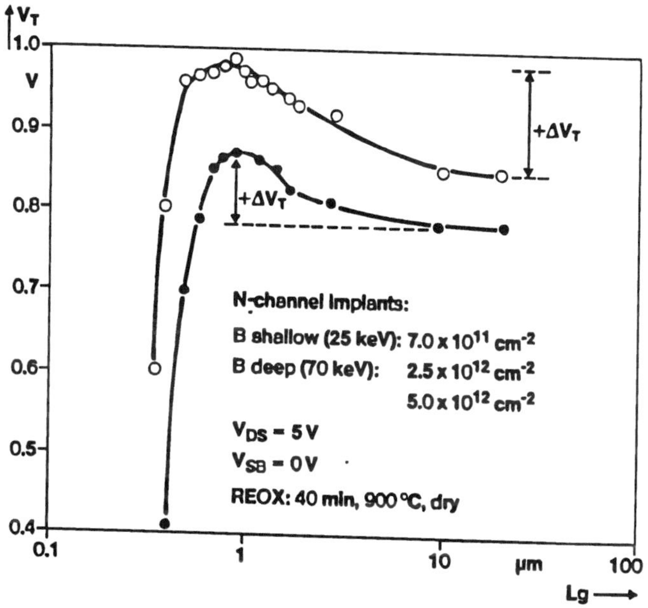

Fig. 2.2.5 A hump can be seen in the V_{th}-L characteristics of a device with *RSCE*. After Orlowski et al. [2.42].

2.2.4 Normal narrow-width effects

Device width has been found to have a significant effect on the device characteristics. Fig. 2.2.6 shows a cross-sectional view of a Localized-oxidation-of-silicon (*LOCOS*) isolated MOSFET along the width direction. It can be seen that the polysilicon gate overlaps the thick oxide on both sides of the thin gate oxide or channel region. The thick oxide on both sides, which is tapered and recessed, is called the field oxide. Under the field oxide, a field implantation may be performed to prevent surface inversion by the gate electrode. For the devices fabricated with the widely used *LOCOS* isolation process, it has been found that V_{th} increases as the channel width decreases, as shown in Fig. 2.2.7. This is considered the "normal" narrow width effect and is explained by the contribution of charges in the depletion layer region or in the edge of the field implant region. As the width of the device decreases, the contribution of these charges to the total depletion region charge becomes greater, leading to an increased V_{th} [2.52, 2.53]. Many different approaches have been proposed to model V_{th} changes caused by the narrow width effect [2.54, 2.55]. They will be discussed in Chapter 3.

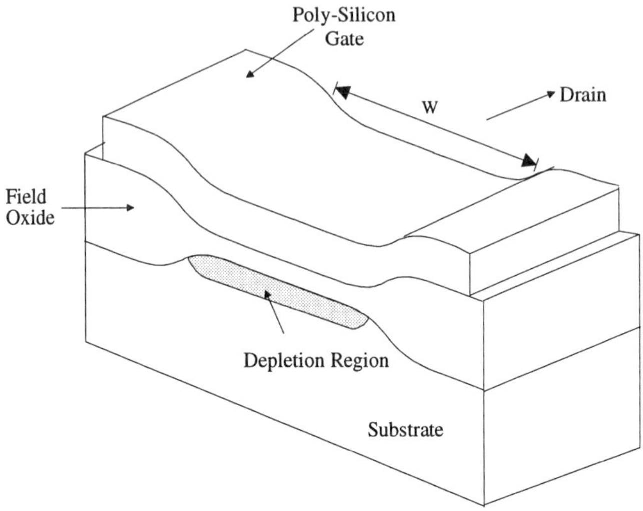

Fig. 2.2.6 A cross section of a *LOCOS* isolated MOSFET along the width direction.

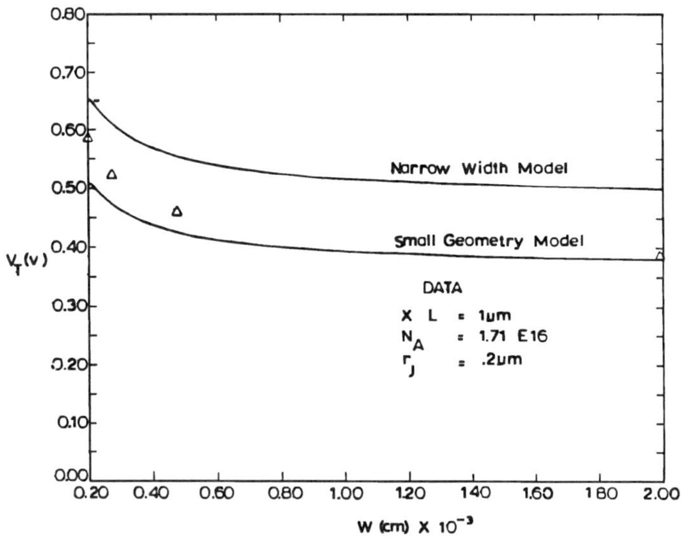

Fig. 2.2.7 Increase of V_{th} caused by the narrow width effect as device channel width decreases. After Akers [2.55].

2.2.5 Reverse narrow-width effects

The effect discussed above is called the normal narrow width effect, in which V_{th} increases as the channel width decreases. Another narrow width effect, in which V_{th} decreases as the channel width decreases as shown in Fig. 2.2.8 [2.56], has been observed in devices with certain new isolation technologies such as fully recessed or trench isolation technologies shown in Fig. 2.2.9. For a device with trench isolation, the field oxide is buried in the substrate and the field lines from the gate electrode are focused by the sharp geometry of the channel edge. Thus, at the edges of the channel an inversion layer is formed at a lower voltage than at the center. As a result, V_{th} is lower in devices with smaller *W*'s. This behavior is called the reverse narrow-width effect, in deference to the "normal" narrow width effect discussed in section 2.2.4.

The reverse narrow width effect is sensitive to several process and device factors such as the doping concentration in the sidewalls of the silicon, the trench isolation spacing, and the shape of the corner region at the edge of the gate.

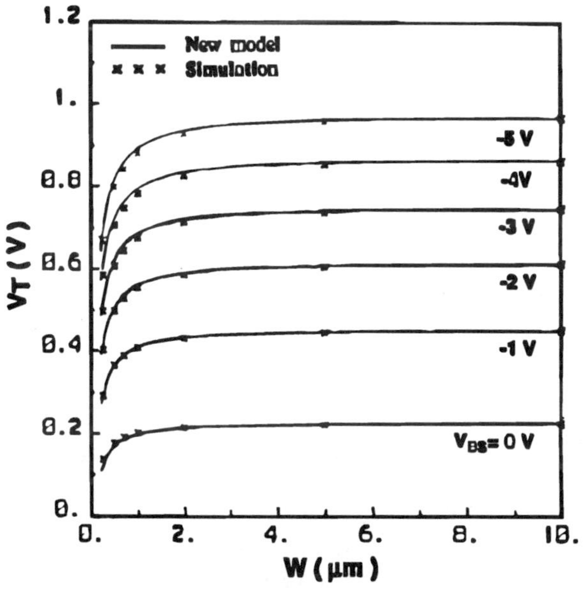

Fig. 2.2.8 V_{th} may decrease as channel width decreases. After Chung and Li [2.57].

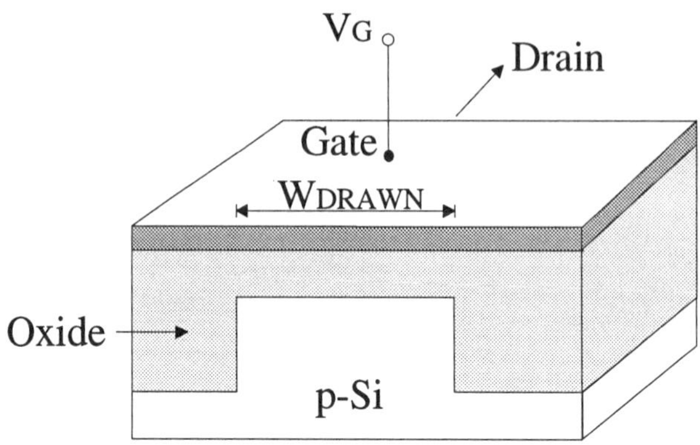

Fig. 2.2.9 A MOSFET device with trench isolation.

2.2.6 Body bias effect and bulk charge effect

In circuit applications, MOSFETs are biased at the drain, gate, and body (often with the source as the reference). In that case, the voltage between the body and the source (V_{bs}) and the voltage between the drain and the source (V_{ds}) are not zero. The body effect refers to the influence of V_{bs} on V_{th}. The body effect results in an increase in the threshold voltage of a MOSFET (with reference to the source) when a reverse bias V_{bs} is applied. The bulk charge effect is closely related to the body bias effect and refers to the changing threshold voltage along the channel when V_{ds}>0. V_{th} is not constant along the channel because the width of the depletion region along the channel is not uniform in the presence of a nonzero V_{ds}. In that case, V_{th} will be a function of the position, that is, a function of *V(y)* along the channel. *V(y)* is the channel potential voltage.

1. Body effect due to V_{bs}

We begin by assuming that the surface potential at the onset of inversion in a MOSFET is $2\phi_B$ when V_{bs}=0. With a reverse bias V_{bs}, the surface potential at the onset of the strong inversion becomes [2.58]

$$\phi_s(inv) = 2\phi_B - V_{bs} \tag{2.2.7}$$

The channel depletion layer becomes wider due to the V_{bs} bias, and can be expressed approximately as:

$$X_{dep} = \sqrt{\frac{2\varepsilon_{si}(2\phi_B - V_{bs})}{qN_{ch}}} \tag{2.2.8}$$

The depletion layer charge (per unit area) is

$$Q_{b\max} = -qN_{ch}X_{dep} \tag{2.2.9}$$

The inversion charge (per unit area) Q_{inv} in strong inversion in terms of V_{gb} and V_{bs} is:

$$Q_{inv} = -C_{ox}(V_{gb} - V_{FB} - 2\phi_B + V_{bs} - \gamma\sqrt{2\phi_B - V_{bs}}) \tag{2.2.10}$$

Assuming again that Q_{inv} is zero at the onset of inversion, and also noting that $V_{gs} = V_{gb} + V_{bs}$, we can find the threshold voltage when $V_{bs} \neq 0$ from Eq. (2.2.10),

$$V_{th} = V_{FB} + 2\phi_B + \gamma\sqrt{2\phi_B - V_{bs}} \tag{2.2.11}$$

The factor γ is the body factor or body effect coefficient, which determines how much V_{th} will increase when V_{bs} is increased. By looking at the expression of γ, we know that the body effect is stronger for devices with heavier substrate doping and thicker gate oxides.

2. Body effect due to V_{ds}

Because of the existence of the drain voltage V_{ds}, the depletion width along the channel is not uniform, as shown in Fig. 2.2.10. The non-uniform depletion width creates a non-uniform threshold voltage along the channel.

The threshold voltage will be a function of the position along the channel and can be expressed as [2.58]

$$V_{th}(y) = V_{th}(0) + \gamma(\sqrt{\phi_s - V_{bs} + V(y)} - \sqrt{\phi_s - V_{bs}}) \tag{2.2.12}$$

where y is the distance from the source. $V_{th}(0)$ means the threshold voltage at the source.

Using Taylor expansion, a linear expression can be found to describe the bulk charge effect due to V_{ds} [2.58],

$$V_{th}(y) = V_{th}(0) - aV(y) \tag{2.2.13}$$

where $a = 1 + \frac{g\gamma}{2\sqrt{\phi_s - V_{bs}}}$ and $g = 1 - \frac{1}{1.744 + 0.836(\phi_s - V_{bs})}$ [2.58].

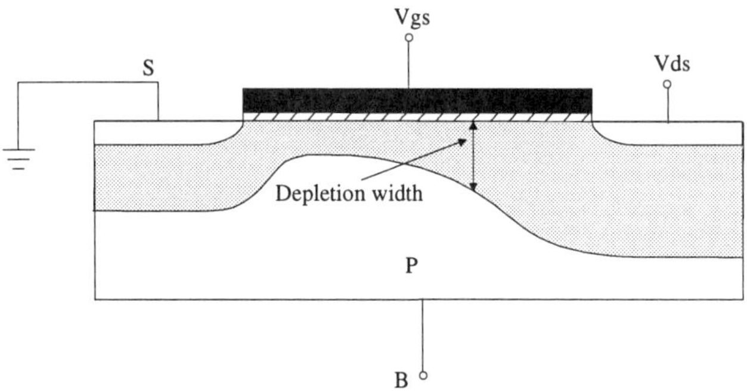

Fig. 2.2.10 Non-uniform nature of the depletion width and hence of V_{th} along the channel due to the drain voltage.

2.3 Channel Charge Theory

Fig. 2.3.1 illustrates an n-channel MOSFET with an applied V_g bias and V_{bs}=V_{ds}=0. The gate bias V_g is divided among the voltage across the oxide V_{tox}, the surface potential ϕ_s, and the work function difference ψ_{ms} between the gate material and the substrate [2.4],

$$V_g = V_{tox} + \phi_s + \psi_{ms} \tag{2.3.1}$$

Charge is induced in the semiconductor substrate and meets the charge neutrality condition as given in the following equations:

$$Q_g + Q_{int} + Q_s = 0 \tag{2.3.2a}$$

$$Q_s = -\varepsilon_{si} E_s \tag{2.3.2b}$$

$$Q_g = -\varepsilon_{si} E_{ox} = V_{tox} C_{ox} \tag{2.3.2c}$$

where Q_g is the charge on the gate, Q_{int} is the effective interface charge in the SiO_2-Si interface, and Q_s is the charge in the semiconductor under the oxide.

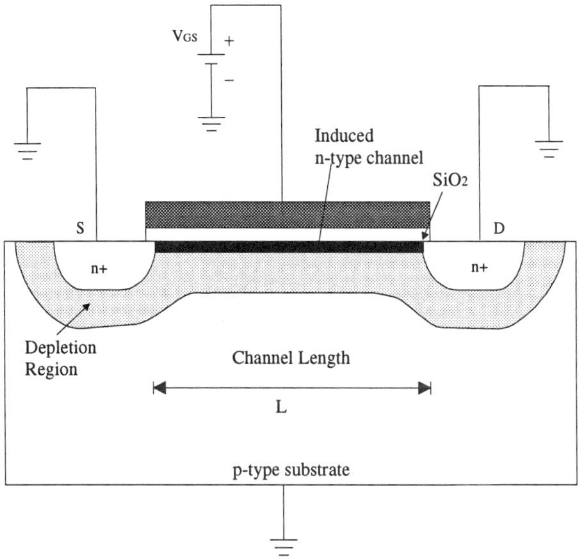

Fig. 2.3.1 N-channel MOSFET with a V_g bias (V_s=V_d=0).

Combining Eq. (2.3.1) and Eq. (2.3.2), we have,

$$V_g = V_{FB} + \phi_s - \frac{Q_s}{C_{ox}} \tag{2.3.3a}$$

$$V_{FB} = \psi_{ms} - \frac{Q_{int}}{C_{ox}} \tag{2.3.3b}$$

The potential distribution along the vertical dimension of the channel depletion layer can be found by solving the Poisson equation:

$$\frac{d^2\phi}{dx^2} = -\frac{q}{\varepsilon_{si}}(N_D - N_A + p - n) \tag{2.3.4}$$

where N_D and N_A are the donor and acceptor concentrations, n and p are the electron and hole concentrations in the semiconductor.

The surface field E_{si}(x=0) can be solved from Eq. (2.3.4) by using the boundary conditions, ϕ(x=0)= ϕ_s and E_s(x=0)=-$\partial\phi/\partial x|x$=0, as a function of the surface potential ϕ_s [2.59, 2.60]:

$$E_s(x=0)=\sqrt{\frac{2qN_A}{\varepsilon_{si}}[v_t\exp(-\phi_s/v_t)+\phi_s-v_t} \\ \overline{+\exp(-2\phi_B/v_t)(v_t\exp(-\phi_s/v_t)-\phi_s-v_t)]^{1/2}} \quad (2.3.5)$$

The charge Q_s in the semiconductor can be obtained from Q_s=-$\varepsilon_{si}E_s$(x=0):

$$|Q_s|=\sqrt{2q\varepsilon_{si}N_A}[[v_t\exp(-\phi_s/v_t)+\phi_s-v_t \\ +\exp(-2\phi_B/v_t)(v_t\exp(\phi_s/v_t)-\phi_s-v_t)]^{1/2} \quad (2.3.6)$$

Depending on the applied gate bias, there are three different charge conditions: (1) Accumulation; (2) Depletion; (3) Inversion. Eq. (2.3.6) is applicable in all three regimes. Fig. 2.3.2 shows Q_s vs. surface potential in an n-channel MOSFET, from accumulation through depletion to the strong inversion regions. Next we will discuss the charge characteristics in each of the three regions.

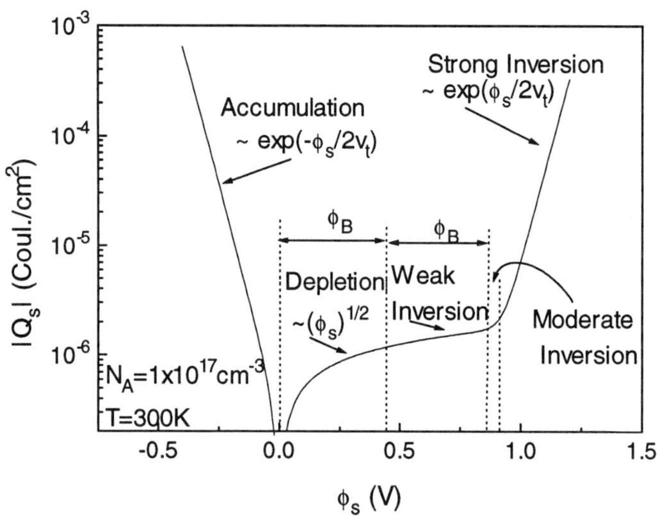

Fig. 2.3.2 Charge in semiconductor of an n-channel MOSFET at different gate bias conditions calculated with Eq. (2.3.6).

2.3.1 Accumulation

If the applied gate bias V_g is less than the flat band voltage V_{FB} (usually V_{FB} is negative for an n-channel device), the negative voltage at the gate will create negative charge on the gate and induce positive charge at the silicon surface, which means excess holes are accumulated at the surface. The charge region corresponding to the bias condition of $V_{gs}<V_{FB}$ is called accumulation [2.60]. When the accumulation of charge occurs, $\phi_s<0$ and $Q_s>0$ in an n-channel MOSFET.

In the accumulation region, ϕ_s <0 and the term with $\exp(-\phi_s / v_t)$ in Eq. (2.3.6) is dominant. Thus, Eq. (2.3.6) can be approximated with

$$Q_s \approx \sqrt{2q\varepsilon_{si}N_A v_t}\exp(-\phi_s / 2v_t) \tag{2.3.7}$$

2.3.2 Depletion

Consider the case of $V_{FB}<V_{gs}<V_{th}$. Because $V_{gs}>V_{FB}$, the gate bias repels positive charges (holes) from the silicon surface and leaves negatively charged acceptors behind. This charge condition is called depletion because the holes are depleted by the applied gate bias [2.60]. In the depletion region, it is assumed that concentrations of holes and electrons are zero, which is the well known depletion approximation [2.60]. Thus, in an n-type MOSFET in which $N_A>>N_D$, Eq. (2.3.4) can be simplified into

$$\frac{d^2\phi}{dx^2}=\frac{q}{\varepsilon_{si}}N_A \tag{2.3.8}$$

Assuming that the width of the depletion layer in the semiconductor is X_{dep}, we can use the following boundary conditions to solve Eq. (2.3.8),

$$\phi=\phi_s \text{ and } \frac{d\phi}{dx}=-E_s \text{ at x=0} \tag{2.3.9a}$$

and

$$\phi=\frac{d\phi}{dx}=0 \text{ at } x=X_{dep} \tag{2.3.9b}$$

The solution of Eq. (2.3.8) is

$$\phi = \phi_s(1 - \frac{x}{X_{dep}})^2 \quad (2.3.10)$$

$\phi = 0$ at X_{dep}, the width of the depletion layer,

$$X_{dep} = \sqrt{\frac{2\varepsilon_{si}\phi_s}{qN_A}} \quad (2.3.11)$$

The depletion charge can be obtained easily,

$$Q_{dep} = -qN_A X_{dep} = -\sqrt{2\varepsilon_{si} q N_A \phi_s} \quad (2.3.12)$$

Combining Eq. (2.3.3a) and Eq. (2.3.12), the relationship between the surface potential ϕ_s and the gate bias V_{gs} can be found,

$$\phi_s = \left[(\frac{\gamma^2}{4} + V_g - V_{FB})^{1/2} - \frac{\gamma}{2} \right]^2 \quad (2.3.13)$$

where γ is the body effect coefficient, and has been given previously.

Therefore, the relationships between the width of the depletion layer (and hence the depletion charge density) and the gate bias are given as follows,

$$X_{dep} = \sqrt{\frac{2\varepsilon_{si}}{qN_A}} \left[(\frac{\gamma^2}{4} + V_g - V_{FB})^{1/2} - \frac{\gamma}{2} \right] \quad (2.3.14)$$

$$Q_b = -\sqrt{2\varepsilon_{si} q N_A} \left[(\frac{\gamma^2}{4} + V_g - V_{FB})^{1/2} - \frac{\gamma}{2} \right] \quad (2.3.15)$$

Eqs. (2.3.12) - (2.3.15) are used frequently in the compact modeling of MOSFETs.

2.3.3 Inversion

If the positive gate bias continues to increase and is much larger than V_{FB}, the device will enter the inversion region. The inversion region can be further divided into the weak inversion and strong inversion regions depending on the applied gate voltage. The inversion regime corresponding to a surface potential between ϕ_B and $2\phi_B$ is called weak inversion, and the region in which the surface potential is large than $2\phi_B$ is called strong inversion. It has become a classic definition that the surface potential is $2\phi_B$ at the onset of strong inversion, at which the electron concentration at the surface in the channel is equal to N_A [2.60]. In compact modeling, the condition of $\phi_s=2\phi_B$ has also been used as the somewhat arbitrary separating point between weak inversion and strong inversion. Recently the necessity of defining a new region between weak inversion and strong inversion has arisen because of the interest in low power circuits. The new region between the weak inversion and strong inversion has been called the moderate inversion region [2.13]. In the moderate inversion region, the surface potential is within a few v_t's of $2\phi_B$.

In the inversion region, because $\phi_s>\phi_B$, some terms in Eq. (2.3.6) can be ignored. The total charge in the semiconductor including the charges in both the inversion layer and the depletion layer becomes

$$Q_s = -\sqrt{2q\varepsilon_{si}N_A}\{\phi_s + v_t \exp[(\phi_s - 2\phi_B)/v_t)\}^{1/2} \tag{2.3.16}$$

As discussed before, the charge in the depletion layer is given by (2.3.12). Thus we can get the inversion charge by subtracting the depletion charge from the total charge,

$$Q_{inv} = -\sqrt{2q\varepsilon_{si}N_A}\left[\{\phi_s + v_t \exp[(\phi_s - 2\phi_B)/v_t]\}^{1/2} - \phi_s^{1/2}\right] \tag{2.3.17}$$

a. Strong inversion

In strong inversion, $|Q_{inv}|>>|Q_b|$. As a result, the exponential term in Eq. (2.3.17) is the dominant term. The inversion charge can be written as,

$$Q_{inv} \approx -\sqrt{2q\varepsilon_{si}N_A v_t}\exp\left(\frac{\phi_s - 2\phi_B}{2v_t}\right) \tag{2.3.18}$$

In Eq. (2.3.18), the inversion layer charge is an exponential function of the surface potential. Therefore, a small increase in the surface potential induces a large change in Q_{inv}. In that case, the surface potential does not change much as the gate bias increases, which is the basis of the assumption that the surface potential is pinned at $2\phi_B$ when a device is in strong inversion. However, we should point out that the surface potential does change even in the strong inversion region, as shown in Fig. 2.3.3. Depending on the oxide thickness and doping concentration, the surface potential in strong inversion can be 200mV higher than $2\phi_B$. Therefore, the $2\phi_B$ approximation used in modeling the charge and threshold voltage has underestimated the actual surface potential value. For devices with thin gate oxides, this fact will impact the accuracy of the models based on this approximation in predicting the actual device characteristics without any correction through the parameter extraction.

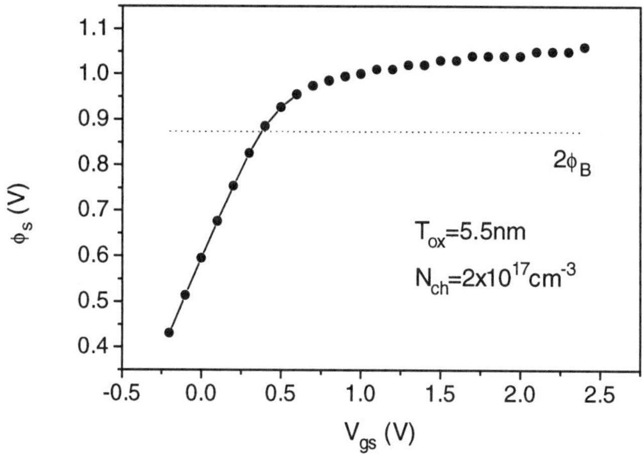

Fig. 2.3.3 The surface potential ϕ_s increases even in strong inversion as gate bias increases.

b. Weak inversion

In weak inversion, $\phi_B<\phi_s<2\phi_B$, and the inversion charge $|Q_{inv}|$ is much less than the depletion charge $|Q_b|$. The term $v_t \exp[(\phi_s - 2\phi_B)/v_t]$ in Eq. (2.3.17) is small compared with the term ϕ_s, and we can approximate the first square root in the parentheses in Eq. (2.3.17) with the first two terms of its Taylor expansion,

$$\sqrt{v_t \exp[(\phi_s - 2\phi_B)/v_t] + \phi_s} \approx \sqrt{\phi_s} + \frac{v_t \exp[(\phi_s - 2\phi_B)/v_t]}{2\sqrt{\phi_s}} \tag{2.3.19}$$

Combining Eq. (2.3.19) with Eq. (2.3.17), we obtain

$$Q_{inv} \approx -\frac{\sqrt{2q\varepsilon_{si}N_A}}{2\sqrt{\phi_s}} v_t \exp(\phi_s - 2\phi_B / v_t) \qquad (2.3.20)$$

c. Moderate inversion

The moderate inversion region is a narrow operation region. Usually a device operates in the moderate inversion regime only over a range of surface potential of several v_t's. An accurate definition of the boundary between the moderate inversion and strong inversion regions is quite complicated [2.13]. As the low power market develops, this moderate inversion region becomes more important.

In this operation region, both drift and diffusion current are important and need to be included in modeling the device characteristics. This increases the difficulty of getting an analytical expression for the current-voltage characteristics in this region. Different approaches including both analytical and numerical solutions have been proposed to model the device behavior in this important operation region. This region is discussed more in Chapter 4.

2.4 Carrier Mobility

Mobility is a key parameter in MOSFET modeling. It is a measure of the ease of carrier motion in semiconductor materials. Carrier transport in a MOSFET mainly takes place along the interface between silicon and SiO_2. The carrier mobility at the interface is lower than in the bulk and depends on both the transverse and longitudinal electric fields [2.62]. Two decades ago, the gate oxide was thick, and the transverse electric field induced by gate bias was low so that the influence due to the transverse electric field could be ignored. But in today's short channel devices, oxide thicknesses are very thin. The transverse field becomes high so that its influence on the carrier mobility cannot be ignored. In this case, the carrier mobility is not constant as the gate bias changes.

The relationship between carrier mobility and the electric field in MOSFETs is a subject well studied since the 1970's [2.62-2.67]. At least three scattering mechanisms have been proposed to account for the dependence of mobility on the electric field: phonon scattering, coulomb scattering due to the charge centers at or close to the interface, and surface roughness scattering [2.66]. Each mechanism may be dominant under some specific conditions of the doping concentration, temperature and bias. For good quality interfaces, phonon scattering is generally the dominant scattering mechanism at room temperature, as shown in Fig. 2.4.1 [2.66].

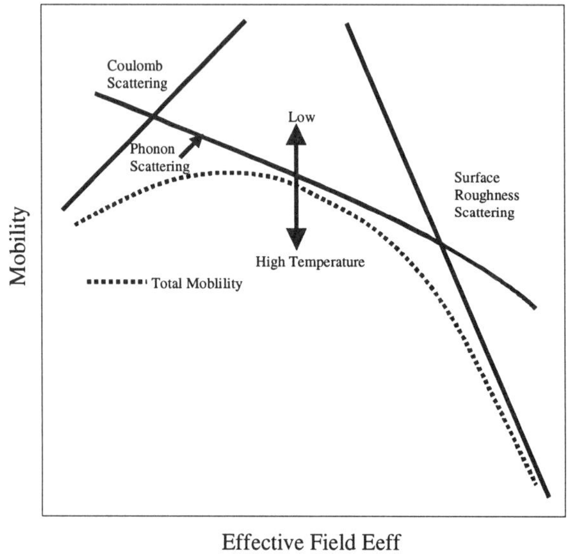

Fig. 2.4.1 Different scattering mechanisms may dominate the mobility behavior depending on the bias and temperature conditions. After Takagi et al. [2.66].

Based on these scattering mechanisms, physical mobility models have been proposed [2.68, 2.69, 2.70]. In general, mobility depends on many process parameters and bias conditions such as the gate oxide thickness, doping concentration, threshold voltage, gate voltage and substrate voltage, etc.

Many studies of the mobility in MOSFETs have been reported [2.63- 2.73]. One universal model has been proposed recently and verified by experiments with both N- and P- channel devices [2.67], as given in the following:

For NMOS electrons:

$$\mu_{eff}\left(\frac{cm^2}{V.\text{sec.}}\right)=\frac{540}{1+\left(\frac{V_{gs}+V_{th}}{5.4T_{ox}}\right)^{1.85}} \tag{2.4.1}$$

For PMOS holes:

$$\mu_{eff}(\frac{cm^2}{V\,\text{sec.}}) = \frac{185}{1 + (\frac{V_{gs} + 1.5V_{th} - \delta}{3.38T_{ox}})} \quad (2.4.2)$$

Here δ is zero for p^+ poly-gate and 2.3V for n^+ poly gate [2.67]. Comparisons of the above expressions with measured data are shown in Fig. 2.4.2.

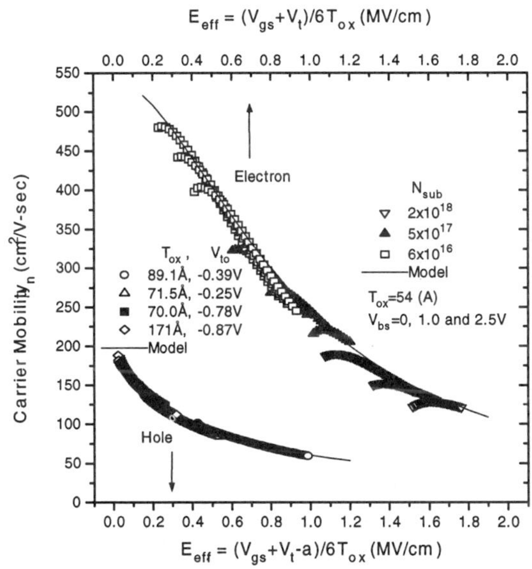

Fig. 2.4.2 The universal mobility model fits experimental data of different technologies. After Chen et al. [2.67].

2.5 Velocity Saturation

Carrier saturation velocity is another important parameter that affects the characteristics of short channel MOSFETs. Previously, people used the concept of channel pinch-off to explain the current saturation characteristics in MOSFETs [2.60]. The current saturates when the conducting channel is pinched-off at the drain side under the gate. In the pinch-off region, the carrier charge density is assumed to be zero and hence the carrier velocity must be infinite to ensure current continuity. It can describe the experimental results for long channel devices, however, it is not physically accurate and cannot model the *I-V* characteristics of short channel devices. The concept of velocity saturation must be introduced [2.62].

If the lateral electrical field is small, the carrier drift velocity is given by [2.73]

$$v = \mu_{eff} E \tag{2.5.1}$$

where μ_{eff} is the effective electric field and is independent of the lateral field.

However, as the lateral field becomes higher, velocity deviates from Eq. (2.5.1), and saturates at high fields [2.73]-[2.76]. The only time when v is a linear function of E is when E is small.

An accurate model for the drift velocity is [2.77]

$$v = \frac{\mu_{eff} E}{[1 + (E/E_{sat})^m]^{1/m}} \tag{2.5.2}$$

where m= 2 for electron and m=1 for holes [2.78, 2.79]. E_{sat} is the critical field at which carrier velocity becomes saturated, and is linked with the saturation velocity, v_{sat}, by $E_{sat} = v_{sat}/\mu_{eff}$. For electrons v_{sat} varies between 6 ~ 10×10^4 m/s, and is between 4 ~ 8×10^4 m/s for holes [2.80].

The model expressed in Eq. (2.5.2) fits experimental data very well [2.66], but yields a very complicated current equation if used in modeling *I-V* characteristics [2.77].

To obtain a simpler analytical expression of the *I-V* model, compact modeling assumes m=1 in Eq.(2.5.2) for both electrons and holes with the following form [2.62]:

$$v = \frac{\mu_{eff} E}{1 + E/E_{sat}} \tag{2.5.3}$$

where E_{sat} is a fitting parameter extracted from the measured data.

Eq. (2.5.3) differs significantly from Eq. (2.5.2) for n-channel MOSFETs [2.79]. As an improved solution, a piece-wise velocity-field relationship has been suggested and adopted in compact modeling [2.62]

$$v = \frac{\mu_{eff} E}{1 + E/(2E_{sat})} \qquad E < 2E_{sat} \tag{2.5.4a}$$

$$v = v_{sat} \qquad E \geq 2E_{sat} \tag{2.5.4b}$$

where E is the lateral electric field along the channel direction.

As shown in Fig. 2.5.1, Eq. (2.5.4) matches Eq. (2.5.2) more closely than Eq. (2.5.3). But, Eq. (2.5.4) saturates earlier than Eq. (2.5.3), which may result in a lower V_{dsat} in an *I-V* model based on Eq. (2.5.4). In reality, adjustments can be made through the parameter extraction process so that Eq. (2.5.4) is adequate for compact modeling.

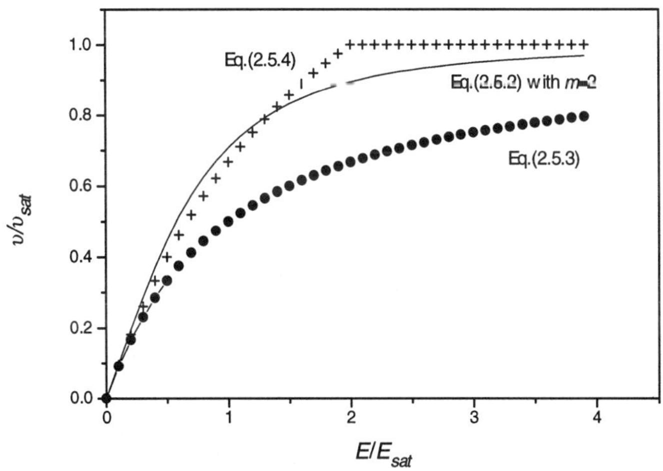

Fig. 2.5.1 Velocity-field relationships for electrons and holes in silicon. The curves are calculated from Eq.(2.5.2) with m=2, Eq. (2.5.3), and Eq. (2.5.4).

2.6 Channel Length Modulation

When devices operate in the saturation region, the channel of the device can be divided to two parts, as shown in Fig. 2.6.1. One part is from the source to the velocity saturation point, and another part is from the velocity saturation point to the drain. As the drain bias increases, the velocity saturation region grows slightly. The device behaves as if the effective channel length has been reduced so that the drain current can increase with increasing V_d, especially for short channel devices. This effect is called channel length modulation (*CLM*). *CLM* should be modeled accurately in a compact model because it affects both the device current and output resistance, which is an important parameter in analog circuit design [2.81].

Analytical models have been developed to model *CLM* by solving the 2 dimensional Poisson equation near the drain [2.82]:

$$\frac{\partial V^2(x,y)}{\partial x^2}+\frac{\partial V^2(x,y)}{\partial y^2}=-\frac{\rho(x,y)}{\varepsilon_{si}} \tag{2.6.1}$$

where ρ is the charge density in the velocity saturation region.

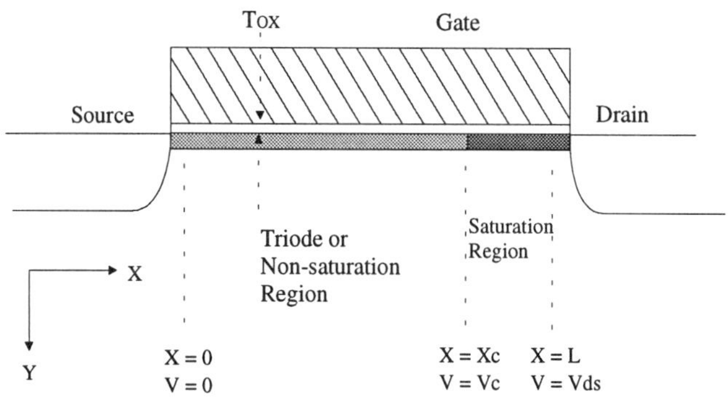

Fig. 2.6.1 Schematic representation of a MOSFET in the saturation region. X_c is the point at which the carrier velocity becomes saturated and V_c is the critical voltage at the saturation point.

This equation can only be solved numerically. However, some analytical solutions for the length of the L_d region have been obtained by using approximations. For example [2.82],

$$L_d=\sqrt{\frac{2\varepsilon_{si}(V_{ds}-V_{dsat})}{qN_{ch}}+\left(\frac{\varepsilon_{si}E_y(L-L_d)}{qN_{ch}}\right)^2}-\frac{\varepsilon_{si}E_y(L-L_d)}{qN_{ch}} \tag{2.6.2}$$

where V_{ds} is the drain bias, N_{ch} is the doping concentration (assumed uniform) in the channel, V_{dsat} is the saturation voltage, and $E_y(L\text{-}L_d)$ is the lateral electric field at the beginning of the velocity saturation region.

The analysis to obtain Eq. (2.6.2) ignores the mobile carriers in the velocity saturation region, and also does not account for the influence of the vertical oxide field. A more physical approach to model *CLM* was suggested by using a quasi-two dimensional approach [2.83, 2.84, 2.85]. Fig. 2.6.2 illustrates the schematic of the quasi-two dimensional analysis of *CLM*. The length of the velocity saturation region can be derived with consideration for the mobile carriers in the L_d region and for the vertical oxide electric field [2.83, 2.84, 2.85]:

$$\Delta L = l\ln\left(\frac{(V_{ds} - V_{dsat}}{lE_{sat}} + \frac{E_m}{E_{sat}}\right) \qquad (2.6.3)$$

$$E_m = \sqrt{\left(\frac{V_{ds} - V_{dsat}}{l}\right)^2 + E_{sat}^2} \qquad (2.6.4)$$

where $l = \sqrt{\frac{\varepsilon_{si}}{\varepsilon_{ox}} T_{ox} X_j}$, T_{ox} is the oxide thickness, and X_j is the source/drain junction depth. E_{sat} and V_{dsat} are the electric field and voltage at the saturation point.

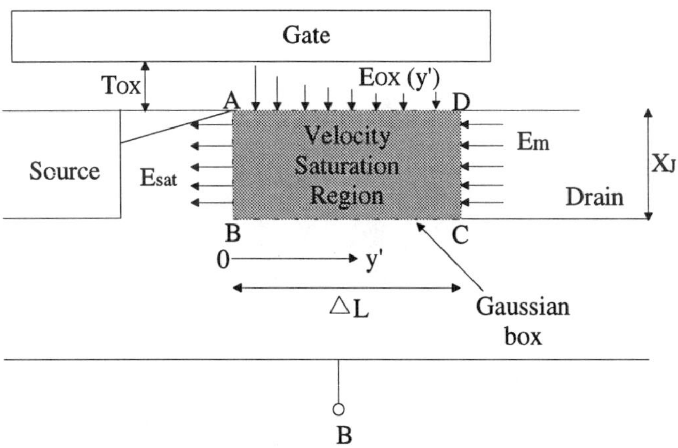

Fig. 2.6.2 Schematic diagram illustrating the velocity saturation region. After Ko [2.83].

2.7 Substrate Current Due to Impact Ionization

As the scaling of MOSFETs proceeds, impact ionization of carriers in the high field region (velocity saturation or pinch-off region) becomes serious. The holes generated during the process of impact ionization flow through the substrate and result in a substrate current. Typical substrate current characteristics of an n-channel MOSFET with channel length of 1μm are shown in Fig. 2.7.1. The substrate current characteristics can be divided into three main parts. In region I, band-to-band tunneling (*BTBT*) dominates [2.88]. In region II, the impact ionization current I_{ht} is dominant [2.89, 2.90]. In region III between the above two regions, leakage current of the reverse biased p-n junction diode may be the major component of the substrate current [2.91].

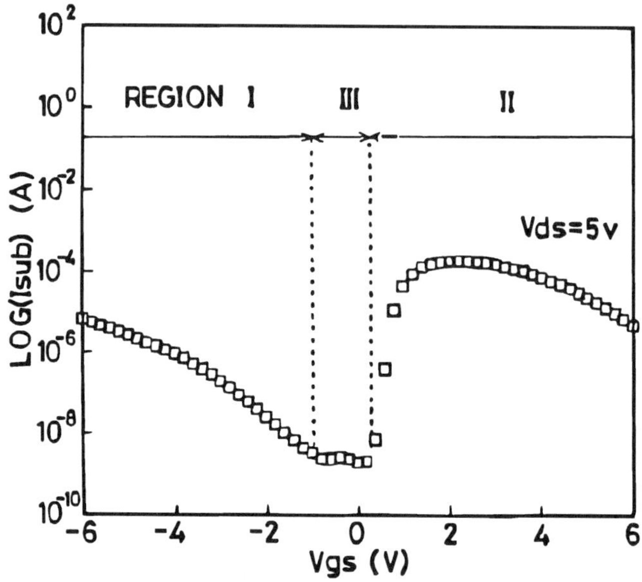

Fig. 2.7.1 Typical characteristics of substrate current vs. gate bias. Region I, II, and III are dominated by band-to-band tunneling, impact ionization, and thermal generation. After Tanizawa et al. [2.94].

Impact ionization has been studied extensively as an important topic in VLSI development [2.92, 2.93]. If electrons in the channel of an n-MOSFET acquire more than about 1.5eV of energy, impact ionization can occur. During the process of impact ionization, electron-hole pairs are generated. The electrons produced in this manner are either attracted to the drain (contributing to additional drain current) or, if possessing sufficient energy, injected into the oxide. The generated holes, on the other hand, enter into the substrate and constitute a parasitic substrate current I_{sub}.

Hot carriers can also generate interface traps or oxide trapped charges, and result in the degradation of device performance [2.95, 2.96, 2.97]. The substrate current can be used to predict the device reliability and has been used in some circuit simulators to analyze the reliability of circuits [2.98, 2.99]. Recently, substrate current models have become important for the simulation of analog circuits [2.61, 2.90, 2.100].

Fig. 2.7.2 shows the physical mechanism of the generation of substrate current in an n-channel MOSFET. For the MOS transistor operating in the saturation mode, the electric field near the drain region causes impact ionization of carriers. The generated electrons are swept into the drain whereas the holes flow into the substrate.

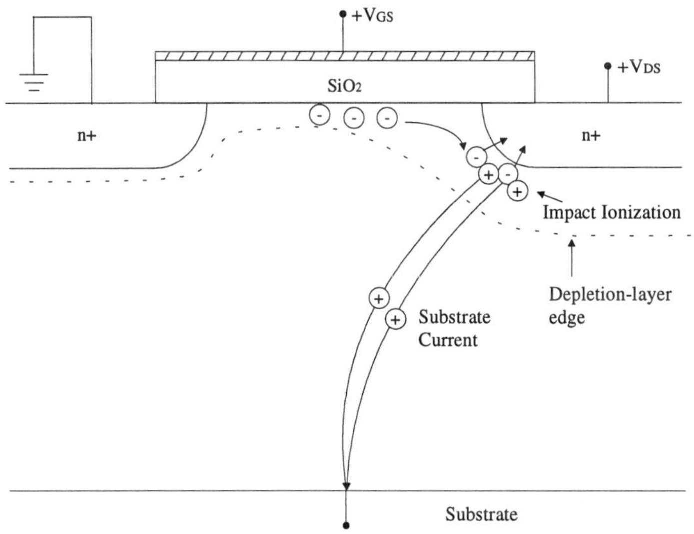

Fig. 2-7.2 Generation of the substrate current due to impact ionization.

The substrate current causes an IR drop in the substrate, resulting in a body bias that causes V_{th} to drop. The substrate current induced body bias effect (*SCBE*) results in a current increase that is many times larger than I_{sub} itself as shown in Fig. 2.7.3.

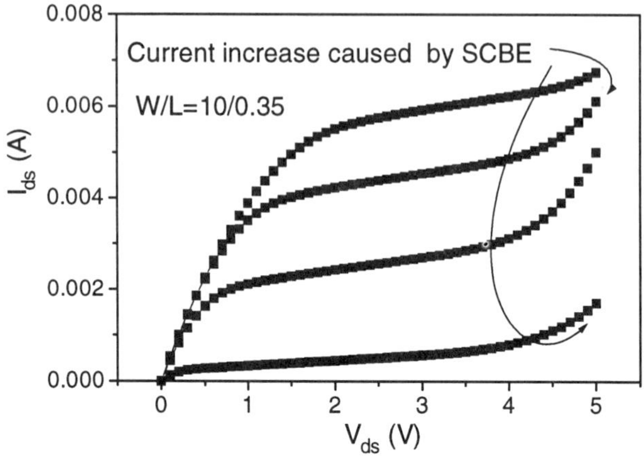

Fig. 2.7.3 Drain current increase caused by the substrate current induced body effect can be seen in the high drain bias region.

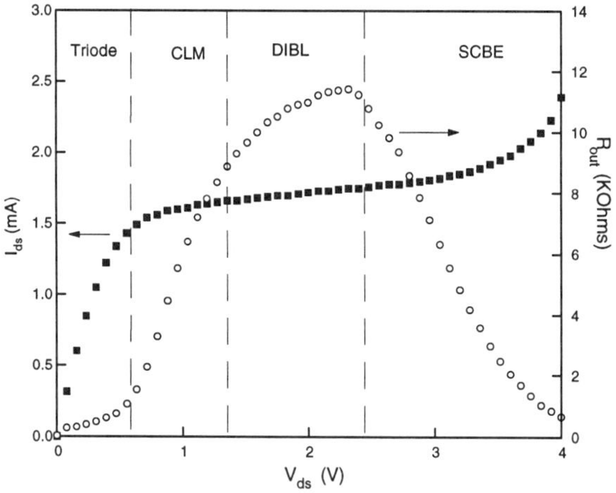

Fig. 2.7.4 General behavior of the MOSFET output resistance. After Huang et al. [2.86].

It has been found that the accurate and scalable modeling of *CLM* and the substrate current induced body effect caused by hot carrier effects (*HCE*) is very important to describe the output resistance characteristics in a compact model [2.86, 2.87]. Fig. 2.7.4 qualitatively shows the regions where different mechanisms dominate the output resistance. The *SCBE* results in a dramatic decrease in output conductance in the high V_{ds} region.

Band to band tunneling (*BTBT*) current, which occurs at the surface of the depletion layer under the gate-drain overlap region, is notable in the subthreshold region in thin-gate-oxide MOSFETs when the gate is grounded and the drain is biased at high voltage [2.94, 2.95]. A large electric field exists across the oxide (E_{ox}), which must be supported by charge in the drain region. This charge is provided by the formation of a deep depletion region in the drain. An inversion layer cannot form at the silicon surface of the n-type drain region. As the thermally generated holes arrive at the surface to form the inversion layer, they are immediately swept laterally into the substrate (which is a region of lower potential for holes). In the deep-depletion mode the band bending can exceed 1.2eV. This allows the energy states in the valence and conduction bands to line up (see Fig. 2.7.5). Thus, band to band tunneling occurs, generating electrons and holes. Electrons flow into the drain and holes into the body. The resulted drain to body current is called *BTBT* current or gate-induced drain leakage (*GIDL*) current, as shown in Fig. 2.7.4. Like other forms of leakage current, *BTBT* current will contribute to standby power, or power consumption.

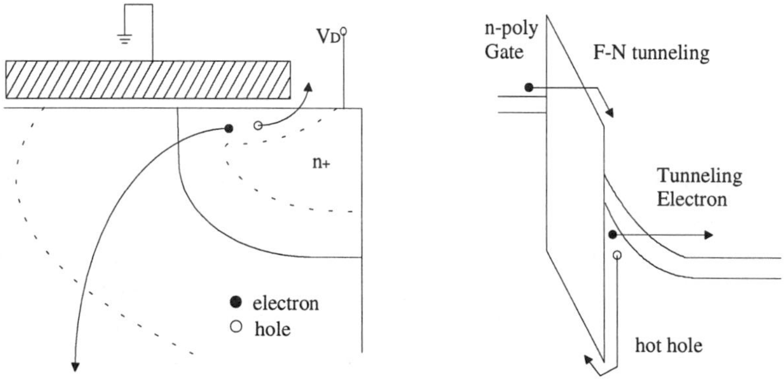

Fig. 2.7.5 Schematic illustration of *BTBT* mechanism in thin oxide MOSFETs. After Chen et al. [2.88].

A simple analytical equation for the *BTBT* current has been derived [2.96]:

$$I_{bt} = AE_s \exp(-B / E_s) \tag{2.7.1}$$

where A and B are material parameters involving the band gap energy and effective mass. E_s is the electric field at the Si surface and can be given by

$$E_s = \frac{V_{dg} - qE_g}{3T_{ox}} \tag{2.7.2}$$

where E_g is the energy band gap of silicon and is also the voltage drop (band bending) in the substrate necessary for tunneling to occur, q is the electron charge, and V_{dg} is the voltage between the drain and gate.

2.8 Polysilicon Gate Depletion

As a gate voltage is applied to a heavily doped poly-Si gate, e.g. NMOS with n+ polysilicon (poly-Si) gate, a thin depletion layer in the poly-Si can be formed at the interface between the poly-Si and the gate oxide. This depletion layer is very thin because of the high doping concentration in the poly-Si gate. But its effect cannot be ignored for devices with gate oxides thinner than 10nm.

Fig. 2.8.1 shows an NMOSFET device with a depletion region in the n^+ poly Si gate. The doping concentration in the n^+ poly-Si gate is ***NGATE*** and the doping concentration in the substrate is ***NSUB***. The gate oxide thickness is ***Tox***. If we assume the doping concentration in the gate is infinite, then no depletion region will exist in the gate, and there would be one sheet of positive charge at the interface between the poly-Si gate and gate oxide. In reality, the doping concentration is finite. The positive charge near the interface of the poly-Si gate and the gate oxide is distributed over a finite depletion region with thickness X_{poly}. In the presence of the depletion region, the effective gate voltage is the voltage at the top of the gate oxide. In other words, part of the gate voltage is lost across the depletion region in the gate. That means the effective gate voltage will be reduced by a fraction of a volt.

The effective gate voltage can be calculated as follows. Assume that the doping concentration in the poly gate near the interface is ***NGATE***. The voltage drop in the poly-Si gate (V_{poly}) can be calculated as [2.101]

$$V_{poly} = \frac{1}{2} X_{poly} E_{poly} = \frac{qN_{GATE} X_{poly}^2}{2\varepsilon_{si}} \tag{2.8.1}$$

where E_{poly} is the electric field at the poly-gate/oxide interface. The boundary condition at the interface is

$$\varepsilon_{ox} E_{ox} = \varepsilon_{si} E_{poly} = \sqrt{2q\varepsilon_{si} N_{GATE} V_{poly}} \tag{2.8.2}$$

where E_{ox} is the electric field in the gate oxide. The gate voltage satisfies

$$V_{gs} - V_{FB} - \phi_s = V_{poly} + V_{ox} \tag{2.8.3}$$

where V_{ox} is the voltage drop across the gate oxide and $V_{ox}=E_{ox}T_{ox}$.

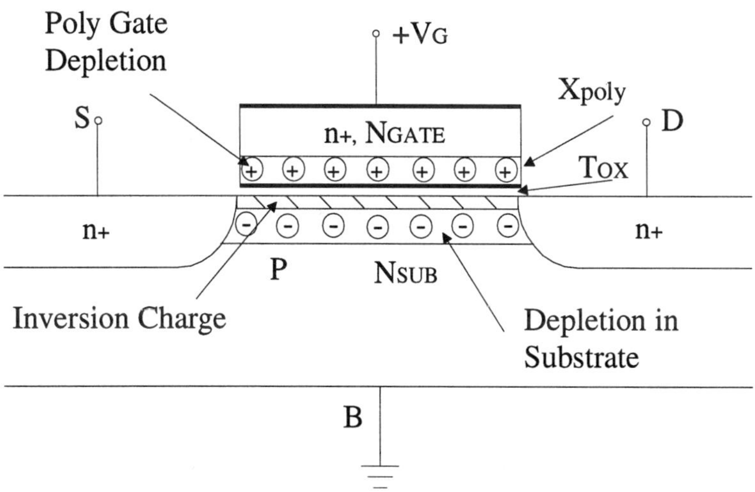

Fig. 2.8.1 A MOSFET with the poly gate depletion region is shown. The device is in the strong inversion region.

From Eqs. (2.8.1) to (2.8.3), we obtain

$$a(V_{gs} - V_{FB} - \phi_s - V_{poly})^2 - V_{poly} = 0 \tag{2.8.4}$$

$$a = \frac{\varepsilon_{ox}^2}{2q\varepsilon_{si} N_{GATE} T_{ox}^2} \tag{2.8.5}$$

By solving Eq. (2.8.4), we find the effective gate voltage (V_{gs_eff})

$$V_{gs\_eff} = V_{FB} + \phi_s + \frac{q\varepsilon_{si} NGATE T_{ox}^2}{\varepsilon_{ox}^2}\left(\sqrt{1 + \frac{2\varepsilon_{ox}^2 (V_{gs} - V_{FB} - \phi_s)}{q\varepsilon_{si} NGATE T_{ox}^2}} - 1\right) \tag{2.8.6}$$

Fig. 2.8.2 shows V_{gs_eff} / V_{gs} versus the gate voltage. The threshold voltage is assumed to be 0.4V. If T_{ox}=30 Å, the effective gate voltage can be reduced by 10% due to the poly gate depletion effect.

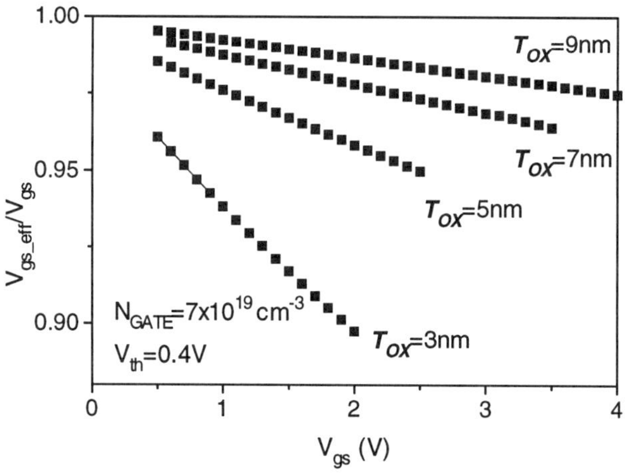

Fig. 2.8.2 The effective gate voltage versus applied gate voltage at several T_{ox}.

We can estimate the drain current reduction in the linear region as a function of the gate voltage. Assume that the drain voltage is very small, e.g. 50mV. The linear drain current is proportional to $C_{ox}(V_{gs} - V_{th})$. The ratio of the linear drain current with and without poly gate depletion is equal to

$$\frac{I_{ds}(V_{gs\_eff})}{I_{ds}(V_{gs})} = \frac{(V_{gs\_eff} - V_{th})}{(V_{gs} - V_{th})} \tag{2.8.7}$$

Fig. 2.8.3 shows $I_{ds}(V_{gs\_eff}) / I_{ds}(V_{gs})$ versus the gate voltage using Eq. (2.8.7). The drain current can be reduced by more than 10 percent due to gate depletion.

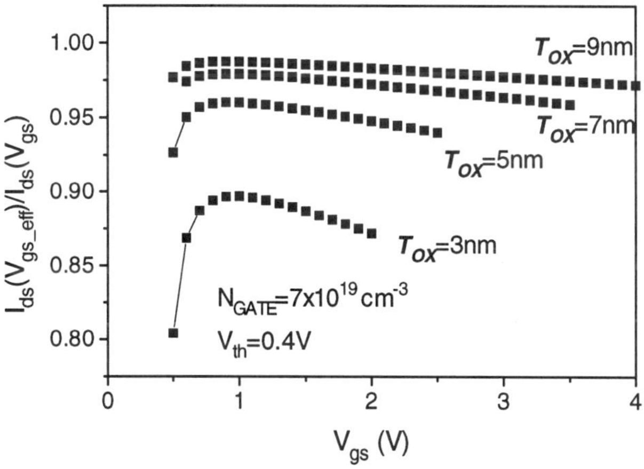

Fig. 2.8.3 Ratio of the linear region current with poly gate depletion effect to the current without poly gate depletion effect.

The polysilicon gate depletion effect can also influence the capacitance characteristics in the strong inversion region [2.101]. A significant capacitance reduction has been observed in MOSFETs with oxide thickness less than 5nm [2.101]. Thus the polysilicon gate depletion effect has to be accounted for in modeling the capacitance characteristics of devices with very thin oxide thickness [2.101, 2.102].

2.9 Velocity Overshoot Effects

As MOSFET dimensions shrink toward 0.1μm, velocity overshoot is expected to become a significant issue [2.103, 2.104]. Velocity overshoot arises when the carrier transit time becomes comparable to the energy relaxation time (i.e. when mobile carriers do not have enough time to reach equilibrium with the applied electric field). Because of the velocity overshoot effect, the steady-state field-dependent mobility model, Eq. (2.4.1), which assumes equilibrium with the applied electric field by scattering of carriers, may no longer be sufficient for modeling the MOS device characteristics accurately.

Recent work shows that the energy balance equation should be incorporated in the drift-diffusion equation to consider the spatial variation of carrier energy in order to account for the velocity overshoot phenomenon in deep submicron MOS devices. It has been shown that an electric field gradient can cause electrons to overshoot the saturation velocity [2.103]. There is an insufficient number of phonon-scattering events experienced by the electron during its flight, with the result that electrons can be accelerated to velocities higher than the saturation velocity.

It has been known that the velocity overshoot effect can improve drain current and transconductance [2.104, 2.105]. Several theoretical models of velocity overshoot have been reported to give physical explanation and understanding of this effect [2.106, 2.107, 2.108, 2.109]. Some analytical models of *I-V* including velocity overshoot have been reported [2.110, 2.111]. The general modeling approach can be described as follows.

The drift velocity in an inhomogenous electric field can be expressed approximately as [2.110],

$$v = \mu E + \lambda \frac{dE}{dx} \tag{2.9.1}$$

where μE is the drift velocity in a homogenous field. E is the longitudinal electric field in the device. The λ parameter is a function of the electric field; however, it can be treated as a constant for the usual longitudinal field range found in short channel MOSFETs to get an analytical expression for the *I-V* model with the consideration of the velocity overshoot effect.

The velocity expression given above can be put in the drain-current equation, and an integral can be taken from the source to the drain along the channel to get an analytical *I-V* expression including the velocity overshoot effect.

However, no compact models implemented in widely used circuit simulators have included this effect yet, even through some detailed research work related to compact modeling has been reported recently [2.111, 2.112, 2.113].

2.10 Self-heating Effect

Heat dissipation is a critical issue in circuit design. The temperature rise caused by the power consumption of devices in a chip may need to be considered because the device density has become huge in modern VLSI circuits. Generally the self-heating effect on the electrical characteristics can be ignored in a device model because the thermal conductivity of silicon substrates is high. However, this may not be true in some cases such as SOI CMOS, high voltage and current devices used in power IC's, and future very high device density (0.1μm or less) circuits where the devices may have significant self-heating effects (*SHE*). It has been reported that *SHE* can decrease the drain drive current and result in negative differential conductance at low frequency and high power operation regions [2.114, 2.115].

The *SHE* was first reported in MOSFETs in 1987 [2.116]. When *SHE* occurs, the temperature in the device increases and temperature sensitive parameters such as threshold voltage, mobility, carrier velocity, and impact ionization rate can change. Negative conductance can also be observed if *SHE* becomes significant at high V_{ds} and high I_{ds}, as shown in Fig. 2.10.1 [2.120]. *SHE* can be observed easily in the measured characteristics of SOI MOSFETs with short channel lengths. Recently, it is also observed in bulk deep submicron MOSFETs, in which a significant increase of the output resistance is seen in the high drain voltage regime as shown in Fig. 2.10.2. The normal output resistance characteristics are given in Fig. 2.7.3, which shows a decrease in R_{out} as the drain bias becomes large.

SHE was studied using numerical two-dimensional device simulators incorporating heat flow [2.117]. Usually, the temperature rise induced by *SHE* in a device can be described by the expression [2.118, 2.119]:

$$T_c = T_o + R_{th} I_d V_{ds} \tag{2.10.1}$$

where I_d and V_{ds} are drain current and the drain-source voltage, respectively, R_{th} is the thermal resistance (for the dynamic case, R_{th} will be replaced by the thermal impedance of the device). T_o is the ambient temperature.

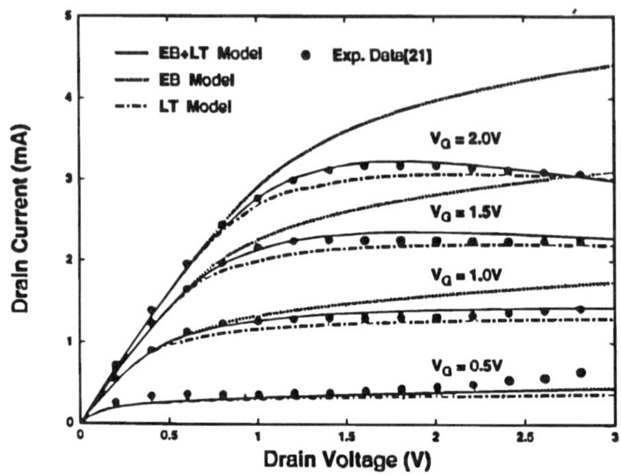

Fig. 2.10.1 A reduction of I_{ds} (and hence negative output conductance) in a SOI MOSFET is shown at high V_g as the drain voltage increases. After Chen et al. [2.120].

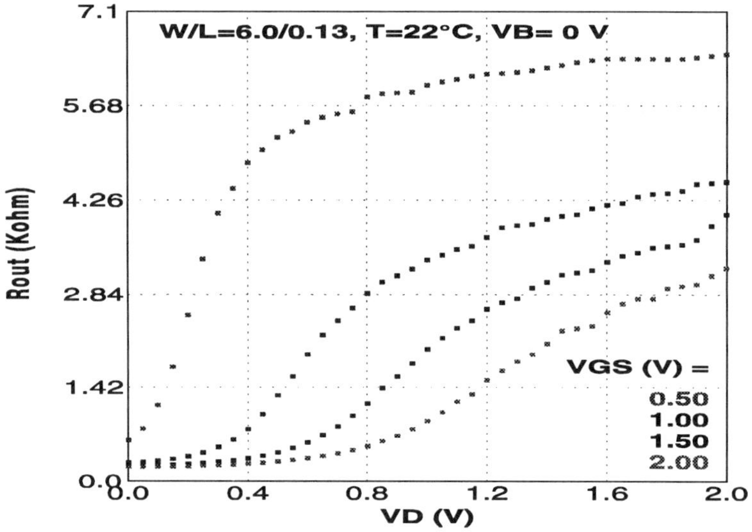

Fig. 2.10.2 The output resistance increases in this bulk device with channel length of 0.13μm (T_{ox}=4.5nm) as the drain bias increases.

Recently, some analytical *SHE* models have been developed for SOI MOSFETs by considering the influence of power dissipation and temperature rise on the electrical characteristics self-consistently [2.120, 2.121]. In general,

$$I_{ds} = \frac{I_{ds0}}{1 + \chi R_{th} I_{ds0} V_{ds}} \tag{2.10.2}$$

Here I_{dso} is the current at the ambient temperature ($T_c = T_o$). χ is a function of buried oxide thickness, gate oxide thickness, silicon film thickness, and the applied bias voltages [2.121].

2.11 Inversion Layer Quantization Effects

MOSFETs continue to be scaled to deep submicron channel lengths. For the gate to continue to control the channel effectively, the gate dielectric thickness needs to be reduced to the low nm level and the channel doping concentration needs to be increased towards 10^{18}cm^{-3}. This results in very high surface electric field so that the conduction energy band is split into discrete energy level of a 2-D electron gas leading to quite different inversion charge densities than that predicted by classical theory. For example, the quantum mechanical nature of the inversion layer is such that the charge distribution is not peaked at the interface as the classical theory predicts, but is displaced by a finite distance as shown in Fig 2.11.1. As a result, the threshold voltage is underestimated by the classical theory. In addition, the C_{ox} in the Q_{inv} given in Eq. (2.2.20) must be modified to include the finite inversion charge thickness. Therefore, the thickness of the inversion layer and hence its effect on the inversion charge density for a given V_g-V_{th} cannot be ignored and need to be accounted for in compact modeling.

Another important impact of the quantization effect on the MOSFET characteristics is the necessary modification in the extraction of the oxide thickness from the *C-V* measurements. Because the charge distribution is no longer peaked at the Si/SiO$_2$ interface, but at a distance Δd away, the effective oxide thickness inferred from *C-V* data is larger than the physical oxide thickness and is given by,

$$T_{ox,\,effective} = T_{ox,\,physical} + \frac{\varepsilon_{ox}}{\varepsilon_{si}} \Delta d \tag{2.11.1}$$

For the ultra-thin oxides used in some deep submicron devices, the second term in Eq. (2.11.1) can be significant and lead to large discrepancies between experimental *C-V* curves and the theoretical curves based only on the physical oxide thickness if the inversion layer quantization effects are ignored.

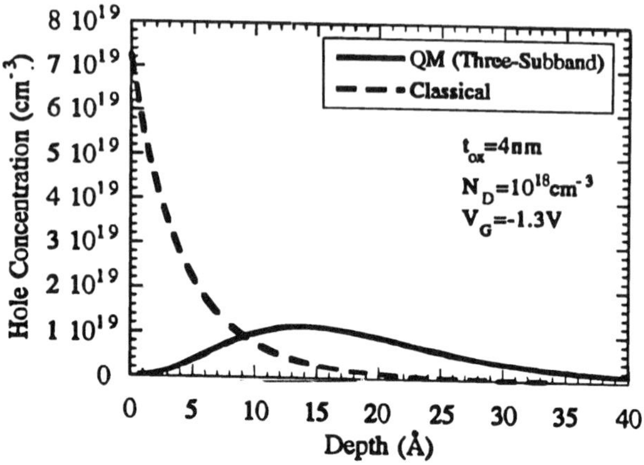

Fig. 2.11.1 Comparison of the charge distribution predicted classically and quantum mechanically in a MOSFET. After Hareland et al. [2.130].

Quantization of electrons and its impact on device behavior has been studied using the effective-mass approximation [2.122, 2.123, 2.124]. The effects of quantization phenomena on the threshold voltage have also been studied [2.125, 2.126]. A model for the influence of the quantization effects on the device characteristics has been developed by accounting for the finite thickness of the inversion layer in modeling the inversion charge and capacitance [2.127, 2.128]. Other models account for the quantization effects on V_{th} and surface potential with a correction for the intrinsic carrier density by using an effective band-gap widening [2.102, 2.129]:

$$ni^{QM} = ni^{CL} e^{-\frac{\Delta Eg}{2KT}} \tag{2.11.2}$$

$$\Delta Eg = \frac{13}{9}\beta\left(\frac{\varepsilon si}{4KT}\right)^{1/3} Eseff^{2/3} \tag{2.11.3}$$

where β is a constant, for example, β=4.1x10^{-8}eV-cm [2.129], n_i^{CL} is the classical intrinsic carrier density. E_{seff} is the effective surface electric field. Eq. (2.11.2) has been implemented in some device simulators to describe the quantization effects in MOSFETs.

References

[2.1] R. M. Swanson and J. D. Meindl, "Ion-implanted complementary MOS transistors in low-voltage circuits," *IEEE Journal of Solid-State Circuits*, vol. SC-7, pp. 146-153, 1972.

[2.2] J. Y. Chen, "CMOS - The emerging technology," *IEEE Circuit and Devices Magazine*, p. 16, 1986.

[2.3] O. J. McCarthy, *MOS Devices and Circuit Design*, John Wiley & Sons, New York, 1981.

[2.4] S. M. Sze, *Semiconductor Devices: Physics and Technology*, John Wiley & Sons, New York, 1985.

[2.5] R. S. Muller and T. I. Kamins, *Device Electronics for Integrated Circuits*, 2nd. Ed., John Wiley & Sons, New York, 1986.

[2.6] S. Wolf, *Silicon Processing for the VLSI Era, Volume 3- The Submicron MOSFET,* Lattice Press, Sunset Beach, CA, 1995.

[2.7] R. F. Pieret, *Field Effect Devices, Vol. IV Modular series on Solid-State Devices*, Addison-Wesley Publishing Co. Reading, MA, 1983.

[2.8] J. H. Huang et al., *BSIM3 Manual (Version 2.0),* University of California, Berkeley, March 1994.

[2.9] J. W. Slotboom, G. Streutker, G. J. T. Davids, and P. B. Hartog, "Surface impact ionization in silicon devices," *IEDM, Technical Digest,* pp. 494-497, 1987.

[2.10] W. Shochley, "Problems related to pn junction in silicon," *Solid-state Electronics*, vol. 2, pp. 35-67, 1961.

[2.11] F. C. Hsu, R. S. Muller, and C. Hu, "A simplified model of short channle MOSFET characteristics in the breakdown model," *IEEE Trans. on Electron Devices*, vol. ED-30, pp. 571-576, 1983.

[2.12] R. R. Troutman and S. N. Chakravarti, "Subthreshold characteristics of insulated-gate field-effect transistors," *IEEE Trans. on Circuit Theory*, vol. CT-20, pp. 659-665, 1973.

[2.13] Y. P. Tsividis, "Moderate inversion in MOS devices," *Solid-State Electronics*, vol. 25, pp. 1099-1104, 1982.

[2.14] Y. P. Tsividis, *Operation and Modeling of the MOS Transistor*, McGraw-Hill, New York, 1987.

[2.15] S. Odanaka et al., "A new half-micron p-channel MOSFET with efficient punchthough stops," *Symp. VLSI Tech. Digest of Tech Papers*, pp.62-63, 1985.

[2.16] M. Stinson et al., "Effects of ion implantation on deep-submicrometer drain-engineering MOSFET technologies," *IEEE Trans. on Electron Devices*, vol. ED-38, p. 487, 1991

[2.17] J. E. Chung et al., "Performance and reliability design issues of deep submicron MOSFETs," *IEEE Trans. on Electron Devices,* vol. ED-38, p.545, 1991.

[2.18] R. R. Troutman, "VLSI limitations from drain-induced barrier lowering," *IEEE Trans. on Electron Devices*, vol. ED-26, p.461, 1979.

[2.19] A. Hori et al., "A 0.1μm CMOS with a step channel profile formed by ultrahigh vacuum CVD and in situ doped ions," *IEDM Tech. Dig.*, pp. 909-911, 1993.

[2.20] K. Noda et al., "A 0.1 μm delta-doped MOSFET fabricated with post-low-energy implanting selective epitaxy," *IEEE Trans. on Electron Devices*, vol. ED-45, pp.809-814, 1998.

[2.21] Y. Okumura et al., "A novel source-to-drain non-uniformity doped channel (NUDC) MOSFET for high-current drivablity and threshold voltage controllability," in *IEEE IEDM Tech. Dig.*, 1990, p 391.

[2.22] T. Hori et al., "A 0.1um CMOS technology with tilt-implanted punchthrough stoper (TIPS)," in *IEEE IEDM Tech. Dig.* 1994, p. 75.

[2.23] T. Hori et al., "A new p-channel MOSFET with large tilt angle implanted punchthrough stopper (LATIPS)," IEEE Electron Device Letters, EDL-9, p. 641, 1988.

[2.24] C. S. Rafferty et al., "Explanation of reverse short channel effect by defect gradients," *IEDM Tech. Dig.* p.311, 1993.

[2.25] C. Duvvury, "A guide to short channel effects in MOSFETS," *IEEE Circuit and Systems Magazine*, p.6, 1986.

[2.26] P. P. Wang, "Device characteristics of short channel and narrow width MOSFETs," *IEEE Trans. on Electron Devices*, vol. ED-25, pp. 779-786, 1978.

[2.27] T. Y. Chan, J. Chen, P. K. Ko, and C. Hu, "Impact of gate-induced drain leakage current on device scaling," *IEDM Tech. Dig.*, pp. 718-721, 1987.

[2.28] K. Kurimoto, Y. Odake, and S. Odanka, "Drain leakeage current characteristics due to band-to-band tunneling in LDD MOS devices," *IEDM Tech. Dig.*, pp.621-624, 1989.

[2.29] L. D. Yau, "A simple theory to predict the threshold voltage of short channel IGFET's," *Solid-State Electronics*, vol. 17, pp.1059-1063, 1974.

[2.30] G. W. Talor, "The effects of two-dimensional charge sharing on the above threshold characteristics of short channel IGFETs," *Solid-State Electronics*, vol.22, pp. 701-717, 1979.

[2.31] C. R. Viswanathan, B. C. Burkey, G. Lubberts, and T. J. Tredwell, "Threshold voltage in short channel MOS devices," *IEEE Trans. on Electron Devices*, vol. ED-32, pp. 932-940, 1985.

[2.32] T. A. Fjeldly and M. Shur, "Threshold voltage modeling and the subthreshold regime of operation of short channel MOSFETs," *IEEE Trans. on Electron Devices*, vol. ED-40, pp. 137-145, 1993.

[2.33] N. D. Arora, "Semi-empirical model for the threshold voltage of a double implanted MOSFET and its temperature dependence," *Solid-State Electronics*, vol. 30, pp. 559-569, 1987.

[2.34] H. S. Lee, "An analysis of the threshold voltage for short-channel IGFET's," *Solid-State Electronics*, vol.16, p.1407, 1973.

[2.35] G. W. Taylor, "Subthreshold conduction in MOSFET's," *IEEE Trans. on Electron Devices*, vol. ED-25, p.337, 1978.

[2.36] Z. H. Liu et al., "Threshold voltage model for deep-submicron MOSFET's," *IEEE Trans. on Electron Devices*, vol. ED-40, pp.86-98, 1993.

[2.37] S. G. Chamberlain and S. Ramanan, "Drain-induced barrier-lowering analysis in VLSI MOSFET devices using two-dimensional numerical simulations," *IEEE Trans. on Electron Devices*, vol. ED-33, pp. 1745-1753, 1986.

[2.38] E. Eitan and D. Frohman-Bentchkowsky, "Surface conduction in short channel MOS devices as a limitation to VLSI scaling," *IEEE Trans. on Electron Devices*, vol. ED-29, pp. 254-266, 1982.

[2.39] S. C. Jain and P. Balk, "A unified analytical model for drain-induced barrier Lowering and drain-induced high electric field in a short-channel MOSFET," *Solid-State Electronics,* vol. 30, pp, 503-511, 1987.

[2.40] C. G. Sodini, S. S. Wong, and P. K. Ko, "A framework to evaluate technology and device design enhancements for MOS IC's," *IEEE J. Solid-State Circuits*, SC-24, pp. 118-12 , 1989.

[2.41] M. Nishida and H. Onodera, “An anomalous increase of threshold voltage with shortening of the channel lengths for deeply boron-implanted n-channel MOSFETs”, *IEEE Trans. Electron Devices*, ED-28, p1101, 1981.

[2.42] M. Orlowski, C. Mazure, and F. Lau, “Submicon short-channel effects due to gate re-oxidation induced lateral diffusion,” *IEDM Tech. Dig*. pp. 632-635, 1987.

[2.43] C. Y. Lu, and J. M. Sung, "Reverse short-channel effects on threshold voltage in submicron salicide devices," *IEEE Electron Device letters*. EDL-10, p.446, 1989.

[2.44] H. Jacobs et al., "MOSFET reverse short-channel effect due to silicon interstitial capture in gate oxide," *IEDM Tech. Dig*. p. 307, 1993.

[2.45] L. A. Akers, and J. J. Sanchez, "Threshold voltage models of short, narrow, and small geometry MOSFET," *IEEE Trans. Electron Devices,* vol. ED-29, pp. 776-778, 1982.

[2.46] S. T. Hsu et al., "Physical mechanism of the reverse short channel effect in MOS transistors," *Solid-state Electronics*, vol.34, pp.605-608, 1991.

[2.47] H. Hanafi et al., "A model for anomalous short channel behavior in MOSFETs," *IEEE Electron Device letters,* EDL-14, p.575, 1993.

[2.48] D. Sadana et al., "Enhanced short channel effects in NMOSFETs due to boron redistribution induced by arsenic source and drain implant," *IEDM Tech. Dig.,* p.849, 1992.

[2.49] Y. Cheng, T. Sugii, K. Chen, and C. Hu, “Modeling of small size MOSFETs with reverse short channel and narrow width effects for circuit simulation”, *Solid State Electronics*, vol. 41, (9), pp. 1227-1231, 1997.

[2.50] T. Kunikiyo et al., "Reverse short channel effects due to lateral diffusion of point-defect induced by source/drain ion implantation," *IEEE Trans. on Computer-aided Design of Integrated Circuits and Systems*, vol. 13, pp. 507-514, 1994.

[2.51] N. D. Arora and M. S. Sharma, "Modeling the anomolous threshold voltage behavior of submicrometer MOSFETs," *IEEE Electron Device letters,* EDL-13, p.92, 1992.

[2.52] C. R. Ji and C. T. Sah, "Analysis of the narrow gate effect in submicrometer MOSFET's," *IEEE Trans. Electron Devices,* vol. Ed-30, pp.1672-1677, 1983.

[2.53] G. Merckel, "A simple model of the threshold voltage of short and narrow channel IGFETs," *Solid-State Electronics,* vol.23, pp.1207-1213, 1980.

[2.54] S. S. Chuang and C. T. Sah, "Threshold voltage models of the narrow-gate effect in micron and submicron MOSFETs," *Solid-State Electronics*, vol. 31, pp. 1009-1021, 1988.

[2.55] L. A. Akers, "An analytical expression for the threshold voltage of a small geometry MOSFET," *Solid-State Electronics*, vol. 24, pp.621-627, 1981.

[2.56] K. K. Hsueh, J. J. Sanchez, T. A. Demassa, and L. A. Akers, "Inverse-narrow-width effects and small geometry MOSFET threshold voltage model," *IEEE Trans. Electron Devices*, ED-35, pp. 325-338, 1988.

[2.57] S. S. Chung and T. C. Li, "An analytical threshold voltage model of trench isolated MOS devices with nonuniformly doped substrate," *IEEE Trans. Electron Devices*, vol. 39, no.3, pp.614-622, 1992.

[2.58] B. J. Sheu et al., *Compact short channel IGFET model (CSIM)*, Electronics Research Laboratory Memo No. UCB/ERL-M84/20, University of California, Berkeley, 1984.

[2.59] M. S. Shur, *Physics of Semiconductor Devices*, Prentice Hall, New Jersey, 1990.

[2.60] E. H. Nicollian and J. R. Brews, *MOS Physics and Technology*, Wiely-Interscience, New York, 1982.

[2.61] Y. Cheng et al., *BSIM3 version 3.1 User's Manual*, University of California, Berkeley, Memorandum No. UCB/ERL M97/2, 1997.

[2.62] C. G. Sodini, P. K. Ko, and J. L. Moll, "The Effects of high fields on MOS device and circuit performance," *IEEE Trans. Electron Devices*, ED-31, p. 1386, 1984.

[2.63] M. S. Liang et al., "Inversion layer capacitance and mobility of very thin gate oxide MOSFETs," *IEEE Trans. Electron Devices*, ED-33, p. 409, 1986.

[2.64] K. Lee et al., "Physical understanding of low field carrier mobility in silicon inversion layer," *IEEE Trans. Electron Devices*, ED-38, p. 1905, 1991

[2.65] E. A. Talkhan, I. R. Mansour, and A. I. Baroor, "Investigation of the effect of drift-filed-dependent mobility on MOSFET characteristics," *Part I and II, IEEE Trans. Electron Devices*, ED-19, p. 899, 1972

[2.66] S. Takagi, A. Toriumi, M. Iwase, and H. Tango, "On the universality of inversion layer mobility in Si MOSFET's: part I - effects of substrate impurity concentration", *IEEE Trans. Electron Devices*, Vol. ED-41, p. 2357, 1994.

[2.67] K. Chen et al., "MOSFET carrier mobility model based on gate oxide thickness, threshold and gate voltages", *Solid-State Electronics*, pp.1515-1518, Vol. 39, No. 10, October 1996.

[2.68] A.G. Sabnis and J.T. Clemens, "Characterization of electron velocity in the inverted <100> Si surface," *IEDM Tech. Dig.*, pp. 18-21, 1979.

[2.69] G.S. Gildenblat, *VLSI Electronics: Microstructure Science*, vol. 18, p. 11, 1989.

[2.70] K.Y. Toh, P.K. Ko, and R.G. Meyer, "An engineering model for short-channel MOS devices," *IEEE J. Solid-State Circuits,* vol. 23, No. 4, 1988.

[2.71] K. Lee, M. Shur, T. A. Fjeldly, and T. Ytterdal, *Semiconductor Device Modeling for VLSI*, Prentice Hall, Englewood Cliffs, New York, 1993

[2.72] N. Arora, *MOSFET Models for VLSI Circuit Simulation*, Springer-Verlag, Wien NewYork, 1994

[2.73] E. A. Talkhan, I. R. Manour and A. I. Barboor, "Investigation of the effect of drift-field-dependent mobility on MOSFET characteristics," Parts I and II. *IEEE Trans. on Electron Devices*, ED-19(8), 899-916, 1972.

[2.74] R. Coen and R.S. Muller, "Velocity of surface carriers in inversion layers on silicon," *Solid-State Electron*, 23, 35-40, 1980.

[2.75] J. A. Cooper and D.F. Nelson, "Measurement of the high-field drift velocity of electron in inversion layer in silicon," *IEEE Electron Device Letters*, EDL-2(7), 169-173, 1983.

[2.76] F. Assaderaghi, J. Chen, P. K. Ko, and C. Hu, "Measurement of electron and hole saturation velocity in silicon inversion layers using SOI MOSFETs", *IEEE International SOI Conference*, p112, 1992.

[2.77] D. M. Caugley and R. E. Thomas, *Proc. IEEE*, vol. 55, pp. 2192-2193, 1967.

[2.78] B. J. Moon, C. K. Park, K. Lee, and M. Shur, "New short channel n-MOSFET current-voltage model in strong inversion and unified parameter extraction technique," *IEEE Trans. on Electron Devices*, ED-38, No. 3, pp. 592-602, 1991

[2.79] B. J. Moon, C. K. Park, K, Rho, K. Lee, M. Shur, and T. A. Fjeldly, "Analytical model for p-channel MOSFETs," *IEEE Trans. on Electron Devices*, ED-38, pp. 2632-2646, 1991

[2.80] F. Fang and X. Fowler, "Hot-electron effects and saturation velocity in silicon inversion layer," *J. Appl. Phys.*, 41, 1825, 1969.

[2.81] P. Gray and R. Meyer, *Analysis and Design of Analog Integrated Circuits*, 3ed edition, New York: John Wiley & Sons, 1993.

[2.82] D. Frohman-Bentchknowsky and A. S. Grove, "Conductance of MOS transistors in saturation," *IEEE Trans. on Electron Devices*, vol.ED-16, pp. 108-113, 1969.

[2.83] P. K. Ko, "Approaches to scaling," *Chap. 1, in advanced MOS device Physics*, N. G. Einspruch and G. Gildenblatt, Eds., Vol. 18 *VLSI Electronics*. Academic Press 1989.

[2.84] Y. A. El-Mansy and A. R. Boothroyd, "A simple two dimensional model for IGFET." *IEEE Trans. Electron Devices*, ED-24, pp. 254-262, 1977.

[2.85] M. E. Banna and M. E. Nokali, “A pseudo-two-dimensional analysis of short channel MOSFETs,” *Solid-state Electronics*, Vol. 31 pp.269-274, 1988.

[2.86] J. H. Huang et al., “A physical model for MOSFET output resistance”, *IEDM, Technical Digest*, Dec. of 1992.

[2.87] Y. Cheng et al., “A physical and scalable BSIM3v3 I-V model for analog/ digital circuit simulation”, *IEEE Trans. Electron Devices*, Vol. 44, pp.277-287, Feb. 1997.

[2.88] J. Chen et al., "Subbreakdown drain leakage current in MOSFETs," *IEEE Electron device Letters*, vol. EDL-8, pp. 515-517, 1987.

[2.89] C. Hu et al., “Hot-electron-induced MOSFET degradation - model, monitor, and improvement,” *IEEE Trans. Electron Devices*, ED-32, p375, 1985.

[2.90] P. K. Ko, "Hot carrier effects in MOSFETs," Ph. D dissertation, Dept. of Electrical Engineering and Computer Sciences, University of California, Berkeley, 1982.

[2.91] K. R. Mistry et al., “Circuit design guidelines for n-channel MOSFET hot carrier robustness,” *IEEE Trans. Electron Devices*, ED-40, p. 1284, 1993.

[2.92] E. Takeda, "Hot carrier effects in submicrometer MOS VLSI," *IEE proceedings*, vol. 131, Pt. I, pp. 153-164, 1984.

[2.93] C. Hu, "Hot carrier Effects," in *Advanced MOS Devices Physics, VLSI Electronics*, vol. 18, pp.119-139, Academic Press Inc., New York, 1989.

[2.94] M. Tanizawa et al., “A complete substrate current model including band-to-band tunneling current for circuit simulation,” *IEEE Trans. on Computer-aided Design of Integrated Circuits and Systems*, vol. 12, no. 11, pp. 1749-1757, 1993.

[2.95] I. C. Chen et al., “Interface-trap enhanced gate-induced leakage current in MOSFET,” *IEEE Electron device Letters*, EDL-10, p216, 1989

[2.96] M. Koyanagi, et al., "Hot electron induced punchthrough (HEIP) effect in submicron PMOSFETs," *IEEE Trans. Electron Devices*, vol. ED-34, pp. 839-844, 1987.

[2.97] K. K. Ng, and G. W. Taylor, "Effects of hot carrier trapping in n- and p-channel MOSFETs," *IEEE Trans. Electron Devices*, vol. ED-30, pp. 871-876, 1983.

[2.98] P. M. Lee, M. M Kuo, P. K. Ko, and C. Hu, *BERT - A circuit aging simulator*, Memo. No. UCB/ERL M90/2, Electron Research Lab, University of California, Berkeley, 1990.

[2.99] T. S. Hobol, and L. A. Glasser, “Relic: A reliability simulator for integrated circuits”, *IEEE Proc. Int. Conf. Computer-Aided Design*, pp517-520, 1986.

[2.100] N. D. Arora, and M. Sharama, “MOSFET substrate current model for circuit simulation,” *IEEE Trans. Electron Devices*, ED-38, pp. 1392-1398, 1991.

[2.101] R. Rios, et al., “An analytical polysilicon depletion effect model for MOSFETs,” *IEEE Electron Device Letters*, Vol. 15, No. 4, pp. 129-131, 1994.

[2.102] Y. Cheng et al., "ICM--An analytical Inversion charge model for accurate modeling of thin gate oxide MOSFETs," *1997 International Conference on Simulation of Semiconductor Processes and Devices*, Sept. 1997, Boston.

[2.103] F. Assaderaghi, et al., "High-Field Transport of inversion layer electrons and holes including velocity overshoot," *IEEE Trans. Electron Devices*, vol. ED-44, p. 664, 1997.

[2.104] J. B. Roladan et al., "Modeling effects of electron-velocity overshoot in a MOSFET," *IEEE Trans. Electron Devices*, vol. ED-44, p. 841, 1997.

[2.105] G. Sai-Halasz et al., "High transconductance and velocity overshoot in NMOS devices at the 0.1-um gate-length level," *IEEE Electron Device Letters*, Vol. 9, pp. 464-466, 1988.

[2.106] G. G. Shahidi, D. A. Antoniadis, H. I, Smith, "Electron velocity overshoot at room and liquid nitrogen temperature in silicon inversion layers," *IEEE Electron Device letters*, Vol. 9, pp. 94-96, 1988.

[2.107] M. Fischetti, and S. Laux, "Monto Carlo analysis of electron transport in small semiconductor devices including band-structure and space-charge effects," *Phys. Rev. B*, Vol. 38, pp. 9721-9745, 1988.

[2.108] Baccarani and M. Wordeman, "An investigation of stead-state velocity overshoot in slilicon," *Solid-state Electronics*, Vol. 28, pp.407-416, 1985.

[2.109] D. Chen et al., "An analytical formulation of the length coefficient for the augment drift-diffusion model including velocity overshoot," *IEEE Trans. Electron Devices*, vol.38 no. 6, 1991.

[2.110] J. H, Sim, "An analytical deep submicron MOS device model considering velocity overshoot behavior using energy balance equation," *IEEE Trans. Electron Devices*, Vol. 42, pp. 864-869, 1995.

[2.111] J. B. Roladan, F. Gamiz, J. A. Lopez,-Villanueva, and J. E. Carceller, "Modeling effects of electron-velocity overshoot in a MOSFET," *IEEE Trans. Electron Devices*, vol. ED-44, p. 841, 1997.

[2.112] P. A. Blakey and K. Joardar, "An analytic theory of the impact of velocity overshoot on the drain characteristics of field-effect transistors," *IEEE Trans. Electron Devices*, vol. ED-39, pp. 740-742, 1992.

[2.113] F. Assaderaghi et al., "High-field transport of inversion layer electrons and holes including velocity overshoot," *IEEE Trans. Electron Devices*, vol. ED-44, p. 664, 1997.

[2.114] K. E. Goodson and M. I. Flik, "Effects of microscale thermal conduction on packing limit of silicon-on-insulator electronic devices," *IEEE Trans. Comp. Hybrids and Manufacturing Tech.*,Vol.15, 715-722, 1992.

[2.115] L. J. McDaid et al., "Physical origin of negative differential resistance in SOI transistors", *Electron. Lett.*, Vol.25, 827-828, 1989.

[2.116] D. Takacs and J. Trager, "Temperature increase by self-heating in VLSI CMOS," *ESSDERC* 1987, Bologna, pp. 59-62.

[2.117] M. Koyanagi et al., "Coupled monte carlo-energy transport simulation with quasi-three-dimensional temperature analysis for SOI MOSFET," *IEEE Trans. Electron Devices*, TED 39, p.2640, 1992.

[2.118] N. Yasuda et al., "Analytical device model of SOI MOSFET's including self-heating," *Japan. J. Appl. Phys.*, Vol.30, pp.3677-3684, 1991.

[2.119] L. T. Su et al., "SPICE model and parameters for fully-depleted SOI MOSFET's including self-heating," *IEEE Electron Devices Letters*, Vol. 15, pp. 374-376, 1994.

[2.120] Y. Chen et al., "An analytical drain current model considering both electron and lattice temperature simultaneously for deep submicron ultrathin SOI NMOS devices with self-heating," *IEEE Trans. Electron Devices*, Vol. 42, pp. 899-906, 1995.

[2.121] Y. Cheng and T. A. Fjeldly, "Unified physical *I-V* model including self-heating effect for fully depleted SOI/MOSFET's," *IEEE Trans. Electron Devices*, vol. 43, pp. 1291-1296, 1996.

[2.122] G. Chindalore et al., "Experimental determination of threshold voltage shifts due to quantum mechanical effects in MOS electron and hole inversion layers", *IEEE electron device letters*, vol. 18, no. 5, 1997.

[2.123] M. J. Van Dort, P. H. Woerlee, and A. J. Walker, "A simple model for quantization effects in heavily-doped silicon MOSFET's at inversion conditions," *Solid-state Electronics*, Vol.37, p.411 1994.

[2.124] M. J. Van Dort, P. H. Woerlee, A. J. Walker, C. A. H. Juffermans, and H. Lifka, "Influence of high substrate doping levels on the threshold voltage and the mobility of deep submicrometer MOSFET's," *IEEE Trans. Electron Devices*, Vol. 39, p.932, 1992

[2.125] Y. Ohkura, "Quantum effects in Si n-MOS inversion layer at high substrate concentration," *Solid-state Electronics*, Vol. 33, p. 1581, 1990

[2.126] S. Jallepalli et al., "Effects of quantization on the electrical characteristics of deep submicron p- and n-MOSFET's," *Symp. VLSI Technology*, p. 138, 1996.

[2.127] Y. King et al., "AC charge centroid model for quantization of inversion layer in NMOSFET," *Int. Symp. VLSI Technology, Systems and Applications, Proc. of Tech. Papers,* Taipei, Taiwan, pp. 245-249, June 1997.

[2.128] W. Liu et al., *An accurate MOSFET intrinsic capacitance model considering quantum mechanic effect for BSIM3v3.2,* Memorandum no. UCB/ERL M98/47, University of California, Berkeley, 1998.

[2.129] R. Rios et al., "A physical compact MOSFET, including quantum mechanical effects, for statistical circuit design applications," *IEDM Tech. Dig.*, pp. 937-940, 1995.

[2.130] S. A. Hareland et al., "A Physically-based model for quantization effects in hole inversion layers," *IEEE Trans. on Electron Devices*, vol. 45, no.1, pp.179-182, 1998.

CHAPTER 3

Threshold Voltage Model

Accurate modeling of the threshold voltage (V_{th}) is one of the most important requirements for the precise description of a device's electrical characteristics. By using the threshold voltage, the device operation can be divided into three operational regions. If the gate voltage is much larger than V_{th}, the MOSFETs is operating in the strong inversion region and the drift current is dominant. If the gate voltage is much less than V_{th}, the MOSFET operates in the weak inversion (or subthreshold) region and diffusion current is dominant. If the gate voltage is very close to V_{th}, the MOSFET operates in the transition region called moderate inversion where both diffusion and drift currents are important.

In this chapter, we will discuss in detail the modeling of the threshold voltage, including the short channel and narrow width effects. We then introduce the threshold voltage model of BSIM3v3. Finally we will discuss some topics, which may cause confusion for BSIM3 users, in the Helpful Hints section.

3.1 Threshold Voltage Model for Long Channel Devices

The schematic diagram of a long channel MOSFET with uniformly doped substrate is shown in Fig. 3.1.1. The x direction is the depth into the silicon measured from the SiO_2-Si interface while the y direction is the length along the channel measured from the source towards the drain.

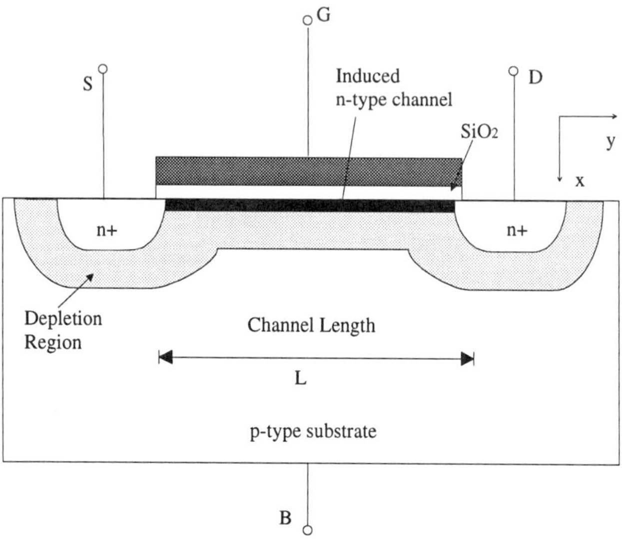

Fig. 3.1.1 A cross-sectional view of an n-channel MOSFET.

When a gate voltage is applied to induce the inversion layer in the channel, drain current flows if a drain bias V_{ds} is applied. The two dimensional Poisson equation is [3.1]

$$\frac{\partial\phi}{\partial x^2} + \frac{\partial\phi}{\partial y^2} = -\frac{\rho(x,y)}{\varepsilon_{si}} \tag{3.1.1}$$

where ϕ is the potential, and ρ(x,y) is the charge density.

For a long channel device with low drain bias, we may use the well known gradual channel approximation (*GCA*) [3.2], which assumes that the electric field gradient in the *y* direction $\partial E_y/\partial y$ is much smaller than the field gradient in the x direction $\partial E_x/\partial x$. In that case, Eq. (3.1.1) can be reduced to:

$$\frac{\partial\phi}{\partial x^2} = -\frac{\rho(x)}{\varepsilon_{si}} \tag{3.1.2}$$

The *GCA* is generally valid when the channel length is much longer than the width of the depletion region and V_{ds} is zero. However, at high drain voltages (saturation region), the *GCA* becomes invalid even for long channel devices, and a quasi-two-dimensional Poisson equation must be solved [3.3, 3.4].

When a gate bias V_g is applied, the relationship between V_{gs} and the surface potential and the charge density induced in the substrate can be written as [3.5]:

$$V_{gs} = V_{FB} + \phi_s - \frac{Q_s}{C_{ox}} \tag{3.1.3}$$

where V_{FB} is the flat band voltage and $Q_s = Q_{inv} + Q_b$. Here Q_{inv} is the inversion charge.

According to the common definition of threshold, i.e. $\phi_s = 2\phi_B$ [3.6], and neglecting Q_{inv} in Q_s since Q_{inv} is small compared with Q_b at the onset point of strong inversion, the threshold voltage can be expressed as

$$V_{th} = V_{FB} + 2\phi_B - \frac{Q_b}{C_{ox}} \tag{3.1.4}$$

where Q_b is the depletion charge at the onset of strong inversion, and can be found by replacing ϕ_s with $2\phi_B - V_{bs}$ (when $V_{bs} \neq 0$) in Eq. (2.3.12). Thus Eq. (3.1.4) becomes,

$$V_{th} = V_{FB} + 2\phi_B - \gamma\sqrt{2\phi_B - V_{bs}} \tag{3.1.5}$$

where γ is the body effect coefficient, and is given by Eq. (2.2.2).

3.2 Threshold Voltage Model with Short Channel Effects

The threshold voltage of a long channel device is independent of the channel length and the drain voltage. Its dependence on the body bias is given by Eq. (3.1.5). However, if the channel length is sufficiently short the measured threshold voltage shows a significant dependence on the channel length and the drain voltage. Also, the dependence of the threshold voltage on the body bias becomes weaker as channel length becomes shorter. These short-channel effects must be included in the threshold voltage expression in order to model submicrometer devices correctly. The long channel theory is based on the one-dimensional theory, that is, it assumes that the space charge under the gate is controlled only by the vertical electric field. However, when the channel length is short (close to the dimension of the depletion region width) this assumption becomes invalid, and the influence of the built-in potential of source/drain and the drain bias has to be accounted for in modeling V_{th}. Two analytical approaches have been used to model the threshold voltage reduction, ΔV_{th}, due to short channel effects:

1). Charge-sharing model, in which analytical expressions for ΔV_{th} are derived with the help of a charge sharing concept [3.7].

2). Quasi 2D model, in which analytical expressions for ΔV_{th} are derived by solving a quasi two-dimensional (2-D) Poisson equation [3.8].

3.2.1 Charge sharing model

In developing the V_{th} model for long channel MOSFETs, an assumption was made that all the space charges under the gate contribute to the vertical electric field, E_x, which determines the voltage across the oxide and V_{th}. When the channel length of a MOSFET is long, this is a reasonable assumption because the influence of the drain and source is small, as shown in Fig. 3.2.1 (a). However, as L approaches the dimensions of the depths of source/drain junctions or the thickness of the depletion region, the space charges in the source/drain junction regions begin to contribute greatly to the formation of the channel depletion region. That is, not all depletion charges in the channel region contribute to E_{ox} or V_{th}. In other words, the charge sharing concept assumes that not all of the electric-field lines emanating from the charge under the gate terminate on the gate charge. Instead, some terminate on the space charge in the source and drain depletion regions. The depletion charge under the gate is actually induced by gate together with the source and drain so that the channel charge can be considered to be "shared" by the gate, source and drain. Thus, less gate charge density (and smaller gate voltage) is needed to induce inversion in short channel MOSFETs than in long channel devices. This means a short channel MOSFET has a smaller V_{th}. Fig. 3.2.1 (b) illustrates the case when V_{ds}=0. As the bias applied to the drain junction is increased, the width of the drain depletion region increases, which makes the portion of the charge shared with the drain increase as shown in Fig. 3.2.2, resulting in an even lower V_{th} [2.31].

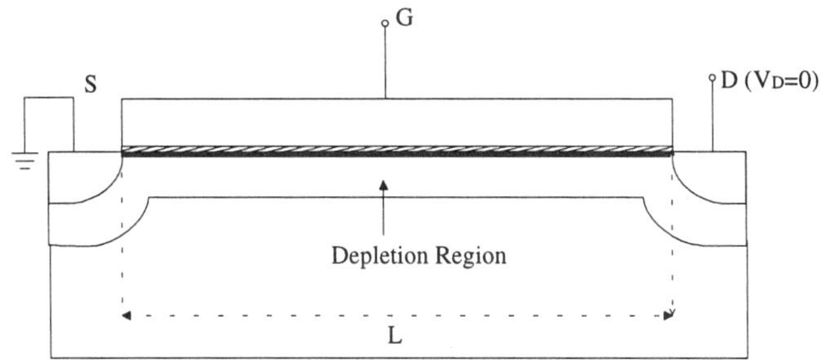

Fig. 3.2.1 (a) Illustration of the depletion region in a long channel MOSFET.

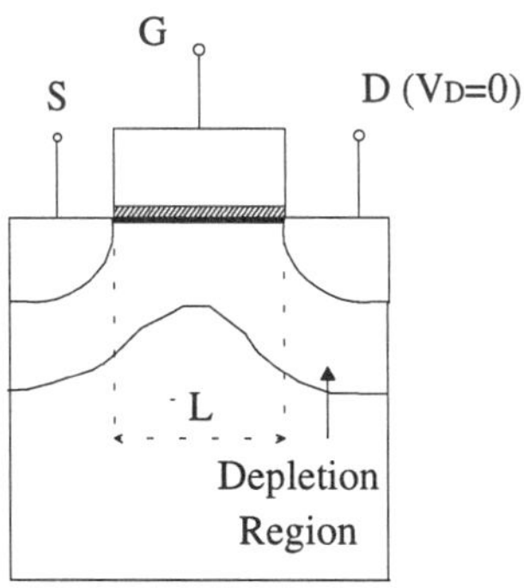

Fig. 3.2.1 (b) The influence of the source/drain regions becomes significant in a short channel MOSFET.

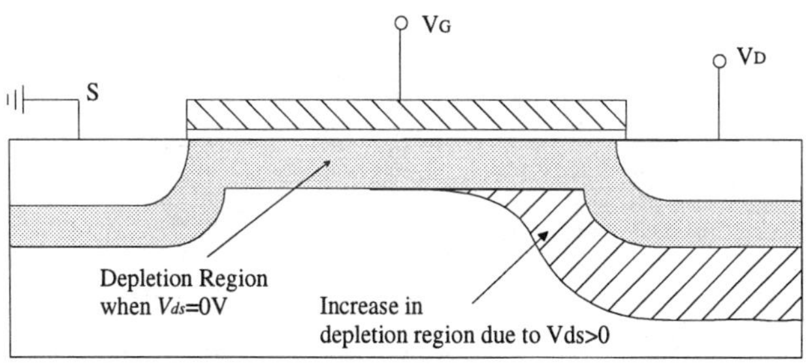

Fig. 3.2.2 The width of the depletion layer is increased when the drain bias is not zero.

Many charge sharing models have been developed using different geometric division approaches [3.7, 3.9, 3.10]. The simplest and most widely used is the one proposed by Yau in 1974 [3.7]. As shown in Fig. 3.2.3, the gate controlled depletion charge is in a trapezoidal area of depth X_{depm}, length L at the surface, and the length L' at the bottom of the depletion region. Therefore, the gate controlled depletion charge can be given by

$$Q_b' = qN_{sub}X_{depm}\left(\frac{L+L'}{2L}\right) \tag{3.2.1}$$

By expressing L' as a function of the source/drain junction depth X_j and X_{depm}, and substituting Q_b in the long channel V_{th} expression, Eq. (3.1.10), with Q_b' given in Eq. (3.2.1), the reduction of V_{th} caused by the short channel effect can be written as,

$$\Delta V_{th} = \frac{qN_{sub}X_{depm}X_j}{C_{ox}L}\left(\sqrt{1+\frac{2X_{depm}}{X_j}}-1\right) \tag{3.2.2}$$

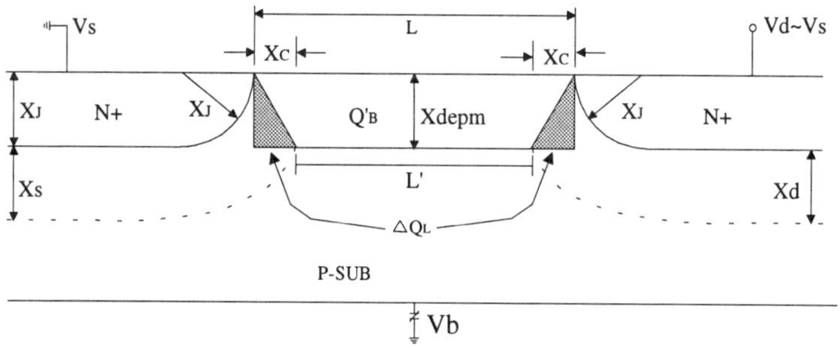

Fig. 3.2.3 Illustration of the charge sharing model. The shaded regions are the depletion charges controlled by the source/drain. After Yau [3.7].

In the above derivation, the following assumptions have been made,

1. The substrate doping N_{sub} is uniform.
2. No potential difference between source and drain, that is V_{ds} =0.
3. The source and drain junctions with depth X_j are cylindrical in shape with radius X_j.
4. The channel depletion width X_{depm} is equal to the source/drain depletion widths.

By modifying Eq. (3.2.2), the above model can also be used to estimate the reduction of V_{th} in the presence of V_{ds},

$$\Delta V_{th} = \frac{qN_{sub}X_{depm}X_j}{2C_{ox}L}[(\sqrt{1+\frac{2X_s}{X_j}}-1)+(\sqrt{1+\frac{2X_d}{X_j}}-1)] \tag{3.2.3}$$

where X_s and X_d are the source and drain depletion widths.

The above simple model qualitatively agrees with observations from experiments. For example, it predicts that (1) the thicker the gate oxide, the larger the ΔV_{th} (the larger the short channel effects); (2) the lower the N_{sub}, the larger the X_{depm} and hence the larger the short channel effects; (3) the larger the X_j, the larger the short channel effects; (4) the larger the V_{ds}, the larger X_d and hence the larger the short channel effects. However, charge sharing models fail to achieve good quantitative agreement with the measured data from devices with shorter channel length (e.g.<1μm) or under large drain voltages [3.11]. This is because the geometric division of the depletion charge is quite arbitrary and has no physical quantitative basis.

Despite the fact that charge sharing models generally cannot provide accuracy in predicting the measured data, they are still useful for providing first order estimates of ΔV_{th} as well as for helping to visualize the short channel effects.

3.2.2 Quasi 2-D models for drain induced barrier lowering effect

The decrease in V_{th} due to the decrease in channel length L and the increase in V_{ds} has been modeled using another approach. As L decreases and V_{ds} increases, the potential profile between S and D is modified and there is a lowering of the potential barrier between S and D. Hence, less gate voltage is needed to bring the surface potential to $2\phi_B$, i.e. the threshold voltage is lowered. This physical description is called drain induced barrier lowering (*DIBL*) [3.12, 3.13, 3.14].

The concept of *DIBL* can be illustrated in Fig. 3.2.4, showing the surface potential plots along the channel for three different (long and short channel) devices [3.12]. The potential, i.e. surface potential, of the long channel device (curve A) is determined by V_g and independent of L or V_{ds}. As the channel length is reduced, the peak surface potential is lower than the long channel case (curve B). Less V_g, i.e. V_{th}, is required to bring the surface potential to $2\phi_B$ and to allow the current to flow. If drain bias V_{ds} increases, the potential peak is further reduced (curve C), resulting in a further decrease in V_{th}.

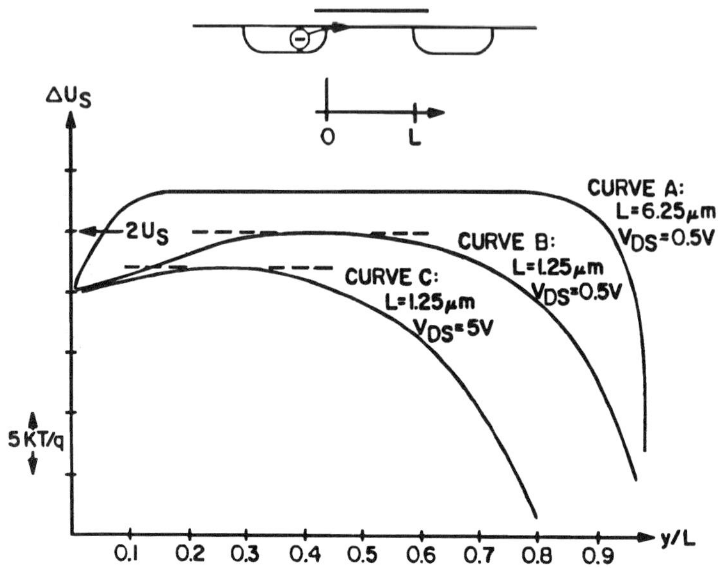

Fig. 3.2.4 Surface potential distribution from the source to the drain. The peak potential decreases with shrinking L and increasing V_{ds}. After Troutman [3.12].

Two-dimensional device simulation has shown that the following factors can influence *DIBL* [3.12, 3.15]:

a. The channel length L: shorter L leads to stronger *DIBL* effects.

b. The gate oxide thickness T_{ox}: thicker T_{ox} leads to stronger *DIBL* effects.

c. The source/drain junction depth X_j: deeper X_j leads to stronger *DIBL* effects.

d. The channel concentration N_{ch}: lower N_{ch} leads to stronger *DIBL* effects.

e. The body bias V_{bs}: higher V_{bs} leads to stronger *DIBL* effects.

According to the above, the reduction of V_{th} in short channel devices can be attributed to the penetration of the junction electric field into the channel region, resulting in a lowering of the potential barrier at the source end. Many

DIBL-based models have been developed [3.13, 3.14, 3.15]. They are mathematically more complex but more accurate than the charge sharing model.

By using a quasi 2-D approach to solve the Poisson equation in the channel of the MOSFET, analytical models for the threshold voltage can be derived [3.13, 3.15]. It has been shown that the threshold voltage developed with this approach can predict the short channel effects accurately [3.13, 3.14, 3.15].

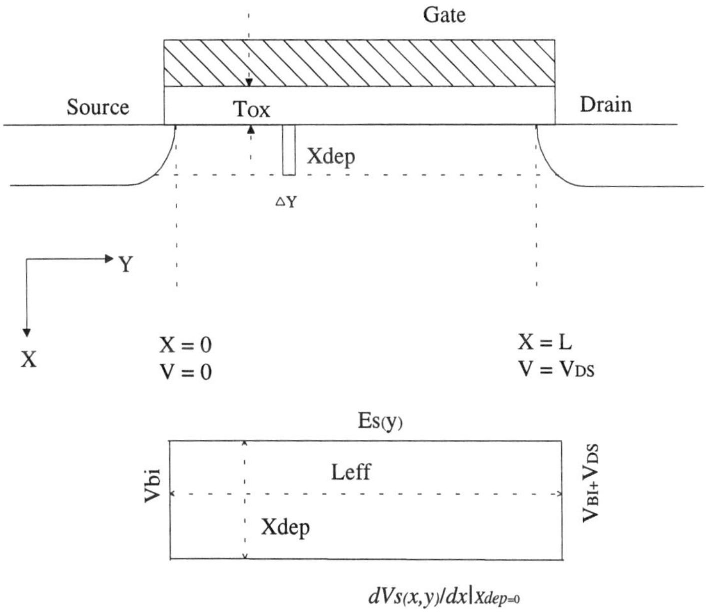

Fig. 3.2.5 The Gaussian box used in the quasi-two-dimensional analysis. After Liu et al. [3.15].

Applying Gauss's law to a rectangular box of height X_{dep} and length Δy in the channel depletion region and neglecting mobile carrier charge, as shown in Fig. 3.2.5 [3.15]:

$$\varepsilon_{si}\frac{X_{dep}}{\eta}\frac{dE_s(y)}{dy}+\varepsilon_{ox}\frac{V_{gs}-V_{FB}-V_s(y)}{T_{ox}}=qN_{sub}X_{dep} \qquad (3.2.4)$$

where E_s(y) is the lateral surface electric field, V_s(y) is the channel potential at the Si-SiO_2 interface, V_{gs} is the gate-source voltage, V_{FB} is the flat-band voltage, N_{sub} is the channel doping concentration, T_{ox} is the gate oxide thickness,

ε_{si} and ε_{ox} are the permittivity of SiO_2 and Si, respectively, and η is a fitting parameter.

The first term on the left hand side of Eq. (3.2.4) is equal to the net electric flux entering the Gaussian box along the y direction. The second term represents the electric flux entering the top surface of the Gaussion box. The solution to Eq. (3.2.4) under the boundary conditions of $V_s(0)=V_{bi}$, and $V_s(L)= V_{ds} + V_{bi}$ (where V_{bi} is the built-in potential between the source-substrate and drain-substrate junctions) is:

$$V_s(y) = V_{sL} + (V_{bi} + V_{ds} - V_{sL})\frac{\sinh(y/l)}{\sinh(L/l)} + (V_{bi} - V_{sL})\frac{\sinh[((L-y)/l]}{\sinh(L/l)} \quad (3.2.5)$$

V_{sL} represents the long channel surface potential, and l is a characteristic length defined as

$$l = \sqrt{\frac{\varepsilon_{si} T_{ox} X_{dep}}{\varepsilon_{ox}\eta}} \quad (3.2.6)$$

X_{dep} is the width of the channel depletion region.

For a given set of V_{gs}, V_{bs} and two V_{ds} values, the channel potential distributions Eq. (3.2.5) for several channel lengths are given in Fig. 3.2.6. This figure shows a large variation in potential along the channel for devices with short channel lengths even when the drain voltage is low. It agrees with the results of 2-D numerical simulation [3.15].

The channel potential V_s(y) has a minimum value, V_{smin}, at y_0 which can be found by solving the equation $dV_s(y)/dy=0$. However, y_0 can be approximated as $L/2$ when $V_{ds} << V_{bi} - V_{sL}$. Thus, V_{smin} can be obtained analytically,

$$V_{s\min} = V_{sL} + [2(V_{bi} - V_{sL}) + V_{ds}]\frac{\sinh(\frac{L}{2l})}{\sinh(L/l)} \quad (3.2.7)$$

The threshold voltage can be defined as the gate voltage which causes V_{smin} to equal $2\phi_B$. Thus, V_{th} can be solved as

$$V_{th}(L) = V_{th0} - \frac{2(V_{bi} - \phi_s) + V_{ds}}{2\cosh(\frac{L}{2l}) - 2} \tag{3.2.8}$$

When $l << L$, the threshold voltage shift ΔV_{th} can be expressed as [3.15]

$$\Delta V_{th}(L) = [2(V_{bi} - 2\phi_B) + V_{ds}](e^{-L/2l} + 2e^{-L/l}) \tag{3.2.9}$$

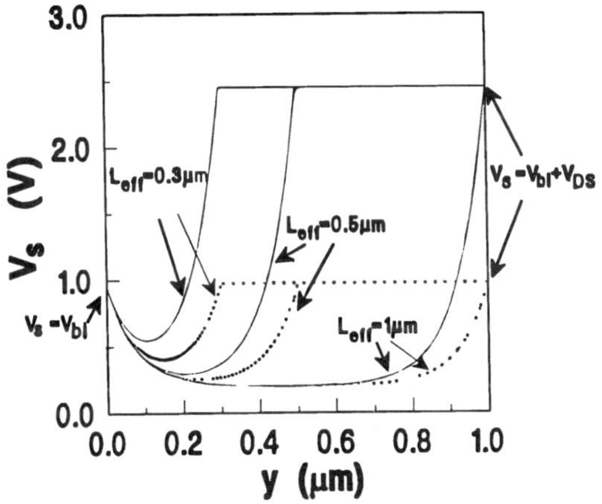

Fig. 3.2.6 Quasi-2D model of surface potential along the channel. After Liu et al. [3.15].

However, Eq. (3.2.9) is only valid for small V_{ds}. When V_{ds} is not small, for $l<<L$, the channel potential can be approximated as:

$$V_s(y) = V_{sL} + (V_{bi} + V_{ds} - V_{sL})e^{(y-L)/l} + (V_{bi} - V_{sL})e^{-y/l} + (V_{bi} + V_{ds} - V_{sL})e^{-L/l} \tag{3.2.10}$$

V_{sL} is a function of the gate bias V_g. The gate voltage that raises the minimum potential to $2\phi_B$ is the threshold voltage. Clearly, Fig. 3.2.6 suggests that short channel devices would have lower V_{th}'s.

The reduction in V_{th} can be shown to be

$$\Delta V_{th} = [3(V_{bi} - 2\phi_B) + V_{ds}]e^{-L/l} + 2\sqrt{(V_{bi} - 2\phi_B)(V_{bi} - 2\phi_B + V_{ds})}\, e^{-\frac{L}{2l}} \tag{3.2.11}$$

Eq. (3.2.11) reduces to Eq. (3.2.9) for large L/l and small V_{ds} as expected.

Fig. 3.2.7 shows the calculated results using Eq. (3.2.9) and Eq. (3.2.11). The results from numerical evaluation of Eq. (3.2.5) are also given in the figures for comparison. When $L>>l$ and $V_{ds}<<V_{bi}-2\phi_B$, Eq. (3.2.9) gives a reasonable estimate of V_{th} shift as shown in Fig. 3.2.7 (a). At high V_{ds}, Eq. (3.2.9) overestimates the V_{th} shift. However, Eq. (3.2.11) can still accurately predict the V_{th} characteristics as shown in Fig. 3.2.7 (b).

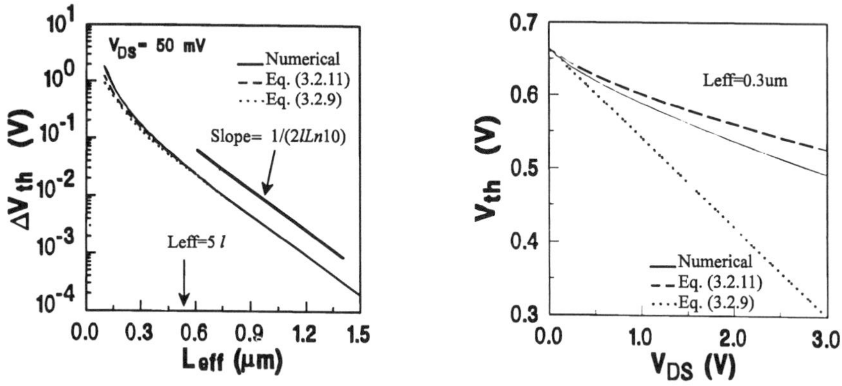

(a) (b)

Fig. 3.2.7 (a) The calculated ΔV_{th} shift versus channel length at V_{ds} =0.05V. The simple analytical models, Eq. (3.2.9) and Eq. (3.2.11), agree with the numerical solutions. (b) Comparison between simple analytical solutions and numerical solution of V_{th} versus V_{ds}. After Liu et al. [3.15].

The above analysis ignored possible voltage drop inside the drain diffusion region. It is valid for both non-*LDD* and *LDD* devices as long as V_{ds} is small. For an *LDD* device, V_{bi} is the built-in potential of the n^-/p junction. When V_{ds} is large (V_{ds} >1V), the voltage drop in the drain region should be subtracted from the $V_{ds} + V_{bi}$ terms in Eq. (3.2.9) and Eq. (3.2.11) for the *LDD* devices.

The characteristic length l is a very important parameter affecting the accuracy of the above quasi 2-D model. Although l calculated from Eq. (3.2.6) has the correct order of magnitude and qualitative dependences, exact values of l need to be extracted from actual devices because of the unknown factor η. The extraction of l can be done by fitting the experimental data of $\log(\Delta V_{th})$ versus L_{eff} in the region of $L_{eff}>5l$. It has been found that for a given technology, a unique l (or η) extracted by the technology can be used for a wide range of L_{eff} and V_{ds} values (for example L=0.2-5μm, V_{ds} =0.05-3.5V) [3.15].

The physical mechanisms behind the short channel effect and the *DIBL* effect can be described using Fig. 3.2.8. The electron (in NMOS) barrier heights along the channel for a long channel and a short channel devices are shown in Fig. 3.2.8. The barrier height of the long channel device is quite insensitive to the drain voltage. However, the barrier height of the short channel device is reduced substantially by the drain voltage. Even when the drain voltage is zero, the barrier height of the short channel device is lower than the long channel device due to the built-in potential of the S/D junctions. Reducing the barrier height will cause the threshold voltage to go down. This is why the threshold voltage of a short channel device is smaller than that of a long channel device; and it is why the threshold voltage of a short channel device is a strong function of the drain voltage.

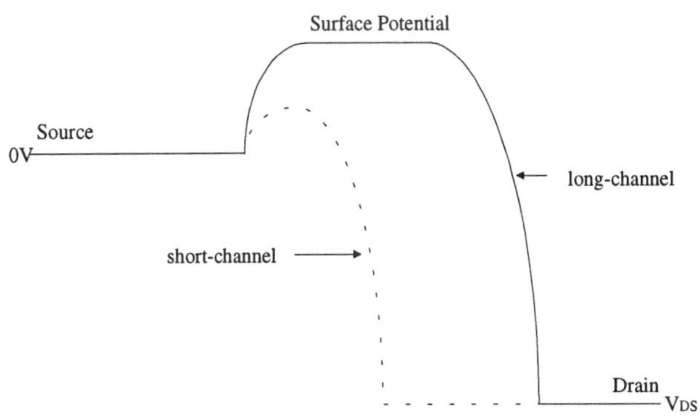

Fig. 3.2.8 Barrier height and potential along the channel for a short and a long *L* devices.

3.3 Narrow Width Effect Model

Fig. 3.3.1 shows a width cross section of a MOSFET with *LOCOS* isolation technology. The narrow width effect is complicated and sensitive to the details of the isolation technology. The following is a simple qualitative description. It can be seen that there is a tapering of the oxide from thin to thick oxide along the width direction from the center of the device to the field oxide region. The polysilicon gate overlaps the thick oxide on both sides of the thin gate oxide. This results in a gate controlled depletion region at the edges of the device. As a result, the gate induced fringing electric field around the device edges controls or is linked with extra depletion charge ΔQ_w with $\Delta Q_w/2$ on each side as shown in Fig. 3.3.2. This additional charge, balanced by the gate charge, cannot be ignored as the channel width decreases. For the case discussed here, the threshold voltage increases as the channel width decreases. The threshold voltage with the narrow width effect can be written as

$$V_{th} = V_{thW} + \frac{\Delta Q_{bW}}{C_{ox}} \tag{3.3.1}$$

where V_{thW} is the threshold voltage in a wide width device, ΔQ_{bW} is the contribution of the extra depletion charge in the thick oxide region, as shown in Fig. 3.3.2.

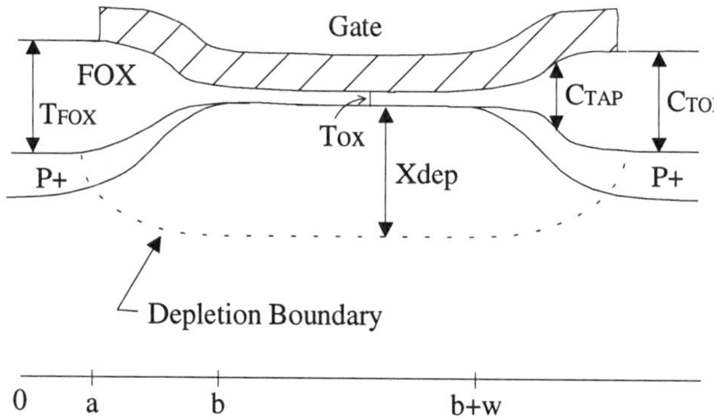

Fig. 3.3.1 A cross-section view of a MOSFET along the gate width direction. After Akers et al. [3.16].

Again, a realistic narrow width model requires a solution from the two dimensional Poisson equation. However, it is not convenient for use in circuit simulation because of the time consuming computations. Many analytical solutions have been suggested [3.16, 3.17]. One of them was proposed by Akers et al. in 1981 [3.16]. The Akers's narrow width model considers three different geometrical approximations of the area enclosing the induced extra charge. The three geometrical shapes considered are a triangle, a quarter circle, and a square.

The extra charge on both sides of the device will contribute a voltage to the threshold voltage with the amount [3.16]:

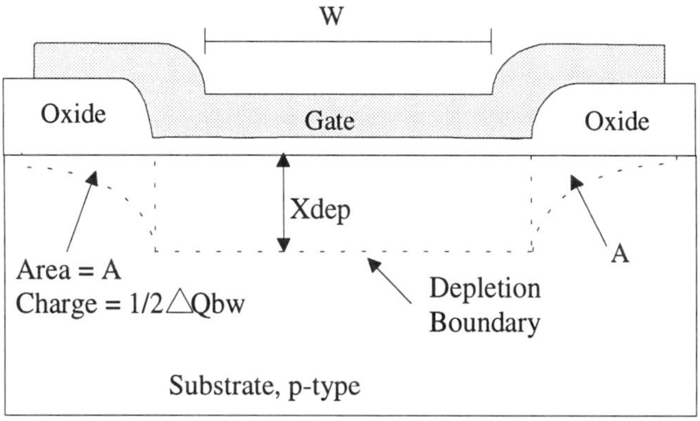

Fig. 3.3.2 The extra depletion charge ΔQ_{bW} induced in the thick oxide region. After Akers et al. [3.16].

$$V_{\Delta QbW} = \frac{qN_A X_{dep}{}^2}{C_{ox} W} \qquad \text{for triangle} \tag{3.3.2}$$

$$V_{\Delta QbW} = \frac{qN_A \pi X_{dep}{}^2}{2C_{ox} W} \qquad \text{for quarter circle} \tag{3.3.3}$$

$$V_{\Delta QbW} = \frac{2qN_A X_{dep}{}^2}{C_{ox} W} \qquad \text{for square} \tag{3.3.4}$$

where X_{dep} is the width of the channel depletion.

A general form of the threshold voltage for a narrow width device can be given as,

$$V_{th} = V_{thW} + \delta \frac{qN_A X_{dep}{}^2}{C_{ox} W} \tag{3.3.5}$$

$$\delta = \begin{cases} 1 & triangle \\ \pi/2 & quarter \quad circle \\ 2 & square \end{cases} \tag{3.3.6}$$

The comparison between the measurements and the models indicated that the square shape, that is δ =2, in Eq. (3.3.2) gave the best fit [3.16]. In compact modeling, δ may be an extracted parameter.

In the above discussion, the substrate doping concentration and gate oxide thickness were assumed to be uniform in the derivation. However, in a real device the substrate doping concentration is not uniform. Also, the capacitances of the tapered oxide and thick oxide are different from the gate oxide capacitance and need to be considered separately in modeling the narrow width effects.

Akers et al. further suggested a more complicated analytical expression to include the effects of the tapered oxide capacitance, the depletion charge under the thick recessed field oxide due to the gate overlap, and field doping encroachment at the channel edges [3.16]. However, the simpler model equation given in Eq. (3.3.5) is more often used in compact modeling.

3.4 Threshold Voltage Model in BSIM3v3

In this section, we discuss the threshold voltage model in BSIM3v3. It accounts for the physical effects discussed in chapter 2 with regard to V_{th}.

3.4.1 Modeling of the vertical non-uniform doping effects

From Eq. (2.2.1), we see that the threshold voltage should depend linearly on $\sqrt{\phi_s - V_{bs}}$, with a slope known as γ. However, experimental data (as shown in Fig. 2.2.2) in general displays a non-linear dependence. The slope γ becomes smaller as the body bias V_{bs} becomes more negative (for NMOS) (V_{bs}, unless stated otherwise, is always a reverse bias of the source/body junction). This non-linearity comes from the non-uniform substrate doping in the vertical direction of the body.

The doping concentration may be higher or lower at the interface of the gate oxide and the body than deep in the body due to the choice of the energies of ion implantation for the channel and well. This non-uniform body doping will make γ in Eq. (2.2.1) a function of the substrate bias. An approximate step-doping profile, as shown in Fig. 3.4.1, can be used to obtain an analytical expression of V_{th} as a function of V_{bs}. If the depletion width is less than X_T as shown in Fig. 3.4.1, N_A in Eq. (3.4.2) is equal to $\boldsymbol{N_{CH}}$[a], otherwise it is equal to $\boldsymbol{N_{SUB}}$. Therefore, we have

$$V_{th} = \boldsymbol{VTH0} + \gamma_1(\sqrt{\phi_s - V_{bs}} - \sqrt{\phi_s}), \quad |V_{bs}| < |V_{bx}|, \tag{3.4.1}$$

$$V_{th} = \boldsymbol{VTH0} + \gamma_1(\sqrt{\phi_s - V_{bs}} - \sqrt{\phi_s}) + \gamma_2(\sqrt{\phi_s - V_{bs}} - \sqrt{\phi_s - V_{bx}}), \quad |V_{bs}| \geq |V_{bx}| \tag{3.4.2}$$

where the definition of $\boldsymbol{V_{TH0}}$ has been given in Chapter 2. ϕ_s[b] is the surface potential at threshold, given in Eq. (2.2.4). γ_1 and γ_2 are given as

$$\gamma_1 = \frac{\sqrt{2q\varepsilon_{si}\boldsymbol{N_{CH}}}}{C_{ox}} \tag{3.4.3}$$

$$\gamma_2 = \frac{\sqrt{2q\varepsilon_{si}\boldsymbol{N_{SUB}}}}{C_{ox}} \tag{3.4.4}$$

V_{bx} is the body bias at which the depletion width is equal to $\boldsymbol{X_T}$. Therefore, V_{bx} satisfies

$$\frac{q\boldsymbol{N_{CH}}\boldsymbol{X_T}^2}{2\varepsilon_{si}} = \phi_s - V_{bx} \tag{3.4.5}$$

To unify the V_{th} expression in Eq. (3.4.1) and Eq. (3.4.2), a general expression of V_{th} under different body bias is proposed as

$$V_{th} = \boldsymbol{V_{TH0}} + \boldsymbol{K_1}(\sqrt{\phi_s - V_{bs}} - \sqrt{\phi_s}) - \boldsymbol{K_2}V_{bs} \tag{3.4.6}$$

$\boldsymbol{K_1}$ and $\boldsymbol{K_2}$ are usually determined by fitting Eq. (3.4.6) to the measured V_{th} data. For theoretical discussions, they can be determined by the criteria that the V_{th} values given by Eq. (3.4.2) and Eq. (3.4.6) and their derivatives versus V_{bs} should be the same at $\boldsymbol{V_{bm}}$, where $\boldsymbol{V_{bm}}$ is the maximum substrate bias voltage. At $V_{bs} = \boldsymbol{V_{bm}}$, let Eqs. (3.4.2) and (3.4.6) be equal:

a. Note: In this book, all BSIM3v3 model parameters are bold-faced Italic characters with capitalized first letters followed by capitalized subscripts.

b. $\phi_s \equiv 2\phi_B$ in the threshold voltage model equations in BSIM3v3.

$$\gamma_1(\sqrt{\phi_s - V_{bx}} - \sqrt{\phi_s}) + \gamma_2(\sqrt{\phi_s - V_{bm}} - \sqrt{\phi_s - V_{bx}})$$
$$= K_1(\sqrt{\phi_s - V_{bm}} - \sqrt{\phi_s}) - K_2 V_{bm} \quad (3.4.7)$$

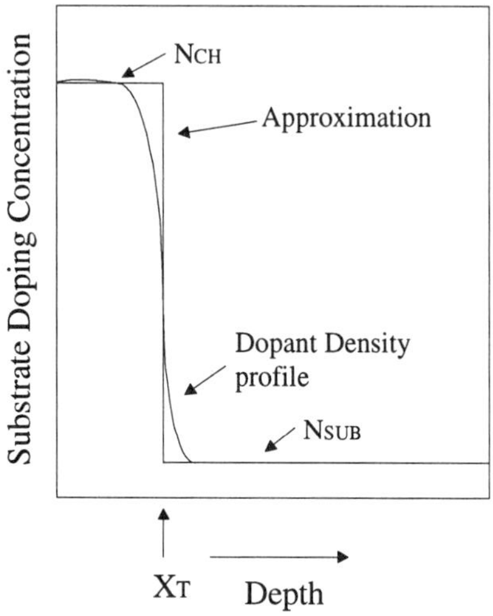

Fig. 3.4.1 One possible body doping profile and approximation.

We can get a second equation by letting the derivative of Eqs. (3.4.2) and (3.4.6) be equal at $V_{bs} = V_{bm}$ and obtaining

$$K_1 = \gamma_2 - 2K_2\sqrt{\phi_s - V_{bm}} \quad (3.4.8)$$

Solving Eqs. (3.4.7) and (3.4.8) gives

$$K_2 = (\gamma_1 - \gamma_2)\frac{\sqrt{\phi_s - V_{bx}} - \sqrt{\phi_s}}{2\sqrt{\phi_s}(\sqrt{\phi_s - V_{bm}} - \sqrt{\phi_s}) + V_{bm}} \quad (3.4.9)$$

K_1 can be obtained from Eq. (3.4.8) with K_2 given in Eq. (3.4.9).

It has been found that using K_1 and K_2 as fitting parameters yields better accuracy than calculating K_1 and K_2 with Eq. (3.4.8) and Eq. (3.4.9). Usually K_1 and K_2 are determined experimentally if measured data is available. However, if device data is not available but the model user knows the doping concentration profile, or the user wants to use the physical nature of K_1 and K_2 for specific purposes such as statistical modeling, the user can input the doping concentrations and other process parameters (e.g. N_{CH}, N_{SUB}, X_T and T_{OX}). K_1 and K_2 can then be calculated using Eqs. (3.4.8) and (3.4.9).

In summary, K_1 and K_2 are the key parameters to model the vertical non-uniform doping effect.

3.4.2 Modeling of the RSCE due to lateral non-uniform channel doping

To account for the lateral non-uniform doping effect due to the higher doping concentration near the drain and the source than in the middle of the channel, as shown in Fig. 2.2.3, a step doping profile along the channel length direction as shown in Fig. 3.4.2 may be used as a first order approximation to obtain a V_{th} expression. As a further approximation, the average channel doping can be calculated as follows:

$$
\begin{aligned}
N_{eff} &= \frac{N_{CH}(L-2L_x)+N_{pocket}\cdot 2L_x}{L} = N_{CH}\left(1+\frac{2L_x}{L}\frac{N_{pocket}-N_{CH}}{N_{CH}}\right) \\
&\equiv N_{CH}\left(1+\frac{NLX}{L}\right)
\end{aligned}
\tag{3.4.10}
$$

Where $N_{LX} = 2L_x \frac{N_{pocket}-N_{CH}}{N_{CH}}$. In BSIM3v3, N_{LX} is treated as a fitting parameter extracted from the measured data.

With the introduction of N_{LX} to account for the lateral non-uniform doping, Eq. (3.4.6) may be modified into

$$
\begin{aligned}
V_{th} &= V_{THO} + K_1\left(\sqrt{\phi_s - V_{bs}} - \sqrt{\phi_s}\right) - K_2 V_{bs} \\
&+ K_1\left(\sqrt{1+\frac{NLX}{L}} - 1\right)\sqrt{\phi_s}
\end{aligned}
\tag{3.4.11}
$$

Eq. (3.4.11) can best be understood by first setting $V_{bs} = 0$. At $V_{bs} = 0$, Eq. (3.4.11) models the dependence of V_{th} on L due to the lateral non-uniform doping. Eq. (3.4.11) shows that the threshold voltage will increase as channel length decreases. This is shown in Fig. 3.4.3.

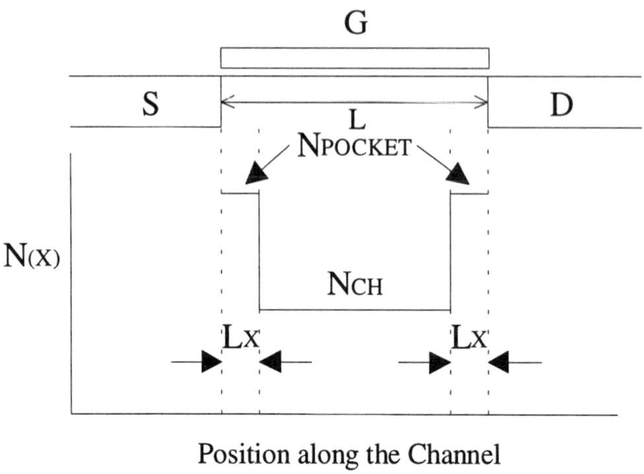

Fig. 3.4.2 A step doping profile can be used to approximate the non-uniform lateral channel doping.

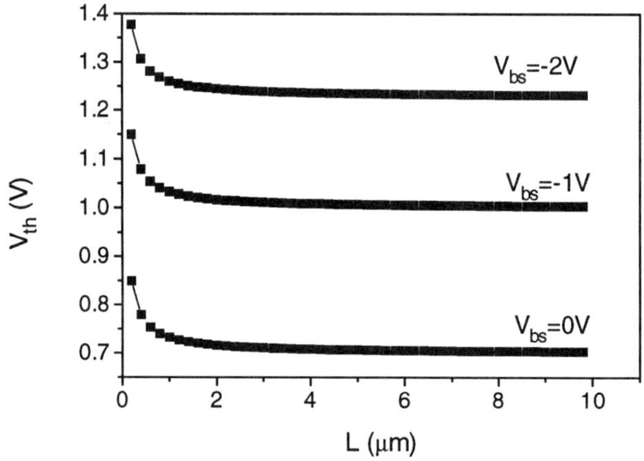

Fig. 3.4.3 Threshold voltage versus channel length calculated by Eq. (3.4.11).

In summary, N_{LX} is the only parameter to represent the lateral non-uniform doping effect. It models the reverse short channel effects (RSCE), as is shown in Fig. 3.4.3.

3.4.3 Modeling of the short channel effect due to drain induced barrier lowering

In BSIM3v3, the drain-induced barrier lowering (*DIBL*) effect is analyzed by solving a quasi two-dimension Poisson equation along the channel as discussed in section 3.2.2. By applying Gauss's law to a rectangular box of height X_{dep} and length Δy in the channel depletion region assuming an equation for the electric potential may be set up and solved, leading to [3.15]

$$\Delta V_{th} = \theta_{th}(L)[2(V_{bi} - \phi_s) + V_{ds}] \tag{3.4.12}$$

where V_{bi} is the built-in voltage of the substrate/source junction. V_{bi} is given by

$$V_{bi} = \frac{K_B T}{q} \ln\left(\frac{NCH NDS}{n_i^2}\right) \tag{3.4.13}$$

where N_{DS} in Eq. (3.4.13) is the source/drain doping concentration and N_{CH} is the channel doping concentration. In Eq. (3.4.12)

$$\theta_{th}(L) = [\exp(-L/2l_t) + 2\exp(-L/l_t)] \tag{3.4.14}$$

l_t is a characteristic length and is given by

$$l_t = \sqrt{\frac{\varepsilon_{si} T_{OX} X_{dep}}{\varepsilon_{ox} \eta}} \tag{3.4.15}$$

X_{dep} is the depletion width in the substrate and is given by

$$X_{dep} = \sqrt{\frac{2\varepsilon_{si}(\phi_s - V_{bs})}{qN_{CH}}} \tag{3.4.16}$$

η in Eq. (3.4.15) is a fitting parameter that accounts for the numerous approximations behind Eq. (3.4.15). For example, X_{dep} is not constant from the source to the drain and is not equal to the quantity in Eq. (3.4.16). Also the electric field is not uniform from the top to the bottom of the depletion region.

As channel length L decreases ΔV_{th} will increase, and in turn V_{th} will decrease. If a MOSFET has a *LDD* structure, N_{DS} in Eq. (3.4.13) is the doping concentration in the lightly doped region. V_{bi} in a *LDD*-MOSFET will be smaller than in a single drain MOSFETs, therefore the threshold voltage reduction due to the short channel effect is smaller in *LDD*-MOSFETs.

Eq. (3.4.12) shows that ΔV_{th} depends linearly on the drain voltage. V_{th} decreases as V_{ds} increases. This is an important aspect of the *DIBL* phenomenon. The severity of the *DIBL* effect has a strong dependence on *L*. If $L>>l_t$, the *DIBL* effect is very weak.

The influences of *DIBL* on V_{th} can be described by Eq. (3.4.14). However, in order to make the model fit a wide range of *L*, V_{ds}, and V_{bs}, several additional parameters such as $\boldsymbol{D_{VT0}}$, $\boldsymbol{D_{VT1}}$, $\boldsymbol{D_{VT2}}$, $\boldsymbol{D_{SUB}}$, $\boldsymbol{E_{TA0}}$ and $\boldsymbol{E_{TAB}}$ are introduced, leading to

$$
\begin{aligned}
V_{th} &= \boldsymbol{V_{TH0}} + \boldsymbol{K_1}(\sqrt{\phi_s - V_{bs}} - \sqrt{\phi_s}) - \boldsymbol{K_2} V_{bs} + \boldsymbol{K_1}\left(\sqrt{1+\frac{\boldsymbol{N_{LX}}}{L_{eff}}} - 1\right)\sqrt{\phi_s} \\
&- \boldsymbol{D_{VT0}}\left(\exp(-\boldsymbol{D_{VT1}}\frac{L_{eff}}{2l_t}) + 2\exp(-\boldsymbol{D_{VT1}}\frac{L_{eff}}{l_t})\right)(V_{bi} - \phi_s) \\
&- \left(\exp(-\boldsymbol{D_{SUB}}\frac{L_{eff}}{2l_{to}}) + 2\exp(-\boldsymbol{D_{SUB}}\frac{L_{eff}}{l_{to}})\right)(\boldsymbol{E_{TA0}} + \boldsymbol{E_{TAB}} V_{bs})\, V_{ds}
\end{aligned}
\tag{3.4.17}
$$

where

$$l_t = \sqrt{\varepsilon_{si} X_{dep} / C_{ox}}\,(1 + \boldsymbol{D_{VT2}} V_{bs}) \tag{3.4.18}$$

$$l_{to} = \sqrt{\varepsilon_{si} X_{dep0} / C_{ox}} \tag{3.4.19}$$

$$X_{dep} = \sqrt{\frac{2\varepsilon_{si}(\phi_s - V_{bs})}{q\boldsymbol{N_{CH}}}} \tag{3.4.20}$$

$$X_{dep0} = \sqrt{\frac{2\varepsilon_{si}\phi_s}{q\boldsymbol{N_{CH}}}} \tag{3.4.21}$$

$\boldsymbol{D_{VT1}}$ replaces $1/(\eta)^{1/2}$ in Eq. (3.4.15). $\boldsymbol{D_{VT2}}$ is introduced to account for the dependence of the doping concentration on substrate bias since the doping concentration in the body is not uniform in the vertical direction. Compared with Eq. (3.4.12), Eq. (3.4.17) allows different *L* dependencies of V_{th} on V_{ds}, and on V_{bs}, i.e. different l_t and l_{to}, in order to achieve better accuracy of matching the V_{th} data. $\boldsymbol{D_{VT0}}$, $\boldsymbol{D_{VT1}}$,$\boldsymbol{D_{VT2}}$, $\boldsymbol{E_{TA0}}$, $\boldsymbol{E_{TAB}}$, and $\boldsymbol{D_{SUB}}$ are determined experimentally. Although Eqs. (3.4.12) and (3.4.17) have many differences, they have the same basic double exponential functional forms. This means that the device physics represented by Eqs. (3.4.12) and (3.4.17) are essentially the same. A philosophy embedded in the BSIM3 model is to find a physically accurate functional form to describes a physical phenomenon and then use fitting parameters and even empirical terms to achieve quantitative match with the device characteristics.

As the body bias becomes more negative, the depletion width increases as shown in Eq. (3.4.20). Thus ΔV_{th} will rise due to the increase in l_t. That is, the *DIBL* effects (i.e. the channel length and V_{ds} dependences of V_{th}) are stronger as V_{bs} is made more negative. This is verified by experimental data shown in Fig. 3.4.3 and Fig. 3.4.4. Although the dependence of V_{th} on V_{ds} is known to be nonlinear [3.15], a linear dependence of V_{th} on V_{ds} is nevertheless sufficient for circuit simulation, as shown in Fig. 3.4.3.

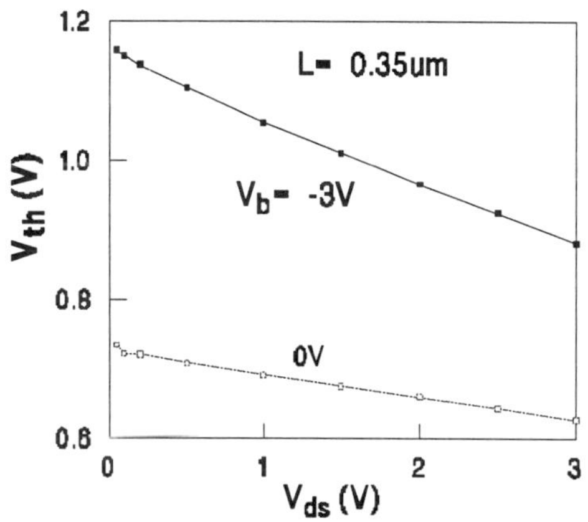

Fig. 3.4.3. Threshold voltage versus the drain voltage at different body biases. After Huang et al. [3.19].

In Eq. (3.4.17), $V_{TH0} + K_1\sqrt{\phi_s - V_{bs}} - K_2 V_{bs}$ and ΔV_{th} due to *DIBL* effect move in opposite directions as V_{bs} varies. Therefore, the changes in $V_{TH0} + K_1\sqrt{\phi_s - V_{bs}} - K_2 V_{bs}$ and in ΔV_{th} will compensate for each other and make V_{th} less sensitive to V_{bs}. This compensation is only evident when the short-channel effect is significant in short-channel devices. Hence, the V_{th} of a short-channel MOSFET is less sensitive to body bias than the V_{th} of a long-channel MOSFET which is seen in Fig 3.4.4.

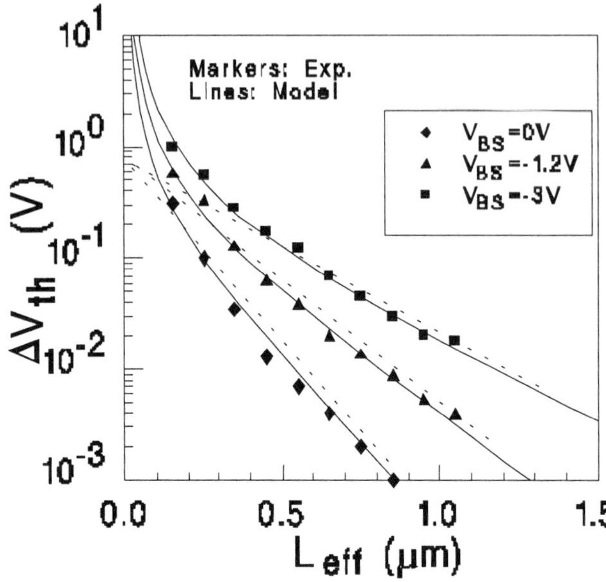

Fig. 3.4.4 Channel length dependence of the threshold voltage at different body biases. After Liu et al. [3.15].

Fig. 3.4.5 shows the threshold voltage versus channel length at different drain voltages and bias voltages and compares the data with Eq. (3.4.14). The additional terms and parameters in Eq. (3.4.17) are necessary to provide a good match between the model and the data.

3.4.4 Modeling of the narrow width effects

Even though several models with narrow width effects were reviewed in Chapter 2, none has been found to be quantitatively accurate. They do not even qualitatively describe the reverse narrow width effect that are encountered in some technologies [3.18]. One must accept the fact that the behavior of the narrow width effect depends on the numerous and changing isolation technologies and no universally accurate physical model of it is available. Models of the threshold voltage shift due to narrow width effect have the general form,

$$\Delta V_{thW} \propto \frac{T_{OX}}{W_{eff}} \phi_s \tag{3.4.22}$$

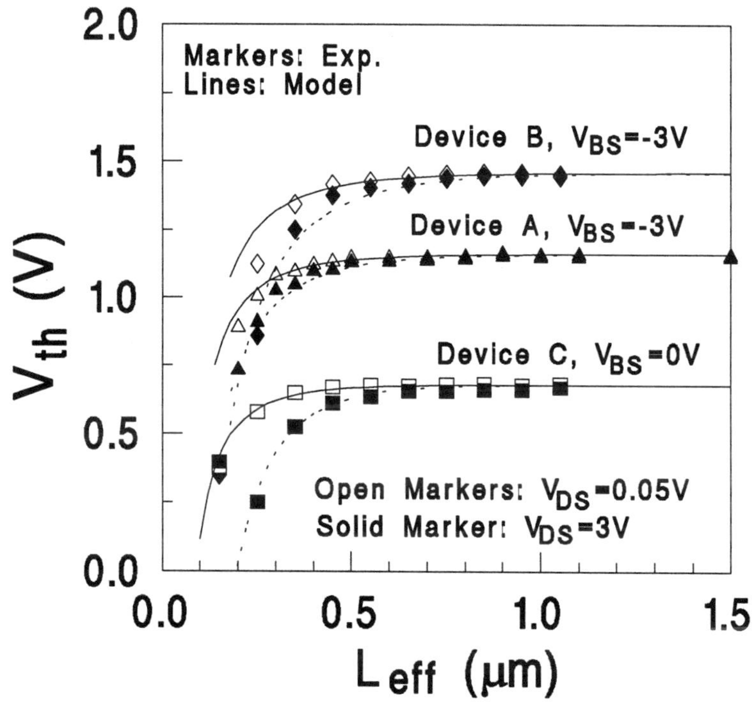

Fig. 3.4.5 Threshold voltage versus channel length at several drain voltages and body-biases. After Liu et al. [3.15].

BSIM3v3 takes an empirical approach to account for the overall narrow width effects (both normal and reverse narrow width effects). Three fitting parameters $\boldsymbol{K_3}$, $\boldsymbol{K_{3B}}$, and $\boldsymbol{W_0}$ are used to model the change in V_{th} due to the narrow W [3.19, 3.20],

$$\Delta V_{thW} = (\boldsymbol{K_3} + \boldsymbol{K_{3B}} V_{bs}) \frac{\boldsymbol{Tox}}{W_{eff}' + \boldsymbol{W_0}} \phi_s \tag{3.4.23}$$

W_{eff}' is the effective channel width [3.20], which will be discussed in detail in Chapter 4.

In the above discussion, it is assumed that the channel length of the device is long enough so that the narrow width effect does not depend on L. To account for the narrow width effect for small channel lengths, BSIM3v3 introduces the following in addition to Eq. (3.4.23) to describe the change in threshold voltage due to narrow W and short L:

$$\Delta V_{thWL} = \boldsymbol{D_{VT0W}}[\exp(-\boldsymbol{D_{VT1W}} \frac{W_{eff}' L_{eff}}{2l_{tw}}) + 2\exp(-\boldsymbol{D_{VT1W}} \frac{W_{eff}' fL_{eff}}{l_{tw}})](V_{bi} - \phi_s) \tag{3.4.24a}$$

$$l_{tw} = \sqrt{\varepsilon_{si} X_{dep} / C_{ox}}(1 + \boldsymbol{D_{VT2W}} V_{bs}) \tag{3.4.24b}$$

where $\boldsymbol{D_{VT0W}}$, $\boldsymbol{D_{VT1W}}$ and $\boldsymbol{D_{VT2W}}$ are parameters extracted from experimental data. Eq. (3.4.24) is a good example of the empirical elements based on experimental observations plus simple physical understanding of the effects, which help make BSIM3 an accurate model.

3.4.5 Complete V_{th} model in BSIM3v3

With the considerations of all the discussed physical effects above, the final and complete V_{th} expression in BSIM3v3 is as follows [3.20]:

$$\begin{aligned}
V_{th} &= \boldsymbol{V_{TH0}} + \boldsymbol{K_1}(\sqrt{\phi_s - V_{bs}} - \sqrt{\phi_s}) - \boldsymbol{K_2} V_{bs} \\
&+ \boldsymbol{K_1}\left(\sqrt{1 + \frac{\boldsymbol{N_{LX}}}{L_{eff}}} - 1\right)\sqrt{\phi_s} + (\boldsymbol{K_3} + \boldsymbol{K_{3B}} V_{bs})\frac{\boldsymbol{T_{OX}}}{W_{eff}' + \boldsymbol{W_0}}\phi_s \\
&- \boldsymbol{D_{VT0}}\left(\exp(-\boldsymbol{D_{VT1}}\frac{L_{eff}}{2l_t}) + 2\exp(-\boldsymbol{D_{VT1}}\frac{L_{eff}}{l_t})\right)(V_{bi} - \phi_s) \\
&- \left(\exp(-\boldsymbol{D_{sub}}\frac{L_{eff}}{2l_{to}}) + 2\exp(-\boldsymbol{D_{sub}}\frac{L_{eff}}{l_{to}})\right)(\boldsymbol{E_{tao}} + \boldsymbol{E_{tab}} V_{bs}) V_{ds} \\
&- \boldsymbol{D_{VT0w}}\left(\exp(-\boldsymbol{D_{VT1w}}\frac{W_{eff}' L_{eff}}{2l_{tw}}) + 2\exp(-\boldsymbol{D_{VT1w}}\frac{W_{eff}' L_{eff}}{l_{tw}})\right)(V_{bi} - \phi_s)
\end{aligned} \tag{3.4.25a}$$

In Eq. (3.4.25a), the second and third terms are used to model the vertical non-uniform doping effect, the fourth term is for the lateral non-uniform doping effect, the fifth term is for the narrow width effect, the sixth and seventh terms are related to the short channel effect due to *DIBL*, and the last term is to describe the small size effect in devices with both small channel length and small width. A simpler model for V_{th} would be preferred if it could offer the same adequate accuracy.

In BSIM3v3.2 [3.35], the $\boldsymbol{T_{OX}}$ dependence of V_{th} is improved by introducing a new model parameter $\boldsymbol{T_{OXM}}$, which can be considered as the nominal gate oxide thickness at which the model parameters are extracted. Eq. (3.4.25a) is modified due to the introduction of $\boldsymbol{T_{OXM}}$

$$
\begin{aligned}
V_{th} = {} & V_{th0OX} + K_{1OX}\sqrt{\phi_s - V_{bs}} - K_{2OX}V_{bs} \\
& + K_{1OX}\left(\sqrt{1+\frac{\boldsymbol{N_{LX}}}{L_{eff}}} - 1\right)\sqrt{\phi_s} + (\boldsymbol{K_3} + \boldsymbol{K_{3B}}V_{bs})\frac{\boldsymbol{T_{OX}}}{W_{eff}' + \boldsymbol{W_0}}\phi_s \\
& - \boldsymbol{D_{VT0}}\left(\exp(-\boldsymbol{D_{VT1}}\frac{L_{eff}}{2l_t}) + 2\exp(-\boldsymbol{D_{VT1}}\frac{L_{eff}}{l_t})\right)(V_{bi} - \phi_s) \\
& - \left(\exp(-\boldsymbol{D_{SUB}}\frac{L_{eff}}{2l_{to}}) + 2\exp(-\boldsymbol{D_{SUB}}\frac{L_{eff}}{l_{to}})\right)(\boldsymbol{E_{TA0}} + \boldsymbol{E_{TAB}}V_{bs})V_{ds} \\
& - \boldsymbol{D_{VT0W}}\left(\exp(-\boldsymbol{D_{VT1W}}\frac{W_{eff}'L_{eff}}{2l_{tw}}) + 2\exp(-\boldsymbol{D_{VT1W}}\frac{W_{eff}'L_{eff}}{l_{tw}})\right)(V_{bi} - \phi_s)
\end{aligned}
\tag{3.4.25b}
$$

$$V_{th0OX} = \boldsymbol{V_{TH0}} - \boldsymbol{K_1}\sqrt{\phi_s} \tag{3.4.25c}$$

$$K_{1OX} = \boldsymbol{K_1}\frac{\boldsymbol{T_{OX}}}{\boldsymbol{T_{OXM}}} \tag{3.4.25d}$$

$$K_{2OX} = \boldsymbol{K_2}\frac{\boldsymbol{T_{OX}}}{\boldsymbol{T_{OXM}}} \tag{3.4.25e}$$

In the implementation of Eqs. (3.4.25) in BSIM3v3, all V_{bs} terms have been substituted with V_{bseff} given in Eq. (3.4.26).

$$V_{bseff} = V_{bc} + 0.5[V_{bs} - V_{bc} - \delta_1 + \sqrt{(V_{bs} - V_{bc} - \delta_1)^2 - 4\delta_1 V_{bc}}\ ] \tag{3.4.26}$$

where δ_1=0.001. When $V_{bs}>V_{bc}$, V_{bseff} approaches V_{bc} which is an upper bound of V_{bseff}. Having an upper bound improves the numerical robustness of the model and enhances simulation convergence.In model implementation, V_{bc} is set to be 0.9 of the V_{bs} at which $dV_{th}/dV_{bs}=0$, where V_{th} is given by Eq. (3.4.6). Thus:

$$V_{bc} = 0.9(\phi_s - \frac{\boldsymbol{K_1}^2}{4\boldsymbol{K_2}^2}) \tag{3.4.27}$$

It should be pointed out that care has been taken in the model implementation to avoid divide-by-zero when K_2=0 in Eq. (3.4.27).

3.5 Helpful Hints

1. Extracting the threshold voltage

One "definition" of the threshold voltage (or onset of strong inversion) is the well known $2\phi_B$ approximation. That is, V_{th} is "defined" as the gate bias at which the surface potential is $2\phi_B$. We have used this in the derivation of V_{th} of the long channel device. However, it may be more useful and meaningful if V_{th} is defined with a measurable behavior of the device, e.g. $I_d = a$W/L μA at $V_g = V_{th}$, where a is a constant, say, 0.1. A number of different definitions for the threshold voltage have been proposed to model or measure V_{th}. Model users should know the differences among them to avoid any confusion when discussing threshold voltages.

The following definitions have been suggested for the threshold voltage. V_{th} is the gate voltage at which (1) the inversion charge density is equal to zero [3.21], (2) the minority carrier density at the surface is equal to the majority carrier density at the boundary of the depletion region [3.22], (3) the change rate in the inversion charge caused by the gate bias is equal to the change rate in the depletion charge density [3.23], (4) the surface potential is equal to $2\phi_B$ [3.24], (5) the extrapolated linear drain current is zero [3.25], (6) the extrapolated square root of the saturation current is zero [3.26], (7) the drain current is at a small constant value [3.27], (8) the gate-channel capacitance is equal to $1/3C_{ox}$ [3.28], (9) the slope of the transconductance g_m vs. V_{gs} is at its maximum [3.29].

Definitions (1)-(4) have been used in analytical model derivations, while definitions (5)-(9) are used to determine the threshold voltage experimentally. It is clear that these threshold voltage definitions are inconsistent and may lead to difficulties and confusion when comparing the results of these definitions. Definition (7) can be made to be more or less consistent with (4), (5) and (6) by careful selection of the drain current value.

Definition (5) has been widely used in the characterization of MOSFETs. However, it is well known that the I_{ds} does not follow a good linear relationship with V_{gs} in modern MOSFETs. Because of the ambiguity of "linear region in the I_{ds}-V_{gs} curves", an engineering methodology has become the standard approach in determining the V_{th} of a MOSFET:

(a). Measure I_{ds}-V_{gs} characteristics at low V_{ds} (<0.1V, typically 50mV),

(b). Determine the maximum slope of the I_{ds}-V_{gs} curve, that is, the maximum g_m point,

(c). Extrapolate I_{ds}-V_{gs} from the maximum g_m point to I_{ds}=0,

(d). Note the corresponding extrapolated V_{gs} value (V_{gs0}) for I_{ds} =0 point.

(e). Calculate V_{th} according to $V_{th} = V_{gs0} - 0.5V_{ds}$.

The reason I_{ds} does not follow a linear relationship with V_{gs} is that the carrier mobility decreases as V_{gs} becomes significantly larger than V_{th}. Extrapolation from the maximum slope point of the I_{ds}-V_{gs} curve (or maximum g_m) arbitrarily eliminates the ambiguity of extrapolating the "linear" I_{ds}-V_{gs} curve [3.30].

The methodology to extract V_{th} discussed last is recommended for extracting threshold voltage for the BSIM3v3 model.

2. Threshold voltages in weak and strong inversion regimes

Usually the *I-V* model in strong inversion can match the measured data well if the "extrapolated" V_{th} is used in the model. However, the *I-V* model in weak inversion usually deviates from the measured *I-V* somewhat if the same V_{th} is used in the weak inversion *I-V* model without any modification [3.31]. The problem is caused by the difference (around 0.1V) between the V_{th} determined with definition (4), which is appropriate for the subthreshold *I-V*, and the V_{th} extrapolated from the I_{ds}-V_{gs} in the strong inversion region. It has been found that $\phi_s = 2\phi_B + 4v_t$ may be a better theoretical criterion for the V_{th} in the strong inversion *I-V* model ($V_g > V_{th}$) while $\phi_s = 2\phi_B$ is a good criterion of the threshold voltage in modeling the weak inversion ($V_g < V_{th}$) regime. The regime between $\phi_s = 2\phi_B$ and $\phi_s = 2\phi_B + 4v_t$ has been called moderate inversion [3.32], which is becoming more and more important as new low power analog applications emphasize the operations of transistors in this regime. In BSIM3v3, a parameter called $\boldsymbol{V_{OFF}}$ is introduced to account for the difference in the two V_{th}'s to achieve good model accuracy in both the strong inversion and the subthreshold regions [3.33, 3.34], which will be discussed again in Chapter 4.

3. The relationship between V_{TH0} and V_{FB} in BSIM3v3

V_{TH0} is an important parameter in the threshold voltage model of BSIM3v3. It is the threshold voltage of a long channel device at zero volt substrate bias.

$$V_{TH0} = V_{FB} + \phi_s + K_1\sqrt{\phi_s} \tag{3.5.1}$$

Where V_{FB} is a constant representing the flat-band voltage. It should be pointed out that V_{FB} is not a user-input parameter in the *I-V* model in BSIM3v3.0 and BSIM3v3.1. However it has become one in BSIM3v3.2 to improve model accuracy for MOSFETs with different gate materials.

The V_{FB} in the V_{th} model of BSIM3v3 can be either calculated or given a fixed value, depending on whether the V_{TH0} parameter is provided by the user.

Usually, V_{TH0} is a parameter specified in the model card. In that case, if V_{FB} is not assigned as a user input parameter, it is calculated in BSIM3v3 as

$$V_{FB} = V_{TH0} - \phi_s - K_1\sqrt{\phi_s} \tag{3.5.2}$$

If V_{TH0} is not specified in the model card, it is calculated using Eq. (3.5.1), and V_{FB} is given a fixed value of -1.0 if it is not given in the model card.

It should be noted that V_{FB} here is different from the v_{fb} that will be used in the capacitance models in Chapter 5, where v_{fb} is calculated to ensure that the same threshold voltage is used in both the *I-V* and capacitance models. The difference between V_{FB} and v_{fb} will be discussed further in Chapter 5.

4. The parameters related to the vertical non-uniform doping

The parameters K_1 and K_2 are introduced to describe the body effect in the presence of non-uniform doping. Usually they are extracted from the measured data. However, they can also be calculated with Eq. (3.4.8) and Eq. (3.4.9) from process parameters such as N_{CH}, T_{OX}, and N_{SUB}, if K_1 and K_2 (and also γ_1 and γ_2) are not provided in the model card. In that case, the γ_1 and γ_2 parameters in Eq. (3.4.8) and Eq. (3.4.9) are calculated with Eq. (3.4.3) and Eq. (3.4.4). For that purpose, process parameters N_{CH}, T_{OX}, N_{SUB} have to be provided as input model parameters. Certainly the γ_1 and γ_2 parameters can also be provided as the input model parameters instead of K_1 and K_2. If γ_1 is given but N_{CH} is not, N_{CH} will be calculated using the following equation,

$$N_{CH} = \frac{\gamma_1^2 C_{ox}^2}{2q\varepsilon_{si}} \tag{3.5.3}$$

If K_1, K_2, γ_1, and γ_2 are not provided as input model parameters, the default value of N_{CH} is used, and γ_1 is calculated using Eq. (3.4.3).

V_{bx}, the critical V_{bs} at which the depletion width is equal to the channel doping depth X_T, is also a model parameter in BSIM3v3. However, it is usually calculated using the following:

$$V_{bx} = \phi_s - \frac{qN_{CH} X_T}{2\varepsilon_{si}} \tag{3.5.4}$$

where X_T is a model parameter with a default value of 1.55x10^{-7} m. N_{CH} is given the default value (1.7x10^{17}cm^{-3}) if it is not calculated using Eq. (3.5.3).

5. V_{th} in buried-channel or depletion-mode MOSFETs

The threshold voltage model in BSIM3v3 was derived for enhancement-mode surface-channel devices, but it is also quite adequate for buried channel devices and even modern depletion-mode devices. Figs. 3.5.1 (a) and (b) shows the characteristics of threshold voltage vs. channel length for p-channel devices with both surface and buried channels [3.36].

6. K_3 and K_{3B} for narrow-width effects

K_3 and K_{3B} describe the narrow width effects for the different isolation technologies. The value of K_3 is usually positive for *LOCOS* field isolation although this is not always the case. This is known as the normal narrow-width effect. For fully recessed *LOCOS* or trench isolation structures, the extracted value of K_3 may be negative. This is known as the reverse narrow-width effect.

Fig.3.5.2 shows the comparison between the model and the measured data for devices using two different isolation technologies [3.37]. Devices from process *A* are n-MOSFETs with p-pocket implant *LDD* fabricated with a dual-gate CMOS shallow trench isolation process. The gate oxide thickness is 4nm. Process *B* uses *LOCOS* technology without a pocket implant. The gate oxide thickness is 12nm. It can be seen in Fig. 3.5.2 that both the normal and reverse narrow width effects can be modeled accurately by BSIM3v3 [3.37].

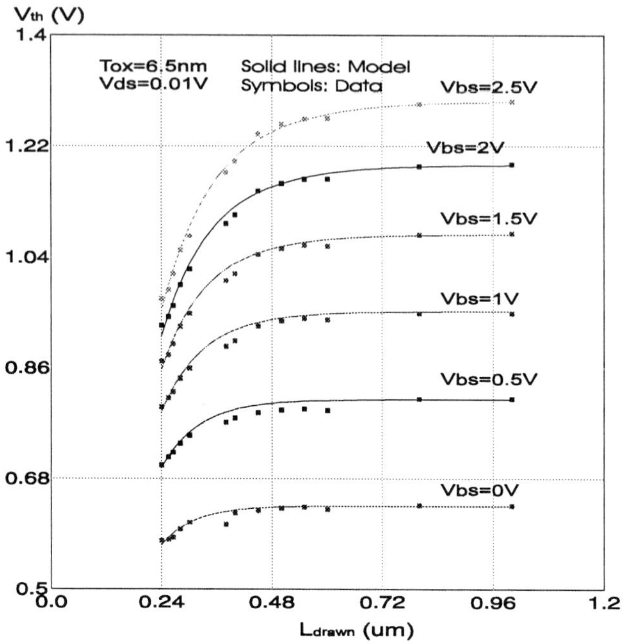

Fig. 3.5.1 (a) Modeled and measured characteristics of V_{th} vs. L for P-MOSFETs with surface channel. After Cheng et al. [3.36].

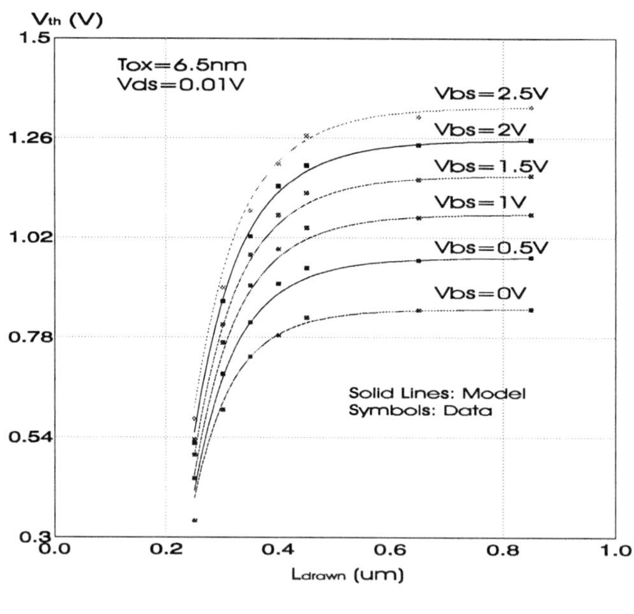

Fig. 3.5.1 (b) Modeled and measured V_{th} vs. L for P-MOSFETs with buried channel. After Cheng et al. [3.36].

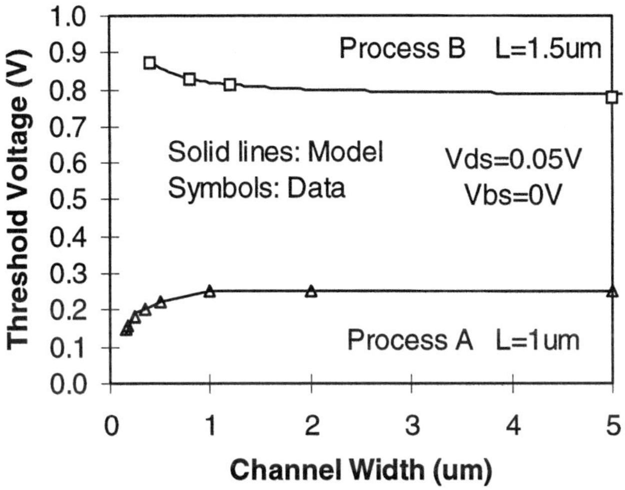

Fig. 3.5.2 Modeled and measured threshold voltage vs. channel width for devices with shallow trench (A) and LOCOS (B) isolation technologies. After Cheng et al. [3.37].

7. N_{LX} parameter for laterally non-uniform doping

Lateral non-uniform doping can produce a reverse short channel effect, which can cause a hump in the V_{th} vs. L characteristics discussed in Chapter 2. This effect can be modeled using the N_{LX} parameter together with the parameters D_{VT0} and D_{VT1}. The value of N_{LX} can be zero if there is no obvious laterally non-uniform effect in the device. However, it should not be negative (the value of N_{LX} may become negative if the global optimization approach is used in parameter extraction)!

Fig. 3.5.3 shows the modeled and measured V_{th} vs. channel length for devices with different pocket implant technologies at V_{ds} =0.05V and several body biases. Process *A* is the same as in Fig. 3.5.2. Process *B* in Fig. 3.5.3 is different from process *B* in Fig. 3.5.2, and is also a technology with p-pocket implant and *LDD* fabricated with a dual-gate CMOS shallow trench isolation process. However the conditions of boron implantation in the channel region and the pocket implantation are different from those in process *A*. The gate oxide thickness is 4nm [3.37].

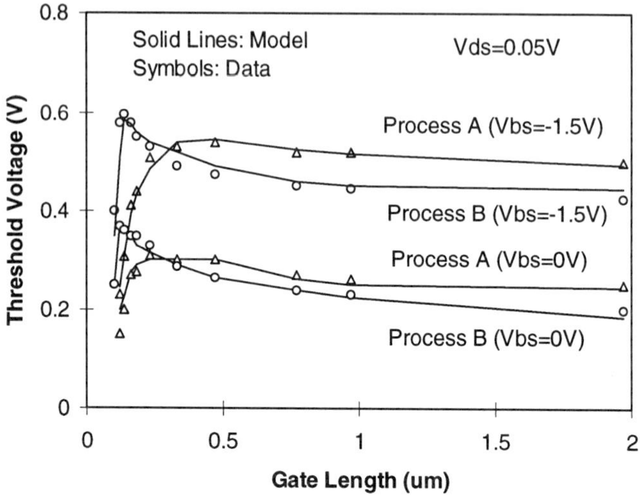

Fig. 3.5.3 Measured and BSIM3v3 modeled threshold voltage vs. channel length. The model can match the data well for the devices with different pocket implant conditions. After Cheng et al. [3.37].

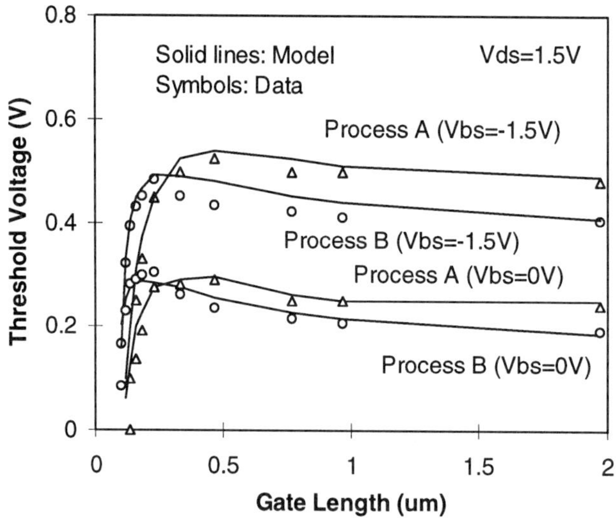

Fig. 3.5.4 Measured and BSIM3v3 modeled threshold voltage vs. channel length at V_{ds} =1.5V for the devices with different pocket implant conditions. After Cheng et al. [3.37].

Because process *A* and process *B* are two different processes, a single set of parameters was extracted for process *A* for all *L*'s while another set of parameters was extracted for process *B*. It can be seen that the data exhibits obvious reverse short channel effects, which are well modeled by BSIM3v3. V_{th} roll-up and roll-off are very strong for devices of process *B* with large-angle-tilt-implantation pocket. Even in this case, BSIM3v3 can model the V_{th} characteristics satisfactorily.

Fig.3.5.4 also shows the measured and modeled V_{th} vs. channel length but at V_{ds} =1.5V. The same sets of parameters used in Fig.3.5.3 are used here. This demonstrates that the model can accurately describe the short channel effects including both *DIBL* (V_{ds} effect) and V_{th} roll-off (*L* effect) for two different conditions of channel engineering. This is critical for circuit simulation because these two effects dominate the V_{th} characteristics in devices with very short channel lengths. Large simulation errors in the current characteristics will occur if the model cannot describe these effects accurately.

8. Understanding the binning approach for threshold voltage parameters

The threshold voltage model in BSIM3v3 accounts for the geometry and process dependencies. Therefore, it not recommended to use any binning approach for any model parameters in Eq. (3.4.25) although any and all of these parameters can be binned if the user desires. Sometimes binning is used even for a parameter like $\boldsymbol{V_{TH0}}$ by users according to their methodologies of parameter extraction. However this practice will lose the built-in geometry dependence in the BSIM3v3 V_{th} model and may cause confusion to circuit designers.

When binning is practiced for the parameter $\boldsymbol{V_{TH0}}$, the value of $\boldsymbol{V_{TH0}}$ in Eq. (3.4.25) will be calculated using the following equation

$$V_{th0,binning} = \boldsymbol{VTH0} + \frac{\boldsymbol{LVTH0}}{L_{eff}'} + \frac{\boldsymbol{WVTH0}}{W_{eff}'} + \frac{\boldsymbol{PVTH0}}{W_{eff}' L_{eff}'} \tag{3.5.5}$$

where $\boldsymbol{L_{VTH0}}$, $\boldsymbol{W_{VTH0}}$, and $\boldsymbol{P_{VTH0}}$ are binning parameters for $\boldsymbol{V_{TH0}}$[a], W_{eff}' and L_{eff} are the effective device channel width (without any bias dependence) and length.

---

a. Note: Users may have some confusion with the use of $\boldsymbol{V_{TH0}}$ in Eq. (3.5.5) and Eq. (3.4.25). The meaning of $\boldsymbol{V_{TH0}}$ in Eq. (3.5.5) is different from that in Eq. (3.4.25) even though the same symbol is assigned due to a historic reason.

Assuming V_{TH0}=0.6V, L_{VTH0}=0.01μmV, W_{VTH0}=0.02μmV, P_{VTH0}=0.03 μm^{2}V, L_{eff}=0.25μm, and W_{eff}'=1μm, the V_{TH0} used in Eq. (3.4.25) to calculate V_{th} for the simulation will be replaced with $V_{th0'binning}$, which here should be $V_{th0'binning}$=0.6+0.01/0.25+0.02/1+0.03/(1x0.25)=0.672V. That is, the value of 0.672V will be used for V_{TH0} in Eq. (3.4.25) instead of 0.6V.

A general description of the binning approach in compact models will be given in Chapter 13. The purpose of the above brief discussion is for the BSIM3v3 users to understand the V_{th} model parameters associated with the binning practice. We do not recommend that BSIM3 users use the binning approach in general since the model may lose its built-in *W* and *L* dependencies which are valuable in circuit optimization and statistical modeling.

9. Parameters in the threshold voltage model

The V_{th} model parameters are listed in Table 3.5.1.

Table 3.5.1 V_{th} model parameters

Symbols in equation	Symbols in source code	Description	Default	Unit
T_{OX}	*tox*	Gate oxide thickness	1.5x10^{-8}	m
T_{OXM}	*toxm*	Nominal T_{ox} at which parameter are extracted	T_{ox}	m
X_J	*xj*	Junction depth	1.5x10^{-7}	m
N_{CH}	*nch*	Channel doping concentration	1.7x10^{17}	1/cm^3
N_{SUB}	*nsub*	Substrate doping concentration	6.0x10^{16}	1/cm^3
V_{TH0}	V_{th0}	Threshold voltage @ V_{bs} =0 for large *L*. Typically V_{th0}>0 for NMOSFET and V_{th0}<0 for PMOSFET	0.7 for nMOS -0.7 for PMOS	V
V_{FB}	*vfb*	Flat band voltage	calculated	V
K_1	*k1*	First-order body effect coefficient	0.53	V$^{1/2}$
K_2	*k2*	Second-order body effect coefficient	-0.0186	none
K_3	*k3*	Narrow width coefficient	80.0	none
K_{3B}	*k3b*	Body effect coefficient of K_3	0.0	1/V

W_0	*w0*	Narrow width parameter	2.5x10^{-6}	m
N_{LX}	*nlx*	Lateral non-uniform doping coefficient	1.74x10^{-7}	m
D_{VT0W}	*dvt0w*	First coefficient of narrow width effect on V_{th} at small L	0	none
D_{VT1W}	*dvt1w*	Second coefficient of narrow width effect on V_{th} at small L	5.3x10^{6}	1/m
D_{VT2W}	*dvt2w*	Body-bias coefficient of narrow width effect on V_{th} at small L	-0.032	1/V
D_{VT0}	*dvt0*	First coefficient of short-channel effect on V_{th}	2.2	none
D_{VT1}	*dvt1*	Second coefficient of short-channel effect on V_{th}	0.53	none
D_{VT2}	*dvt2*	Body-bias coefficient of short-channel effect on V_{th}	-0.032	1/V
V_{BM}	*vbm*	Maximum applied body bias in V_{th} calculation	-3	V

References

[3.1] E. H. Nicollian and J. R. Brews, *MOS Physics and Technology*, Wiely-Interscience, New York, 1982.

[3.2] S. Wolf, *Silicon Processing for the VLSI Era, Volume 3- The Submicron MOSFET*, Lattice Press, Sunset Beach, CA, 1995.

[3.3] Y. A. El-Mansy and A. R. Boothroyd, "A simple two dimensional model for IGFET." *IEEE Trans. Electron Devices*, ED-24, pp. 254-262, 1977.

[3.4] M. E. Banna and M. E. Nokali, "A pseudo-two-dimensional analysis of short channel MOSFETs," *Solid-state Electronics*, Vol. 31 pp.269-274, 1988.

[3.5] S. M. Sze, *Semiconductor Devices: Physics and Technology*, John Wiley & Sons, New York, 1985.

[3.6] J. R. Brews, "The physics of the MOS transistor," in *Silicon Integrated Circuits*, Pt. A, Ed. D. Kahng, New York, Academic Press, 1981.

[3.7] L. D. Yau, "A simple theory to predict the threshold voltage of short channel IGFET's," *Solid-State Electronics*, vol. 17, pp.1059-1063, 1974.

[3.8] D. R. Pool and D. L. Kwong, "Two-dimensional analysis modeling of threshold voltage of short channel MOSFET's," *IEEE Electron Device Letters*, vol. EDL-5, 1984.

[3.9] G. W. Talor, "The effects of two-dimensional charge sharing on the above threshold characteristics of short channel IGFETs," *Solid-State Electronics*, vol.22, pp. 701-717, 1979.

[3.10] H. S. Lee, "An analysis of the threshold voltage for short-channel IGFET's," *Solid-State Electronics,* vol.16, p.1407, 1973.

[3.11] C. R. Viswanathan, B. C. Burkey, G. Lubberts, and T. J. Tredwell, "Threshold voltage in short channel MOS devices," *IEEE Trans. on Electron Devices*, vol. ED-32, pp. 932-940, 1985.

[3.12] R. R. Troutman, "VLSI limitations from drain-induced barrier lowering," *IEEE Trans. on Electron Devices*, vol. ED-26, p.461, 1979.

[3.13] T. A. Fjeldly and M. Shur, "Threshold voltage modeling and the subthreshold regime of operation of short channel MOSFET's," *IEEE Trans. on Electron Devices*, vol. ED-40, pp. 137-145, 1993.

[3.14] J. D. Kendall and A. R. Boothroyd, "A two dimensional analytical threshold voltage model for MOSFET's with arbitrarily doped substrate," *IEEE Electron Device Letters*, vol. EDL-7, p.407, 1986.

[3.15] Z. H. Liu et al., "Threshold voltage model for deep-submicron MOSFET's," *IEEE Trans. on Electron Devices*, vol. ED-40, pp.86-98, 1993.

[3.16] L. A. Akers and J. J. Sanchez, "Threshold voltage models of short, narrow and small geometry MOSFET's: A review," *Solid-State Electronics,* vol. 25, pp.621-641, 1982.

[3.17] S. S. S. Chung and C. T. Sah, "Threshold voltage models of the narrow gate effect in micron and submicron MOSFETs," *Solid-State Electronics,* vol. 31, pp.1009-1021, 1988.

[3.18] C. S. Rafferty et al., "Explanation of reverse short channel effect by defect gradients," *IEDM Tech. Dig*. p.311, 1993.

[3.19] J. H. Huang et al., *BSIM3 Manual (Version 2.0)*, University of California, Berkeley, March 1994.

[3.20] Y. Cheng et al., “A physical and scalable BSIM3v3 I-V model for analog/ digital circuit simulation”, *IEEE Trans. Electron Devices*, Vol. 44, pp.277-287, Feb. 1997.

[3.21] L. Lewyn and G. Masetti, "An IGFET inversion charge model for VLSI system," *IEEE Trans. on Electron Devices*, vol. ED-32, pp.434-440, 1985.

[3.22] R. F. Pirret, *Field Effect Devices*, Addison-Wesley, Reading, 1983.

[3.23] C. G. Sodini, T, W, Eketedt, and J. L. Moll, "Charge accumulation and mobility in thin dielectric MOS transistors," *Solid-State Electronics,* vol. 25, pp.833-841, 1982.

[3.24] J. R. Brews, "A charge-sheet model of the MOSFET," *Solid-State Electronics,* vol. 21, p. 345, 1978.

[3.25] S. C. Sun and J. D. Plummer, *IEEE Trans. on Electron Devices*, vol. ED-27, p. 1497, 1980.

[3.26] P. E. Allen and D. R. Holberg, *CMOS Analog Circuit Design,* Holt, Rinehart and Winston, Inc., New York, 1987.

[3.27] H. G. Lee, S. Y. Oh, and G. Fuller, "A simple and accurate method to measure the threshold voltage of an enhancement-mode MOSFET," *IEEE Trans. on Electron Devices*, vol. ED-29, pp.346-348, 1982.

[3.28] K. Lee, M. Shur, T. A. Fjeldly, and T. Ytterdal, *Semiconductor Device Modeling for VLSI*, Prentice Hall, Englewood Cliffs, New York, 1993.

[3.29] H. S. Wong, M. H. White, T. J. Kritsick, and R. V. Booth, "Modeling of transconductance degradation and extraction of threshold voltage in thin oxide MOSFET's," *Solid-State Electronics,* vol. 30, pp.953-968, 1987.

[3.30] N. Arora, *MOSFET Models for VLSI Circuit Simulation*, Springer-Verlag, Wien New York, 1994.

[3.31] Y. Cheng et al., "A unified MOSFET channel charge model for device modeling in circuit simulation," *IEEE Trans. Computer-aided Design of Integrated Circuits and Systems*, vol.17, pp.641-644, 1998.

[3.32] Y. P. Tsividis, *Operation and Modeling of the MOS Transistor*, McGraw-Hill, New York, 1987.

[3.33] Y. Cheng et al., *BSIM3 version 3.0 User's Manual*, University of California, Berkeley, 1995.

[3.34] Y. Cheng et al., *BSIM3 version 3.1 User's Manual*, University of California, Berkeley, Memorandum No. UCB/ERL M97/2, 1997.

[3.35] W. Liu et al., *BSIM3 version 3.2 User's Manual*, University of California, Berkeley, 1998.

[3.36] Y. Cheng et al., "Quarter-micron surface and buried channel P-MOSFET modeling for circuit simulation", *Semiconductor Science and Technology,* vol. 11, pp. 1763-1769, 1996.

[3.37] Y. Cheng, T. Sugii, K. Chen, and C. Hu, "Modeling of small size MOSFETs with reverse short channel and narrow width effects for circuit simulation", *Solid State Electronics*, vol. 41, (9), pp. 1227-1231, 1997.

CHAPTER 4 *I-V Model*

Accurate modeling of the *I-V* characteristics is a basic requirement for a good compact model. In this chapter we will introduce the essential equations that describe the *I-V* characteristics. We then discuss channel charge and mobility, which are the two key factors influencing the *I-V* characteristics. The single-equation *I-V* model of BSIM3v3 will be introduced after first discussing the piece-wise models. Finally, we discuss some points to understand the *I-V* model in BSIM3v3.

4.1 Essential Equations Describing the I-V Characteristics

The essential equations for describing the current in MOSFETs are

$$J_n = q\mu_n nE + qD_n\nabla n \tag{4.1.1a}$$

$$J_p = q\mu_p pE + qD_p\nabla p \tag{4.1.1b}$$

J_n and J_p are the current densities for electrons and holes respectively. q is the electron charge. μ_n and μ_p are the mobilities of electrons and holes. n and p are the electron and hole concentrations. E is the electric field. D_n and D_p are the diffusion coefficients of electrons and holes, respectively. D_n and D_p are linked to μ_n and μ_p with the following Einstein's relationship:

$$D_n = v_t \mu_n \tag{4.1.2a}$$

$$D_p = v_t \mu_p \tag{4.1.2b}$$

where v_t is the thermal voltage.

The E terms in Eq. (4.1.1) represent the drift current components due to the electric field E. The second terms of Eq. (4.1.1) describe the diffusion current components due to the carrier concentration gradient. In the strong inversion region, as we mentioned in section 2.3, the current is dominated by the drift current. In the subthreshold region, the diffusion current component dominates. However, in the transition region (moderate inversion region) from subthreshold to strong inversion, both drift and diffusion currents are important.

As shown in Eq. (4.1.1), carrier density and the velocity-field relationship (μE) are two fundamental factors determining the *I-V* characteristics. We need to model the channel charge and mobility as well as the velocity-field relationship carefully to describe the current characteristics accurately and physically. We discuss the modeling of channel charge and mobility next before we introduce the modeling of the *I-V* behavior.

4.2 Channel Charge Density Model

In this section, we discuss the modeling of the inversion charge density in the channel, using an n channel MOSFET as an example. It should be pointed out that we ignore the sign of the channel charge to simplify the equations in the following discussion.

4.2.1 Channel charge model in the strong inversion region

In strong inversion the inversion charge density, Q_{inv}, is much larger than the depletion charge density Q_b. As a result, the exponential term in Eq. (2.3.17) is the dominant term, and the inversion charge can be written as [4.1],

$$Q_{inv} \approx \sqrt{2q\varepsilon_{si}N_a v_t}\,\exp[(\phi_s - 2\phi_B)/2v_t] \tag{4.2.1}$$

In Eq. (4.2.1), the inversion layer charge is an exponential function of the surface potential, ϕ_s, with a slope of $1/(2v_t)$. Therefore, a small increase in the surface potential induces a large change in Q_{inv}.

Eq. (4.2.1) shows the dependence of the inversion charge on the surface potential. However, in compact modeling a relationship between the inversion charge density and the gate bias is preferred. According to Eq. (2.3.3a), the inversion charge density in strong inversion can be written as,

$$Q_{inv} = C_{ox}(V_{gs} - V_{FB} - \phi_s) - Q_b \tag{4.2.2}$$

Substituting Q_b in Eq. (4.2.2) with Eq. (2.3.15) and recalling the threshold voltage expression given in Eq. (2.2.1), we obtain,

$$Q_{inv} = C_{ox}(V_{gs} - V_{th}) \tag{4.2.3}$$

Eq. (4.2.3) has been widely used in the compact modeling of MOSFETs.

4.2.2 Channel charge model in the subthreshold region

Eq. (2.3.20) gives the charge expression in subthreshold region, but is still not acceptable for use in compact modeling because of presence of the function $\phi_s(V_g)$. In the following, we will derive a relationship between the surface potential and the gate bias which is suitable for compact modeling.

Because $\phi_B < \phi_s < 2\phi_B$ in the weak inversion region, we can express V_{gs} as

$$V_{gs} \approx V_{gs}|_{\phi s=1.5\phi B} + \frac{\partial V_{gs}}{\partial \phi_s}|_{\phi s=1.5\phi B}(\phi_s - 1.5\phi_B) \tag{4.2.4}$$

We have defined that $V_{gs} = V_{th}$ when $\phi_s=2\phi_B$ so we have the following,

$$\begin{aligned} V_{th} &\approx V_{gs}|_{\phi s=1.5\phi B} + \frac{\partial V_{gs}}{\partial \phi_s}|_{\phi s=1.5\phi B}(2\phi_B - 1.5\phi_B) \\ &= V_{gs}|_{\phi s=1.5\phi B} + \frac{\partial V_{gs}}{\partial \phi_s}|_{\phi s=1.5\phi B}0.5\phi_B \end{aligned} \tag{4.2.5}$$

Eq. (4.2.5) can be rewritten as,

$$V_{gs}|_{\phi_s=1.5\phi_B} = V_{th} - 0.5\phi_B \frac{\partial V_{gs}}{\partial \phi_s}|_{\phi_s=1.5\phi_B} \tag{4.2.6}$$

Combining Eq. (4.2.4) and Eq. (4.2.6) yields,

$$V_{gs} \approx V_{th} + \frac{\partial V_{gs}}{\partial \phi_s}|_{\phi_s=1.5\phi_B}(\phi_s - 2\phi_B) \tag{4.2.7}$$

Combining Eq. (2.3.3a) and Eq. (2.3.13) (because $Q_{inv} << Q_b$ in weak inversion) we find the relationship between V_{gs} and the surface potential as follows,

$$V_{gs} = V_{FB} + \phi_s + \gamma\sqrt{\phi_s} \tag{4.2.8}$$

Therefore,

$$\frac{\partial V_{gs}}{\partial \phi_s} = 1 + \frac{\gamma}{2\sqrt{\phi_s}} \tag{4.2.9}$$

we then define,

$$C_{dep} = \frac{\varepsilon_{si}}{X_{dep}} \tag{4.2.10}$$

where X_{dep} is the width of the depletion layer under the channel, and is given in Eq. (2.3.11).

Thus, Eq. (4.2.9) can be rewritten as (recalling the expression of γ given in Eq. (2.2.3))

$$\frac{\partial V_{gs}}{\partial \phi_s} = 1 + \frac{C_{dep}}{C_{ox}} \equiv n \tag{4.2.11}$$

n can be approximated as a constant in the subthreshold region. As a result, we can obtain approximately the following simple relationship between the surface potential and the gate bias in the weak inversion region,

$$V_{gs} \approx V_{th} + n(\phi_s - 2\phi_B) \tag{4.2.12}$$

Eq. (4.2.12) has been used widely in compact models in the subthreshold region [2.61].

Substituting Eq. (4.2.12) into Eq. (2.3.20) and approximating the ϕ_s outside the exponential term with $2\phi_B$, which is a reasonable approximation in the subthreshold region, the inversion charge density in the weak inversion regime can be written as

$$Q_{inv} \approx \sqrt{\frac{q\varepsilon_{si}N_a}{4\phi_B}}\, v_t \exp\left(\frac{V_{gs} - V_{th} - V_{OFF}}{n v_t}\right) \tag{4.2.13}$$

V_{OFF} is added to account for the difference between V_{th} in the strong inversion and the subthreshold regions as discussed in Chapter 3 [4.6, 4.7].

4.2.3 Continuous channel charge model of BSIM3v3

Separate expressions for channel charge density have been given in Eq. (4.2.3) and Eq. (4.2.6) for the strong inversion and the weak inversion regimes, respectively. We can combine these expressions in the following form [4.5]:

$$Q_{chs0} = C_{ox} V_{gsteff} \tag{4.2.14}$$

$$V_{gsteff} = \frac{2\, n\, v_t \ln\left[1 + \exp\left(\dfrac{V_{gs} - V_{th}}{2\, n\, v_t}\right)\right]}{1 + 2\, n\, C_{ox}\sqrt{\dfrac{2\phi_s}{q\varepsilon_{si}N_{CH}}}\exp\left(-\dfrac{V_{gs} - V_{th} - 2V_{OFF}}{2\, n\, v_t}\right)} \tag{4.2.15}$$

As shown in Figs. 4.2.1 and 4.2.2, V_{gsteff} becomes V_{gs}-V_{th} in the strong inversion region, and follows $\sqrt{\frac{q\varepsilon_{si}N_{CH}}{4\phi_B}}\frac{v_t}{C_{ox}}\exp\left(\frac{V_{gs} - V_{th} - V_{OFF}}{n v_t}\right)$ in the subthreshold region. To demonstrate the continuity of V_{gsteff}, the curves of V_{gsteff} and its first and second derivatives versus V_{gs} are shown in Fig. 4.2.3.

The form of Eq. (4.2.15) was chosen to obtain a continuous equation for the channel charge and to match the measured Q_{chs}-V_{gs} characteristics in the moderate inversion (transition) region. The channel charge expression, Q_{chs0}, will be used in subsequent sections of this chapter to model the drain current.

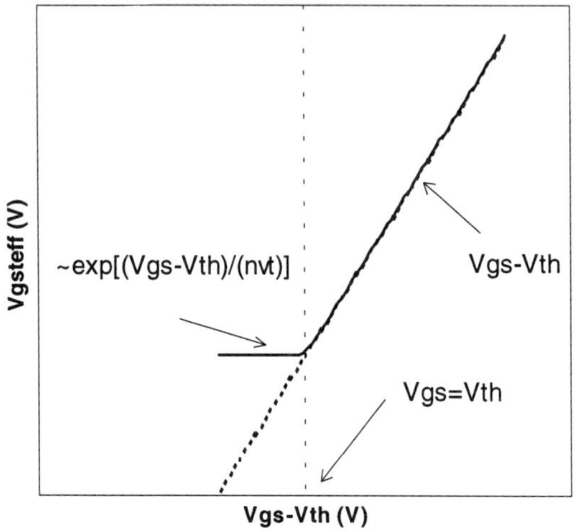

Fig. 4.2.1 V_{gsteff} vs. V_{gs}-V_{th} in a linear plot.

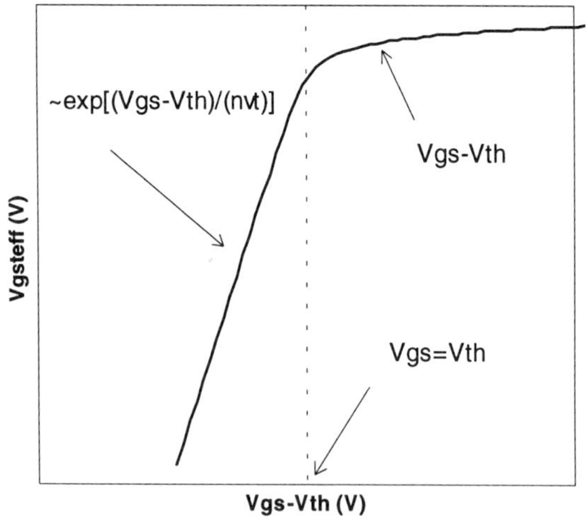

Fig. 4.2.2 V_{gsteff} vs. V_{gs}-V_{th} in a semi-logarithmic plot.

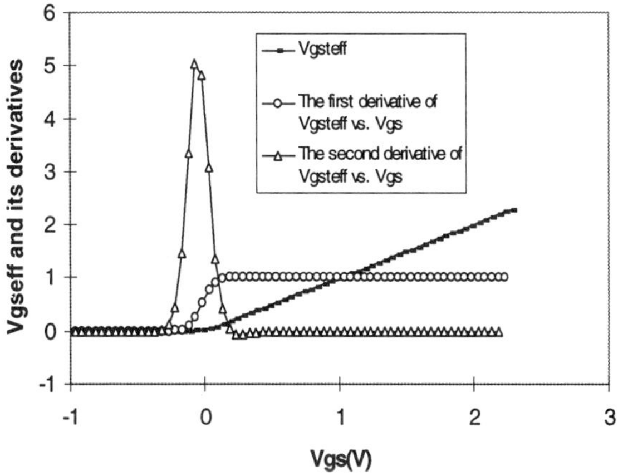

Fig. 4.2.3 V_{gsteff} and its first and second derivatives are continuous function of V_{gs} from the subthreshold through the strong inversion regions.

The charge density model has been verified with experimental data at different bias and design conditions [4.5, 4.8, 4.9]. Fig. 4.2.4 shows that the unified charge expression, Eq. (4.2.14), fits the data and matches Eq. (4.2.3) and Eq. (4.2.13) well in the strong inversion and subthreshold regions, respectively.

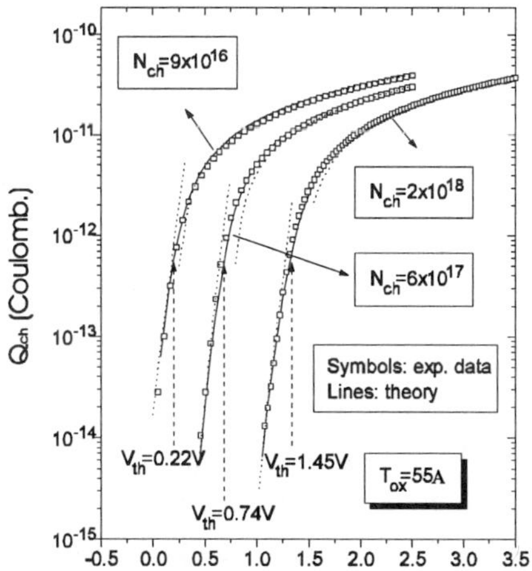

Fig. 4.2.4 The continuous channel charge model fits the measurement data taken from devices with differing channel doping concentrations. It also matches Eq. (4.2.3) and Eq. (4.2.13) (dashed lines) in the strong inversion and subthreshold regions respectively. The unified model covers the weak, moderate, and strong inversion regions of MOSFETs. After Cheng et al. [4.5].

Furthermore, the model accurately predicts the charge in the transition (moderate inversion) region. The continuous and accurate nature of the model makes it very attractive and promising in circuit simulation since the moderate inversion region is becoming more important for low voltage/power circuit applications.

4.2.4 Continuous channel charge model with the effect of V_{ds}

Eq. (4.2.14) is the unified channel charge expression at the source or for V_{ds}=0. For a charge density model to be used in modeling *I-V* characteristics, the influence of V_{ds} on the channel charge must be accounted for. In other words, the channel charge density model has to include a dependency on the channel potential which varies along the channel. For this purpose, consider first the channel charge density for the case of strong inversion in the presence of V_{ds}:

$$Q_{chs(y)} = C_{ox}(V_{gs} - V_{th} - A_{bulk}V_{F(y)}) \tag{4.2.16}$$

where $V_{F(y)}$ stands for the quasi-Fermi potential at any given point y along the channel with respect to the source. A_{bulk} is a parameter accounting for the bulk charge effect due to V_{ds} as discussed in Chapter 2, and its detailed form will be given in section 4.5. Eq. (4.2.16) has been used widely in modeling the *I-V* characteristics in the strong inversion region. This equation can also be written as:

$$Q_{chs(y)} = Q_{chs0} + \Delta Q_{chs(y)} \tag{4.2.17}$$

where Q_{chs0} is given by Eq. (4.2.3), and $\Delta Q_{chs(y)}$ is the increment in the channel charge density induced by the drain voltage and can be given as:

$$\Delta Q_{chs(y)} = -C_{ox}A_{bulk}V_{F(y)} \tag{4.2.18}$$

In the subthreshold region (V_{gs}<<V_{th}), the channel charge density along the channel from source to drain can be written as:

$$Q_{chsubs(y)} = Q_{chsubs0}\exp(-\frac{A_{bulk}V_{F(y)}}{nv_t}) \tag{4.2.19}$$

where $Q_{chsubs0}$ is given by Eq. (4.2.13).

We expand the exponential term in Eq. (4.2.19) using Taylor series and consider the first and second terms:

$$Q_{chsubs(y)} = Q_{chsubs0}\left(1 - \frac{A_{bulk}V_{F(y)}}{nv_t}\right) \tag{4.2.20}$$

Analogous to Eq. (4.2.17), Eq. (4.2.20) can be written as:

$$Q_{chsubs(y)} = Q_{chsubs0} + \Delta Q_{chsubs(y)} \tag{4.2.21}$$

The parameter $\Delta Q_{chsubs}(y)$ is the incremental channel charge density induced by the drain voltage in the subthreshold region. It can be written as:

$$\Delta Q_{chsubs(y)} = -\frac{A_{bulk}V_{F(y)}}{nv_t}Q_{chsubs0} \tag{4.2.22}$$

Note that Eq. (4.2.22) is valid only when $V_{F(y)}$ is very small. This condition is met due to the fact that Eq. (4.2.20) is only used in the linear regime (i.e. $V_{ds} \leq 2v_t$) in the subthreshold region.

Eqs. (4.2.16) and (4.2.20) separately describe the drain voltage dependencies. However, a unified expression for Q_{ch}(y) is needed. To obtain a unified expression along the channel we let:

$$\Delta Q_{ch(y)} = \frac{\Delta Q_{chs(y)}\Delta Q_{chsubs(y)}}{\Delta Q_{chs(y)} + \Delta Q_{chsubs(y)}} \tag{4.2.23}$$

Here, $\Delta Q_{ch}(y)$ is the incremental channel charge density, including both strong inversion and subthreshold regions, induced by the drain voltage. Substituting Eq. (4.2.18) and Eq. (4.2.22) into Eq. (4.2.23), and noticing that the term $C_{ox}V_{gsteff}/Q_{chsubs0}$ is equal to 1 in the subthreshold region and approaches a very small value in strong inversion, we obtain:

$$\Delta Q_{ch(y)} = -\frac{V_{F(y)}}{V_b}Q_{chs0} \tag{4.2.24}$$

where $V_b=(V_{gsteff}+nvt)/A_{bulk}$. In order to remove the complexity caused by the variable n, it is replaced with 2. This is a reasonable approximation because n ranges typically from 1 ~ 2. The expression for V_b now becomes:

$$V_b = \frac{V_{gsteff} + 2v_t}{A_{bulk}} \tag{4.2.25}$$

A unified expression for Q_{ch}(y) from the subthreshold to the strong inversion regimes is now at hand [4.5, 4.7]:

$$Q_{ch(y)} = Q_{chs0}(1 - \frac{V_{F(y)}}{V_b}) \tag{4.2.26}$$

Eq. (4.2.26) will be used in the derivation of the single-equation *I-V* model of BSIM3v3.

4.3 Mobility Model

4.3.1 Piece-wise mobility models

A good model for the carrier surface mobility is critical to the accuracy of a MOSFET model. The scattering mechanisms responsible for the surface mobility include phonons, coulombic scattering, and surface roughness scattering [4.10, 4.11]. For good quality interfaces, phonon scattering is the dominant scattering mechanism at room temperature. In general, mobility depends on many process parameters and bias conditions. For example, mobility depends on the gate oxide thickness, doping concentration, threshold voltage, gate voltage and substrate voltage, etc. Sabnis and Clemens [4.12] proposed an empirical unified formulation based on the concept of an effective field, E_{eff}, which lumps many process parameters and bias conditions together. E_{eff} is defined by

$$E_{eff} = \frac{Q_B + (Q_{inv}/2)}{\varepsilon_{si}} \tag{4.3.1}$$

For an NMOS transistor with n-type poly-silicon gate or a PMOS transistor with p-type gate, Eq. (4.3.1) can be rewritten in a more useful form that explicitly relates E_{eff} to the device parameters by noting $V_{th} = V_{FB} + 2\phi_B + \frac{Q_B}{C_{ox}}$, $Q_{inv} = C_{ox}(V_{gs} - V_{th})$ and $V_{FB} + 2\phi_B \approx 0$. Substituting the above equations into Eq. (4.3.1) one obtains [4.13]:

$$E_{eff} = \frac{V_{gs} + V_{th}}{6T_{ox}} \tag{4.3.2}$$

The physical meaning of E_{eff} can be interpreted as the average electrical field experienced by the carriers in the inversion layer [4.13]. The unified formulation of the mobility is then empirically given by:

$$\mu_{eff} = \frac{\mu_0}{1 + (E_{eff}/E_0)^{\nu}} \tag{4.3.3}$$

Values for μ_0, E_0, and ν were reported by Liang et al. [4.10], and recently Chen et al. [4.13] given in Table 4.3.1:

Table 4.3.1 Mobility and related parameters for electrons and holes.

Parameter	Electron (surface)	Hole (surface)
μ0 (cm2/V)	540	185
E0 (MV/cm)	0.9	0.45
ν	1.85	1.0

Eq. (4.3.3) fits the experimental data very well [4.10], but it involves a power function which is time consuming for circuit simulators such as SPICE. A Taylor expansion of the exponential function in Eq. (4.3.3) is used, and the first three terms are taken with the coefficients left to be determined by fitting to the linear *I-V* data. Thus, we have [4.14]:

$$\mu_{eff} = \frac{\mu_0}{1 + (U_a + U_c V_{bs})\dfrac{V_{gs} + V_{th}}{T_{ox}} + U_b[(V_{gs} + V_{th})/T_{ox}]^2} \tag{4.3.4}$$

where the term $U_c V_{bs}$ is introduced to improve the model accuracy at high body bias. U_a, U_b, and U_c are parameters to be determined by the *I-V* data.

The discussion given above is for devices in the strong inversion region. It can be seen that the mobility in strong inversion is a function of the gate bias. In the subthreshold region the accuracy of the mobility is not as critical because Q_{inv} varies rapidly with V_g and cannot be modeled accurately. It is usually modeled as a constant.

4.3.2 Mobility models in BSIM3v3

The continuity of mobility model is also required to ensure the continuity of the I-V model. To achieve continuity in the mobility model, BSIM3v3 uses a unified mobility expression based on the V_{gsteff} expression of Eq. (4.2.15) [4.7],

$$\mu_{eff} = \frac{\mu_o}{1 + (U_A + U_C V_{bseff})\left(\frac{V_{gsteff} + 2V_{th}}{Tox}\right) + U_B\left(\frac{V_{gsteff} + 2V_{th}}{Tox}\right)^2} \tag{4.3.5}$$

where V_{bseff} is given in Eq. (3.4.26). It can be seen that Eq. (4.3.5) follows Eq. (4.3.4) in strong inversion, and becomes a constant in the subthreshold region.

Several mobility model options are provided for users to choose in BSIM3v3. A selector parameter called ***mobMod*** is introduced for this purpose. The mobility expression in Eq. (4.3.5) has been designated as ***mobMod***=1.

The following empirical mobility model option (***mobMod***=2) is better suited for depletion mode devices [4.7]:

$$\mu_{eff} = \frac{\mu_o}{1 + (U_A + U_C V_{bseff})\left(\frac{V_{gsteff}}{Tox}\right) + U_B\left(\frac{V_{gsteff}}{Tox}\right)^2} \tag{4.3.6}$$

BSIM3v3 also introduced a third mobility model option (***mobMod***=3) [4.7]:

$$\mu_{eff} = \frac{\mu_o}{1 + \left[U_A\left(\frac{V_{gsteff} + 2V_{th}}{Tox}\right) + U_B\left(\frac{V_{gsteff} + 2V_{th}}{Tox}\right)^2\right](1 + U_C V_{bseff})} \tag{4.3.7}$$

It is clear that all of the mobility models given above approach constant values that are independent of V_g when $V_{gs}<V_{th}$.

It should be pointed out that all of the mobility models given above account for only the influence of the vertical electrical field. The influence of the lateral electrical field on the mobility will be considered when discussing the velocity saturation effect in the next session.

4.4 I-V Model in the Strong Inversion Region

4.4.1 *I-V* model in the linear (triode) region

1. Intrinsic case (R_{ds} =0)

In the strong inversion region, the current equation at any point y along the channel is [4.15]

$$I_{ds} = W_{eff} C_{ox} (V_{gst} - A_{bulk} V(y)) v(y) \tag{4.4.1}$$

where $V_{gst} = (V_{gs} - V_{th})$, W_{eff} is the effective device channel width. C_{ox} is the gate capacitance per unit area. $V(y)$ is the potential difference between the channel and the source. A_{bulk} is the coefficient accounting for the bulk charge effect and $v(y)$ is the velocity of carriers.

BSIM3 *I-V* formulation starts with a simple piece-wise saturation velocity model [4.16],

$$v(y) = \mu E_y \qquad E_y < E_{sat} \tag{4.4.2a}$$

$$v(y) = v_{sat} \qquad E_y > E_{sat} \tag{4.4.2b}$$

where E_y is the magnitude of the lateral electric field and E_{sat} is the critical electric field at which the carrier velocity becomes saturated. μ is the mobility including the influence of the lateral electric field E_y and is given by

$$\mu = \frac{\mu_{eff}}{1 + (E_y / E_{sat})} \tag{4.4.3}$$

In order to have a continuous velocity model at $E_y = E_{sat}$, E_{sat} satisfies

$$E_{sat} = \frac{2 v_{sat}}{\mu_{eff}} \tag{4.4.4}$$

Thus, before the electric field reaches E_{sat} the drain current can be expressed as,

$$I_{ds} = W_{eff} C_{ox} (V_{gs} - V_{th} - A_{bulk} V(y)) \frac{\mu_{eff} E_y}{1 + E_y / E_{sat}} \tag{4.4.5}$$

Eq. (4.4.5) can be rewritten as

$$E_y = \frac{I_{ds}}{\mu_{eff} W_{eff} C_{ox} (V_{gst} - A_{bulk} V(y)) - I_{ds} / E_{sat}} = \frac{dV(y)}{dy} \tag{4.4.6}$$

By integrating Eq. (4.4.6) from $y = 0$ to $y = L_{eff}$, the effective channel length, and $V(y) = 0$ to $V(y) = V_{ds}$, we arrive at

$$I_{ds} = \mu_{eff} C_{ox} \frac{W_{eff}}{L_{eff}} \frac{1}{1 + V_{ds} / E_{sat} L_{eff}} (V_{gs} - V_{th} - A_{bulk} V_{ds} / 2) V_{ds} \tag{4.4.7}$$

The drain current model in Eq. (4.4.7) is valid before the carrier velocity saturates, that is, in the linear or the triode region.

2 Extrinsic case (R_{ds} >0)

The parasitic source/drain resistance is an important device parameter which can affect MOSFET performance significantly in short channel devices. The most straightforward and accurate way of modeling the parasitic resistance effect is to use a circuit with resistors in series with the intrinsic MOSFET. This leads to a complicated drain current expression. In order to make the model efficient, the drain current in the linear region can be modeled by extending Eq. (4.4.7) as [4.7, 4.14]

$$I_{ds} = \frac{I_{ds0}}{1 + R_{ds} I_{ds0} / V_{ds}} \tag{4.4.8}$$

where I_{ds0} is the intrinsic current expression given by Eq. (4.4.7). R_{ds} is a variable to account for the influence of the parasitic resistances at the source and drain.

4.4.2 Drain voltage at current saturation, V_{dsat}

1. Intrinsic case (R_{ds} =0)

If the drain voltage (and hence the lateral electric field) is sufficiently high, the carrier velocity near the drain saturates. The channel may be divided into two portions: one adjacent to the source where the carrier velocity is field-dependent and the others adjacent to the drain where the velocity has saturated. At the boundary between the two portions, the channel voltage is the saturation voltage (V_{dsat}) and the lateral electric field is equal to E_{sat}. We can substitute $v = v_{sat}$ and $V_{ds} = V_{dsat}$ into Eq. (4.4.1) to obtain the saturation current:

$$I_{dsat} = WeffC_{ox}(V_{gst} - AbulkV_{dsat})v_{sat} \tag{4.4.9}$$

By equating Eqs. (4.4.1) and (4.4.9) at $V_{ds} = V_{dsat}$, we can solve for the saturation voltage V_{dsat}:

$$V_{dsat} = \frac{E_{sat}Leff(V_{gs} - V_{th})}{AbulkE_{sat}Leff + (V_{gs} - V_{th})} \tag{4.4.10}$$

2. Extrinsic case (R_{ds} >0)

Due to the parasitic resistance, the saturation voltage V_{dsat} will be larger than what is predicted by Eq. (4.4.10). Equating Eq. (4.4.8) with Eq. (4.4.9), V_{dsat} with parasitic resistance R_{ds} may be found to be [4.7, 4.14]

$$V_{dsat} = \frac{-b - \sqrt{b^2 - 4ac}}{2a} \tag{4.4.11a}$$

$$a = A_{bulk}^2 R_{ds} C_{ox} Weffv_{sat} + (\frac{1}{\lambda} - 1)Abulk \tag{4.4.11b}$$

$$b = -[V_{gst}(\frac{2}{\lambda} - 1) + A_{bulk}E_{sat}Leff + 3A_{bulk}R_{ds}C_{ox}Weffv_{sat}V_{gst}] \tag{4.4.11c}$$

$$c = E_{sat}LeffV_{gst} + 2R_{ds}C_{ox}Weffv_{sat}V_{gst}^2 \tag{4.4.11d}$$

$\lambda = A_1V_{gst} + A_2$ is introduced to account for the non-saturating effect of the device *I-V* which will be discussed in section 4.5 in this chapter.

4.4.3 Current and output resistance in the saturation region

A typical *I-V* curve and its output resistance are shown in Fig. 2.7.4. If we only look at the drain current the *I-V* curve can be divided into two parts (without considering the breakdown region): (1) the linear region in which the drain current clearly increases with the drain voltage; and (2) the saturation region in which the drain current has only a weak dependence on the drain voltage. However, the first-order derivative reveals more detailed information about the physical mechanisms which are involved in the saturation region. The output resistance, which is the reciprocal of the derivative of the *I-V* curve, is shown in Fig. 2.7.4. It can be clearly divided into four regions where the R_{out} - V_{ds} dependencies are different [4.17].

The first region is the linear or triode region in which I_d is not saturated. The output resistance is very small because the drain current has a strong dependence on the drain voltage. The other three regions belong to the saturation region. There are three physical mechanisms which affect the output resistance in the saturation region: channel length modulation (*CLM*) [4.18, 4.19], drain-induced barrier lowering (*DIBL*) [4.20, 4.21, 4.22], and the substrate current induced body effect (*SCBE*) [4.7, 4.14]. All three mechanisms affect the output resistance in the saturation range, but each of them dominates in one of three distinct regions.

The drain current depends on the drain voltage only weakly in the saturation region. A Taylor series can be used to expand the drain current in the saturation region [4.7, 4.14, 4.17].

$$I_{ds}(V_{gs},V_{ds}) = I_{ds}(V_{gs},V_{dsat}) + \frac{\partial I_{ds}(V_{gs},V_{ds})}{\partial V_{ds}}(V_{ds}-V_{dsat})$$
$$\equiv I_{dsat}(1+\frac{V_{ds}-V_{dsat}}{V_A}) \tag{4.4.12}$$

where

$$I_{dsat} = I_{ds}(V_{gs},V_{dsat}) = W_{eff}v_{sat}C_{ox}(V_{gst} - A_{bulk}V_{dsat}) \tag{4.4.13}$$

$$V_A = I_{dsat}(\frac{\partial I_{ds}}{\partial V_{ds}})^{-1} \tag{4.4.14}$$

V_A is called the Early voltage (analogous to the BJT case) and is introduced for the analysis of the output resistance in the saturation region. Only the first order term is kept in the Taylor expansion.

1. V_A due to channel length modulation

Considering the influence of channel length modulation only, the Early voltage can be calculated by [4.17]

$$V_{ACLM} = I_{dsat}\left(\frac{\partial I_{ds}}{\partial Leff}\frac{\partial Leff}{\partial V_{ds}}\right)^{-1} = \frac{A_{bulk}E_{sat}Leff + V_{gst}}{A_{bulk}E_{sat}}\left(\frac{\partial \Delta L}{\partial V_{ds}}\right)^{-1} \quad (4.4.15)$$

where ΔL is the length of the velocity saturation region (the operational channel length is $L_{eff}-\Delta L$).

Based on the quasi-two dimensional approximation, i.e. using Eq. (2.6.3), V_{ACLM} can be derived [4.17]

$$V_{ACLM} = \frac{A_{bulk}E_{sat}Leff + V_{gst}}{A_{bulk}E_{sat}l}(V_{ds} - V_{dsat}) \quad (4.4.16)$$

A parameter $\boldsymbol{P_{CLM}}$ is then introduced into the V_{ACLM} expression to compensate for the uncertainty over the value of the parameter l given in section 2.6. Thus, the V_{ACLM} becomes:

$$V_{ACLM} = \frac{1}{\boldsymbol{P_{CLM}}}\frac{A_{bulk}E_{sat}Leff + V_{gst}}{A_{bulk}E_{sat}l}(V_{ds} - V_{dsat}) \quad (4.4.17)$$

2. V_A due to drain-induced barrier lowering

As discussed in section 2.2, threshold voltage is a linear function of the drain voltage. According to Eq. (4.4.14) and Eq. (3.4.12), the Early voltage due to the *DIBL* effect can be calculated as [4.17].

$$V_{ADIBL} = I_{dsat}\left(\frac{\partial I_{ds}}{\partial V_{th}}\frac{\partial V_{th}}{\partial V_{ds}}\right)^{-1} = \frac{1}{\theta_{th}(L)}\left(V_{gst} - \left(\frac{1}{A_{bulk}V_{dsat}} + \frac{1}{V_{gst}}\right)^{-1}\right) \quad (4.4.18)$$

In the derivation of Eq. (4.4.18), the parasitic resistance is assumed to be 0. As expected, V_{ADIBL} is a strong function of L as seen in Eq. (4.4.18). As channel length decreases, V_{ADIBL} drops very quickly. On the other hand, V_{ADIBL} is independent of V_{ds} because only the first order term in the Taylor expansion is kept in Eq. (4.4.12). The combination of the *CLM* and *DIBL* effects determines the output resistance in the third region, as shown in Fig. 2.7.3.

To model the output resistance in the saturation region more accurately, the coefficient $\theta_{th(L)}$ given by Eq. (3.4.14) is replaced by $\theta_{rout(L)}$. $\theta_{th(L)}$ and $\theta_{rout(L)}$ have the same channel length dependency, but independent coefficients.

$$\theta_{rout}(L) = \boldsymbol{P_{DIBLC1}}[\exp(-\boldsymbol{D_{ROUT}} L / 2l_t) + 2\exp(-\boldsymbol{D_{ROUT}} L / l_t)] + \boldsymbol{P_{DIBLC2}} \tag{4.4.19}$$

Thus, Eq. (4.4.18) becomes

$$V_{ADIBL} = \frac{1}{\theta_{rout}(L)}(V_{gst} - (\frac{1}{A_{bulk}V_{dsat}} + \frac{1}{V_{gst}})^{-1}) \tag{4.4.20}$$

$\boldsymbol{P_{DIBLC1}}$, $\boldsymbol{P_{DIBLC2}}$ and $\boldsymbol{D_{ROUT}}$ in Eq. (4.4.19) are the parameters for the *DIBL* effect in the saturation region. The reason why $\boldsymbol{D_{VT0}} \neq \boldsymbol{P_{DIBLC1}}$ and $\boldsymbol{D_{VT1}} \neq \boldsymbol{D_{ROUT}}$ is that the gate voltage modulates the *DIBL* effect. When the threshold voltage is determined, the gate voltage is equal to the threshold voltage. But in the saturation region where the output resistance is modeled, the gate voltage is much larger than the threshold voltage. Drain induced barrier lowering may not be the same at different gate biases. $\boldsymbol{P_{DIBLC2}}$ is usually very small (may be as small as $8\text{x}10^{-3}$). However it is an important parameter in V_{ADIBL} for long channel devices because $\boldsymbol{P_{DIBLC2}}$ will be dominant in Eq. (4.4.19) when the channel length is long.

3. V_A due to substrate current induced body effect

When the electric field near the drain is large (*>0.1MV/cm*), some electrons coming from the source will be energetic (hot) enough to cause impact ionization, creating electron-hole pairs. The substrate current I_{sub} created during impact ionization will increase exponentially with the drain voltage. A well known I_{sub} model was presented in [4.23]:

$$I_{sub} = \frac{A_i}{B_i} I_{dsat}(V_{ds} - V_{dsat})\exp(-\frac{B_i l}{V_{ds} - V_{dsat}}) \tag{4.4.21}$$

where A_i and B_i are experimentally determined parameters and l is given in section 2.6. As I_{sub} flows through the substrate resistance, R_{sub}, it will induce a source-body bias equal to $I_{sub}R_{sub}$. This will cause V_{th} to decrease and I_d to increase.

Based on the above, the Early voltage due to the substrate current induced body effect, V_{ASCBE}, can be written as:

$$V_{ASCBE} \propto \exp(\frac{B_i l}{V_{ds} - V_{dsat}}) \tag{4.4.22}$$

From Eq. (4.4.22) we can see that V_{ASCBE} is a strong function of V_{ds}. *SCBE* is important in the high drain voltage region. The channel length and gate oxide dependence of V_{ASCBE} comes from V_{dsat} and l. BSIM3 introduces $\boldsymbol{P_{SCBE1}}$ and $\boldsymbol{P_{SCBE2}}$, and models V_{ASCBE} as:

$$\frac{1}{V_{ASCBE}} = \frac{\boldsymbol{P_{SCBE2}}}{L_{eff}} \exp(-\frac{\boldsymbol{P_{SCBE1}} l}{V_{ds} - V_{dsat}}) \tag{4.4.23}$$

$\boldsymbol{P_{SCBE1}}$ and $\boldsymbol{P_{SCBE2}}$ are extracted from the measured *I-V* data.

4. V_A at the saturation point ($V_{ds}=V_{dsat}$)

In order to have continuous drain current and output resistance at the linear-saturation transition point, the V_{Asat} parameter is introduced into the Early Voltage expression. V_{Asat} is the Early voltage at $V_{ds}=V_{dsat}$ and can be obtained by differentiating Eq. (4.4.8) with respect to V_{ds} and using Eq. (4.4.14) [4.17],

$$V_{Asat} = \frac{E_{sat} L_{eff} + V_{dsat} + 2R_{ds} v_{sat} C_{ox} W_{eff} (V_{gst} - A_{bulk} V_{ds} / 2)}{1 + A_{bulk} R_{ds} v_{sat} C_{ox} W_{eff}} \tag{4.4.24}$$

5. The full current expression in the saturation region

The total Early voltage V_A without consideration of SCBE can be written as:

$$V_A = V_{Asat} + (\frac{1}{V_{ACLM}} + \frac{1}{V_{ADIBL}})^{-1} \tag{4.4.25}$$

To improve the model accuracy further, another parameter P_{VAG} is introduced in V_A to account for the gate bias dependence of V_A more accurately. Thus, the Early voltage becomes:

$$V_A = V_{Asat} + \left(1 + \frac{P_{VAG} V_{gst}}{E_{sat} L_{eff}}\right)\left(\frac{1}{V_{ACLM}} + \frac{1}{V_{ADIBL}}\right)^{-1} \quad (4.4.26)$$

With the *SCBE*, the full drain current in the saturation region can be written as:

$$I_{ds} = I_{dsat}\left(1 + \frac{V_{ds} - V_{dsat}}{V_A}\right)\left(1 + \frac{V_{ds} - V_{dsat}}{V_{ASCBE}}\right) \quad (4.4.27)$$

where I_{dsat} is the saturation current which can be calculated with Eq. (4.4.9).

4.5 Subthreshold I-V Model

In the subthreshold region, the potential in the channel exhibits a peak between the source and the drain. At or near the peak of the potential, the lateral electric field can be considered zero because the potential gradient is zero. Thus, the drift current can be ignored in the subthreshold region. According to the current density equation given in Eq. (4.1.1), we have:

$$J_n = q D_n \nabla n \quad (4.5.1)$$

We would like to use the charge sheet density expression Q_{inv} in modeling the *I-V* characteristics. If we integrate Eq. (4.5.1) from the Si-SiO_2 interface to the edge of the depletion layer (X_{dep}) in the bulk, the current in the subthreshold region can be given as

$$I_{ds} = W_{eff} \mu_n v_t \frac{dQ_{inv}}{dy} \quad (4.5.2)$$

The current expression can be obtained by integrating Eq. (4.5.2) along the channel from source to drain,

$$I_{ds} = \frac{W_{eff}}{L_{eff}} \mu_n v_t (Q_{dinv} - Q_{sinv}) \quad (4.5.3)$$

where Q_{dinv} and Q_{sinv} are the channel inversion charge at the drain and source.

The channel charges at the source and drain can be written as,

$$Q_{sinv} = \sqrt{\frac{q\varepsilon_{si}N_{CH}}{4\phi_B}}v_t \exp\left(\frac{V_{gs} - V_{th} - V_{OFF}}{nv_t}\right) \quad (4.5.4)$$

$$Q_{dinv} \approx \sqrt{\frac{q\varepsilon_{si}N_{CH}}{4\phi_B}}v_t \exp\left(\frac{V_{gs} - V_{th} - V_{OFF} - V_{ds}}{nv_t}\right) \quad (4.5.5)$$

Therefore, the current expression becomes

$$I_{ds} = I_{s0}\left(1 - \exp\left(-\frac{V_{ds}}{nv_t}\right)\right)\exp\left(\frac{V_{gs} - V_{th} - V_{OFF}}{nv_t}\right) \quad (4.5.6)$$

$$I_{s0} = \mu_n \frac{W_{eff}}{L_{eff}}\sqrt{\frac{q\varepsilon_{si}N_{CH}}{4\phi_B}}v_t^2 \quad (4.5.7)$$

where n is a factor introduced in section 2.3 that will be discussed again in section 4.5 in this chapter. v_t is the thermal voltage (K_BT/q) and V_{OFF} is the offset voltage discussed in section 4.2.

4.6 Single Equation I-V model of BSIM3v3

The development of separate model expressions for such device operation regimes as subthreshold and strong inversion as well the linear and saturation regions is discussed in the previous sections. Although these expressions can each accurately describe device behavior within their own respective region of operation, problems are likely to occur in a transition region between two well-described regions. In order to address this persistent problem, a unified model should be synthesized to preserve region-specific accuracy and to ensure the continuities of current (I_{ds}) and conductance (G_x) and their derivatives in all transition regions. This was accomplished in BSIM3v3 [4.7, 4.16].

This section will describe the unified BSIM3v3 *I-V* model equations. A complete description of all *I-V* model equations and parameters can be found in Appendices *A* and *B*.

By following similar derivations to those given in section 4.4 and based on the continuous channel charge and mobility models, a single equation *I-V* expression is obtained [4.7, 4.16]:

$$I_{ds} = \frac{I_{dso}}{1 + \dfrac{R_{ds} I_{dso}}{V_{dseff}}} \left(1 + \frac{V_{ds} - V_{dseff}}{V_A}\right)\left(1 + \frac{V_{ds} - V_{dseff}}{V_{ASCBE}}\right) \tag{4.6.1}$$

where

$$I_{ds0} = \frac{W_{eff} C_{ox} \mu_{eff} V_{gsteff} V_{dseff} \left(1 - \dfrac{V_{dseff}}{2V_b}\right)}{L_{eff}\left(1 + \dfrac{V_{dseff}}{E_{sat} L_{eff}}\right)} \tag{4.6.2}$$

$$V_A = V_{Asat} + \left(1 + \frac{\boldsymbol{P_{VAG}} V_{gsteff}}{E_{sat} L_{eff}}\right)\left(\frac{1}{V_{ACLM}} + \frac{1}{V_{ADIBLC}}\right)^{-1} \tag{4.6.3}$$

$$V_{Asat} = \frac{E_{sat} L_{eff} + V_{dsat} + 2 R_{DS} v_{sat} C_{ox} W_{eff} V_{gsteff}\left[1 - \dfrac{A_{bulk} V_{dsat}}{2(V_{gsteff} + 2v_t)}\right]}{2/\lambda - 1 + R_{DS} v_{sat} C_{ox} W_{eff} A_{bulk}} \tag{4.6.4}$$

$$V_{ACLM} = \frac{A_{bulk} E_{sat} L_{eff} + V_{gsteff}}{\boldsymbol{P_{CLM}} A_{bulk} E_{sat}\, l}(V_{ds} - V_{dseff}) \tag{4.6.5}$$

$$V_{ADIBLC} = \frac{(V_{gsteff} + 2v_t)}{\theta_{rout}(1 + \boldsymbol{P_{DIBLCB}} V_{bseff})}\left(1 - \frac{A_{bulk} V_{dsat}}{A_{bulk} V_{dsat} + V_{gsteff} + 2v_t}\right) \tag{4.6.6}$$

$$\theta_{rout} = \boldsymbol{P_{DIBLC1}}\left[\exp\left(-\boldsymbol{D_{ROUT}} \frac{L_{eff}}{2l_{t0}}\right) + 2\exp\left(-\boldsymbol{D_{ROUT}} \frac{L_{eff}}{l_{t0}}\right)\right] + \boldsymbol{P_{DIBLC2}} \tag{4.6.7}$$

$$\frac{1}{V_{ASCBE}} = \frac{\boldsymbol{P_{SCBE2}}}{L_{eff}} \exp\left(\frac{-\boldsymbol{P_{SCBE1}}\, l}{V_{ds} - V_{dseff}}\right) \tag{4.6.8}$$

when R_{ds}=0,

$$V_{dsat} = \frac{E_{sat} L_{eff} (V_{gsteff} + 2v_t)}{A_{bulk} E_{sat} L_{eff} + V_{gsteff} + 2v_t} \tag{4.6.9}$$

For R_{ds}>0,

$$V_{dsat} = \frac{-b - \sqrt{b^2 - 4ac}}{2a} \tag{4.6.10a}$$

$$a = A_{bulk}{}^2 W_{eff} v_{sat} C_{ox} R_{ds} + (\frac{1}{\lambda} - 1) A_{bulk} \tag{4.6.10b}$$

$$b = -\left((V_{gsteff} + 2v_t)(\frac{2}{\lambda} - 1) + A_{bulk} E_{sat} L_{eff} + 3A_{bulk}(V_{gsteff} + 2v_t) W_{eff} v_{sat} C_{ox} R_{ds} \right) \tag{4.6.10c}$$

$$c = (V_{gsteff} + 2v_t) E_{sat} L_{eff} + 2(V_{gsteff} + 2v_t)^2 W_{eff} v_{sat} C_{ox} R_{ds} \tag{4.6.10d}$$

$$\lambda = A_1 V_{gsteff} + A_2 \tag{4.6.10e}$$

The parameters such as E_{sat} and v_{sat} have been introduced in Eq. (4.4.11).

V_{dseff} can be written as:

$$V_{dseff} = V_{dsat} - \frac{1}{2}\left(V_{dsat} - V_{ds} - \delta + \sqrt{(V_{dsat} - V_{ds} - \delta)^2 + 4\delta V_{dsat}} \right) \tag{4.6.11}$$

where δ is a user specified parameter with a default value of 0.01.

The V_{dseff} function is introduced to guarantee continuities of I_d and its derivatives at V_{dsat}. The dependence of V_{dseff} on V_{ds} is given in Fig. 4.6.1 and Fig. 4.6.2, from which it can be shown that $V_{dseff} \approx V_{ds}$ until $V_{ds} \approx V_{dsat}$-50δ and smoothly approaches and remains V_{dsat} for V_{ds}>V_{dsat} +50δ in the saturation region. δ is roughly 1% of the transition range between the two branches of V_{dseff}. The plots of V_{dseff} and it's first and second derivatives versus V_{ds} from linear to saturation regimes are shown in Fig. 4.6.3 to demonstrate the continuity of V_{dseff}.

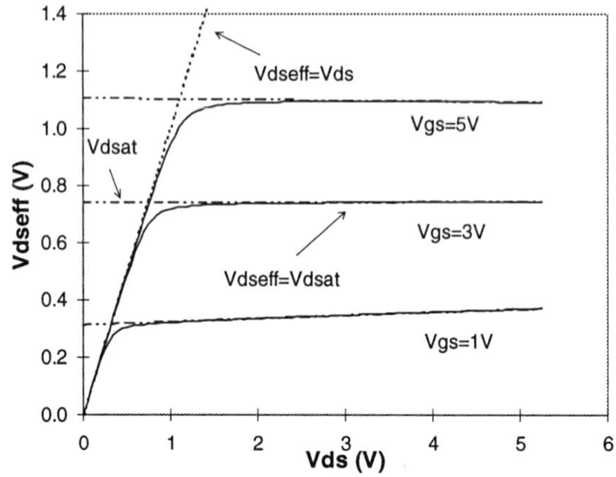

Fig.4.6.1 V_{dseff} vs. V_{ds} for δ=0.01 at several V_{gs}.

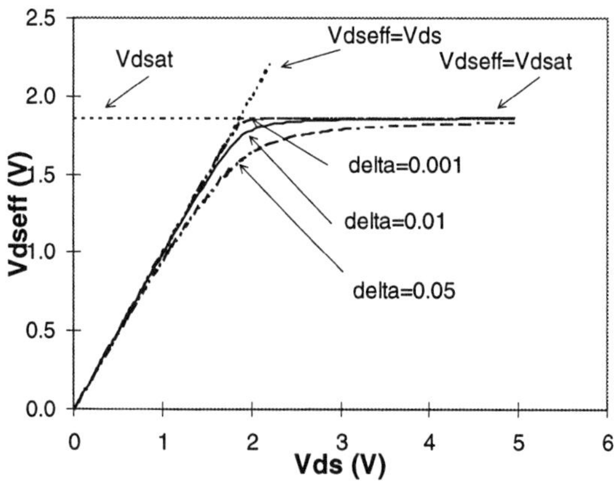

Fig 4.6.2 V_{dseff} for several values of δ, V_{gs}=3V.

It is easy to understand the unification nature of the single equation in Eq. (4.6.1). Eq. (4.6.1) resembles the equation (4.4.7), noticing that $V_{dseff}=V_{ds}$ in the linear region, used to model drain current in the strong inversion regime. However, it can now be used to describe the current characteristics in the subthreshold regime when V_{ds} is very small ($V_{ds}<2v_t$) because of the V_{gsteff} function. Eq. (4.6.1) can also change to Eq. (4.4.28) in the saturation region, where V_{dseff} approaches V_{dsat}. Furthermore, it can also cover both subthreshold and strong inversion due to the introduction of V_{gsteff}.

The single equation *I-V* model in BSIM3v3 has been verified with measured data from different technologies [4.9, 4.16]. Good accuracy and continuity features can be seen in the testing results that will be shown in Chapter 12.

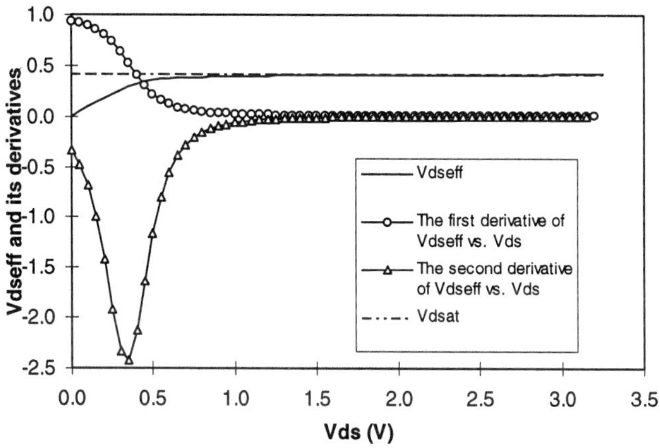

Fig. 4.6.3 The V_{dseff} function and its first and second derivatives are continuous from the linear to saturation regions (δ=0.01).

4.7 Polysilicon Gate Depletion Effect

As discussed in Chapter 2, the polysilicon (poly-Si) gate depletion effect results in a voltage drop V_p across the poly-Si gate which reduces the effective gate voltage. Therefore, when considering the poly-Si depletion effect, the gate voltage V_{gs} is replaced with an effective gate voltage V_{gse} calculated by subtracting the poly-Si gate band bending from V_{gs}, i.e. $V_{gse}= V_{gs}-V_p$ [4.24, 4.25]. V_p is dependent on the gate oxide thickness $\boldsymbol{T_{OX}}$, the voltage across the oxide, and the poly-Si gate doping concentration, denoted by $\boldsymbol{N_{GATE}}$. The formula for V_{gse}, Eq. (2.8.6.), is given here again:

$$V_{gs\_eff} = \boldsymbol{V_{FB}} + \phi_s + \frac{q\varepsilon_{si}\boldsymbol{NGATE}\boldsymbol{T_{OX}^2}}{\varepsilon_{ox}^2}\left(\sqrt{1+\frac{2\varepsilon_{ox}^2(V_{gs}-\boldsymbol{V_{FB}}-\phi_s)}{q\varepsilon_{si}\boldsymbol{NGATE}\boldsymbol{T_{OX}^2}}}-1\right)$$

(4.7.1)

where V_{FB} is the flat band voltage, ε_{si} and ε_{ox} are the dielectric constants of the silicon and SiO_2, respectively. ϕ_s is equal to $2\phi_B$.

In BSIM3v3, the default value of N_{GATE} is 0. Poly-gate depletion effect is modeled in the operation regime where V_{gs} is not less than $V_{FB}+\phi_s$ when N_{GATE} is given in the model card with a value larger than $1\text{x}10^{18}\text{cm}^{-3}$ but less than $1\text{x}10^{25}\text{cm}^{-3}$.

4.8 Helpful Hints

1. The V_{OFF} parameter

The theoretical threshold voltages $V_{th,sub}$ needed to fit the subthreshold current is different from V_{th} that is used to fit strong inversion *I-V*. One explanation is that the surface potential corresponding to the V_{th} in strong inversion is actually higher than $2\phi_B$. The difference between the threshold voltages discussed above is several v_t [4.29]. To account for this fact, a parameter called V_{OFF} is introduced so that

$$V_{th,sub} = V_{th} + V_{OFF} \tag{4.8.1}$$

V_{OFF} is determined experimentally from the measured *I-V* characteristics and is expected to be negative. Due to the physical meaning of V_{OFF}, overly large absolute values of V_{OFF} are not recommended in the model. The recommended range for V_{OFF} is between -0.06 and -0.12 V [4.5].

2. The effective channel length and width:

It has been well known that the electrical channel length and width of a MOSFET are different from the drawn channel length L_{drawn} and width W_{drawn} because of processing related reasons. Usually the effective channel length L_{eff} and channel width W_{eff} are used to characterize the MOSFETs in the compact model for circuit design. There are many different ways to define and extract L_{eff} and W_{eff} [4.1, 4.26]. The effective channel length and width used in BSIM3v3 are modeled in the following manner:

$$L_{eff} = L_{drawn} - 2dL \tag{4.8.2}$$

$$W_{eff} = W_{drawn} - 2dW \tag{4.8.3}$$

where dW and dL are given with the following expressions:

$$dW = W_{INT} + D_{WG}V_{gsteff} + D_{WB}(\sqrt{\phi_s - V_{bseff}} - \sqrt{\phi_s})$$
$$+\frac{W_L}{L^{W_{LN}}} + \frac{W_W}{W^{W_{WN}}} + \frac{W_{WL}}{L^{W_{LN}}W^{W_{WN}}} \quad (4.8.4)$$

$$dL = L_{INT} + \frac{L_L}{L^{L_{LN}}} + \frac{L_W}{W^{L_{WN}}} + \frac{L_{WL}}{L^{L_{LN}}W^{L_{WN}}} \quad (4.8.5)$$

where W_{INT}, D_{WG}, D_{WB}, and L_{INT} are parameters extracted from experimental results. W_L, W_W, W_{LN}, W_{WN}, W_{WL}, L_L, L_W, L_{LN}, L_{WN}, and L_{WL} are additional fitting parameters available to the user to improve the model accuracy.

The formulas for dW and dL are complex but have been found to be necessary to fit the data in some cases. In Eq. (4.8.4), W_{INT} is the traditional "ΔW". It can be extracted from the intercept of a straight line in a $1/R_{ds}$ vs. W_{drawn} plot. The parameters D_{WG} and D_{WB} have been added to account for the contribution of both the gate and substrate biasing effects to the effective channel width. For dL given in Eq. (4.8.5) the parameter L_{INT}, or the traditional "ΔL", can be extracted from the intercept of lines in a R_{ds} vs. L_{drawn} plot.

The other terms in both dW and dL are introduced as fitting parameters for the convenience of the user. They are meant to allow the user to model each parameter as a function of W_{drawn}, L_{drawn}, and their associated product terms.

All of the above geometrical dependencies for dW and dL are set to zero by default. We do not encourage their use unless it is found to be necessary.

In Chapter 3, we have seen another parameter called W_{eff}' in the modeling of V_{th} in BSIM3v3. The definition of W_{eff}' is:

$$W_{eff}' = W_{drawn} - 2dW' \quad (4.8.6)$$

$$dW' = W_{INT} + \frac{W_L}{L^{W_{LN}}} + \frac{W_W}{W^{W_{WN}}} + \frac{W_{WL}}{L^{W_{ILN}}W^{W_{WN}}} \quad (4.8.7)$$

From the above, it can be seen that W_{eff}' is nothing but W_{eff} without the bias dependence.

3. Drain and source parasitic resistance, R_{ds}

In BSIM3v3, the parasitic resistance of drain and source is modeled with the following expression:

$$R_{ds} = \frac{R_{DSW}[1 + P_{RWG}V_{gsteff} + P_{RWB}(\sqrt{\phi_s - V_{bseff}} - \sqrt{\phi_s})]}{(10^6 W_{eff}')^{W_R}} \tag{4.8.8}$$

where W_R is a fitting parameter and R_{DSW} has the units of Ω-μm^{W_R}. P_{RWB} is the body effect coefficient, and P_{RWG} is the gate-bias effect coefficient.

4. The parameters of A_1 and A_2 in BSIM3v3

Holes do not exhibit as prominent or abrupt a velocity saturation effect as electrons. As a result, it is difficult to identify the saturation voltage in the *I-V* curve of a PMOSFET, especially at high gate voltage, because the current continues increasing and saturates slowly. This may be called soft saturation. This specific effect makes the modeling of the PMOSFET difficult, especially in the region of drain voltage close to V_{dsat}.

As discussed in Chapter 2, Eq. (2.5.4) can be used to describe the velocity-field relationship for *n*-channel devices. To use a unified velocity-field relationship for both electrons and holes, a term, $\lambda = A_1 V_{gsteff} + A_2$, is introduced into the velocity-field relationship.

$$v(y) = \frac{E_y \mu_{eff}}{1 + [E_y/(\lambda E_{sat})]} \quad E_y < E_{sat} \tag{4.8.9}$$

$$v(y) = \upsilon_{sat} \quad E_y > E_{sat} \tag{4.8.10}$$

Usually, A_2=1 and A_1=0 for n channel devices, even though users can optimize the A_2 and A_1 parameters to fit the measured *I-V* data around the saturation point (for *n*-channel devices, λ should be kept close to 1 to maintain its physical meaning). For PMOSFETs, the values of A_2 and A_1 should be extracted from measured data. These two parameters should be such as to make λE_{sat} approximately υ_{sat}/μ_{eff} for the P-channel MOSFETs [4.27].

In BSIM3v3 model implementation in simulators, the λ term has been limited to ≤ 1 with a smoothing function as will be discussed in Chapter 11.

5. The n parameter for subthreshold swing

The n parameter can be called the subthreshold swing factor or the subthreshold slope factor because the traditional gate voltage swing or subthreshold slope can be defined as

$$S = \frac{dV_{gs}}{d \log I_{ds}} \approx 2.3 n v_t \tag{4.8.11}$$

The subthreshold swing is the change in the gate voltage V_{gs} required to reduce the subthreshold current I_{gs} by one decade. According to Eq. (4.8.11), the n parameter is the key parameter in determining the subthreshold swing of the device. For long channel devices, n can be modeled as

$$n = 1 + \frac{C_{dep}}{C_{ox}} + \frac{C_{it}}{C_{ox}} \tag{4.8.12}$$

where C_{dep} and C_{it} are the depletion layer capacitance and interface charge capacitance. However, Eq. (4.8.12) does not consider the influence of short channel effects. In short channel devices, the potential at the surface of the channel (point A in Fig. 4.5.1) will be determined by both the gate bias and the drain bias through the coupling of C_{ox} and C_{dsc}, respectively, instead of the gate bias only. The coupling capacitance $C_{dsc}(L)$ is an exponential function of the channel length, as shown by the solution of the quasi 2-D Poisson equation. To reflect this phenomenon in BSIM3v3, the n parameter for the subthreshold swing is described in the following form:

$$n = 1 + N_{FACTOR}\frac{C_{dep}}{C_{ox}} + \frac{C_{IT}}{C_{ox}} + \frac{(C_{DSC} + C_{DSCD}V_{ds} + C_{DSCB}V_{bseff})\left(\exp(-D_{VT1}\frac{L_{eff}}{2l_t}) + 2\exp(-D_{VT1}\frac{L_{eff}}{l_t})\right)}{C_{ox}} \tag{4.8.13}$$

where the parameter N_{FACTOR} is introduced to cover for any uncertainty in the calculation of the depletion capacitance, and is determined experimentally. C_{IT} is called the interface charge capacitance, and accounts for the influence of the interface charge density. C_{DSC}, C_{DSCD}, and C_{DSCB} are parameters to describe the coupling effects between the drain and the channel due to the *DIBL* effect discussed in section 2.2. V_{bseff}, D_{VT1} and l_t have been discussed in section 3.4.

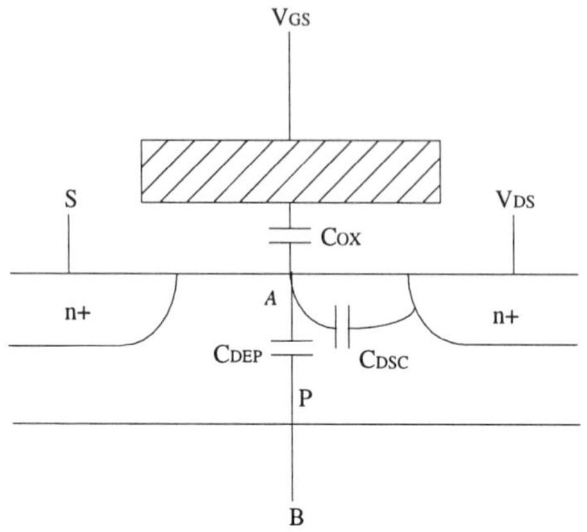

Fig. 4.5.1 MOSFET device with gate oxide capacitance, drain/source, and channel coupling capacitances. All the capacitances have an effect on the channel potential.

6. The A_{bulk} parameter for the bulk charge effect

When the drain voltage is large and/or when the channel length is long, the depletion region "thickness" of the channel is not uniform along the channel length. This will cause the threshold voltage to vary along the channel. This effect is called the bulk charge effect and has been discussed in section 2.2.

In BSIM3v3, the parameter A_{bulk} is used to account for the bulk charge effect, including both the short channel effects and narrow width effects, as shown in Eq. (4.8.14). Several extracted parameters such as $\boldsymbol{A_0}$, $\boldsymbol{A_{GS}}$, $\boldsymbol{B_0}$, and $\boldsymbol{B_1}$ are introduced in BSIM3v3 to account for the channel length and width dependencies of the bulk charge parameter. In addition, the parameter $\boldsymbol{K_{ETA}}$ is introduced to model the change in the bulk charge effect at high body bias conditions.

$$A_{bulk} = (1 + \frac{K_{1OX}}{2\sqrt{\phi_s - V_{bseff}}}\{\frac{\boldsymbol{A_0} L_{eff}}{L_{eff} + 2\sqrt{\boldsymbol{X_J} X_{dep}}}[1 - \boldsymbol{A_{GS}} V_{gsteff}(\frac{L_{eff}}{L_{eff} + 2\sqrt{\boldsymbol{X_J} X_{dep}}})^2] + \frac{\boldsymbol{B_0}}{W_{eff}' + \boldsymbol{B_1}}\})\frac{1}{1 + \boldsymbol{K_{ETA}} V_{bseff}} \tag{4.8.14}$$

In Eq. (4.8.14), $\boldsymbol{A_0}$, $\boldsymbol{A_{GS}}$, $\boldsymbol{B_0}$, $\boldsymbol{B_1}$, and $\boldsymbol{K_{ETA}}$ are extracted from experimental *I-V* data. K_{1OX} is given in Eq. (3.4.25d). It is known that A_{bulk} is close to 1 if the channel length is small, and rises as channel length increases.

7 The gain factor *K* and *K'*

The gain factor K or K' is a widely used parameter in circuit design. According to the classic definition, K is a measure of the gain of the device in the linear region, and can be written as

$$K = \mu C_{ox} \tag{4.8.15a}$$

and K' is a measure of the gain of the device in saturation region and defined as

$$K' = \frac{\mu C_{ox}}{2} \tag{4.8.15b}$$

For MOSFETs with long channel lengths and thick oxide thickness, the carrier mobility is approximately a constant in the whole channel region so that it can be extracted directly. Thus the K or K' factor can be used as a measure of device performance. However, as device technology develops, the channel length becomes shorter, and oxide thickness becomes thinner, the K or K' parameters cannot be obtained simply and need to be defined with some modifications to Eq. (4.8.15a) and Eq. (4.8.15b). This is true since the mobility is not a constant in today's devices but a function of gate bias and drain bias. We need to choose the appropriate value for the mobility in order to use Eq (4.8.15) for K and K'. For example, it may be reasonable if choosing $0.85\mu_0$ in the linear region and $0.8\mu_0$ in saturation region for μ in Eq. (4.4.15a) for K and in Eq. (4.4.15b) for K' where μ_0 is the low field mobility value in BSIM3v3.

8. BSIM3v3 model equations for hand calculations

Circuit designers like to use hand calculations to get some general idea about their design or to verify some design methodology. The following equations were developed two decades ago and used widely for hand calculation:

In the linear region:

$$I_{ds} = \mu_n C_{ox} \frac{W}{L}(V_{gs} - V_{th} - V_{ds}/2)V_{ds} \tag{4.8.16}$$

In the saturation region,

$$Idso = \frac{\mu_n W C_{ox}}{2L}(V_{gs} - Vth)^2 \tag{4.8.17}$$

It is not easy to perform any direct hand calculations with some advanced compact MOSFET models like the BSIM3v3 equation given in Eq. (4.6.1) because it is a single equation covering all of the operation regimes. However, the single equation can be simplified into several piece-wise equations in different operation regions for the purpose of simple hand calculations.

In strong inversion, the *I-V* equation can be simplified to

$$I_{ds} = \mu_{eff} C_{ox} \frac{Weff}{Leff} \frac{1}{1 + V_{ds}/(E_{sat} L)} (V_{gs} - V_{th} - Abulk\, V_{ds}/2) V_{ds} \quad V_{ds} < V_{dsat} \tag{4.8.18}$$

$$Ids = Weff v_{sat} C_{ox} (V_{gs} - Vth - A_{bulk} V_{dsat})(1 + \frac{V_{ds} - V_{dsat}}{V_A}) \qquad V_{ds} > V_{dsat} \tag{4.8.19}$$

$$V_{dsat} = \frac{E_{sat} Leff (V_{gs} - V_{th})}{AbulkE_{sat} Leff + (V_{gs} - V_{th})} \tag{4.8.20}$$

The definitions of all parameters in Eq. (4.8.18) and Eq. (4.8.19) have been given previously. Please note that the μ_{eff} in Eq. (4.8.18) is a function of the gate bias and the body bias. Some parameters, such as V_{th} and V_A, are to be extracted from measurements. Some approximate values can be used for *Abulk*, say, a value around 1 for hand calculations.

9. The equations related to the *DIBL* effect

DIBL causes the reduction of Vth due to a lowering of the potential barrier in the channel. It is a physical effect usually considered only when $V_{gs} \leq Vth$. However, when $V_{gs} > Vth$, there is still a reduction of Vth due to the electrostatic coupling between the drain and the channel as if the drain is serving as an unwanted extra gate. Because of this, the influence of *DIBL* in Vth is not absent when $V_{gs} > Vth$. Instead, in BSIM3v3, a correction term is introduced in V_A to account for the influence of this *DIBL* effect in the saturation region. The *DIBL* effect in strong inversion is in fact usually stronger than the *DIBL* effect in the subthreshold region. It usually determines the peak output resistance in the R_{out} versus V_{ds} plot [4.16, 4.17].

10. Parameters in the *I-V* model

The parameters in the *I-V* model are listed in Table 4.8.1.

Table 4.8.1 *I-V* Model parameters

Symbols in equation	Symbols in source code	Description	Default	Unit
mobMod	mobmod	Mobility model selector	1	none
μ_0	uo	Mobility at $T = T_{NOM}$ NMOSFET PMOSFET	670.0 250.0	$cm^2/V/sec$
U_A	ua	First-order mobility degradation coefficient	2.25×10^{-9}	m/V
U_B	ub	Second-order mobility degradation coefficient	5.87×10^{-19}	$(m/V)^2$
U_C	uc	Body-effect of mobility degradation coefficient	***mobMod***= 1,2: -4.65×10^{-11} ***mobMod***= 3: -0.0465	m/V^2 1/V
V_{SAT}	vsat	Saturation velocity at $T = T_{NOM}$	8.0×10^4	m/sec
A_0	a0	Bulk charge effect coefficient for channel length	1.0	none
A_{GS}	ags	Gate bias coefficient of the bulk charge effect	0.0	1/V
B_0	b0	Bulk charge effect coefficient for channel width	0.0	m
B_1	b1	Bulk charge effect width offset	0.0	m
K_{ETA}	keta	Body-bias coefficient of the bulk charge effect	-0.047	1/V
A_1	a1	First non-saturation parameter	0.0	1/V
A_2	a2	Second non-saturation parameter	1.0	none
R_{DSW}	rdsw	Parasitic resistance per unit width	0.0	$\Omega\text{-}\mu m^{Wr}$

P_{RWG}	prwg	Gate bias effect coefficient of R_{ds}	0	V^{-1}
P_{RWB}	prwb	Body bias effect coefficient of R_{ds}	0	$V^{-1/2}$
W_R	wr	Width offset from W_{eff} for R_{ds} calculation	1.0	none
W_{INT}	wint	Width offset fitting parameter without bias effect	0.0	m
L_{INT}	lint	Length offset fitting parameter without bias effect	0.0	m
D_{WG}	dwg	Coefficient of W_{eff}'s gate dependence	0.0	m/V
D_{WB}	dwb	Coefficient of W_{eff}'s body bias dependence	0.0	$m/V^{1/2}$
V_{OFF}	voff	Offset voltage in the subthreshold region at large W and L	-0.08	V
N_{FACTOR}	nfactor	Subthreshold swing factor	1.0	none
E_{TA0}	eta0	*DIBL* coefficient in subthreshold region	0.08	none
E_{TAB}	etab	Body-bias coefficient for the subthreshold *DIBL* effect	-0.07	1/V
P_{CLM}	pclm	Channel length modulation parameter	1.3	none
P_{DIBLC1}	pdiblc1	First output resistance *DIBL* effect correction parameter	0.39	none
P_{DIBLC2}	pdiblc2	Second output resistance *DIBL* effect correction parameter	0.0086	none
P_{DIBLCB}	pdiblcb	Body effect coefficient of *DIBL* correction parameters	0	1/V
D_{ROUT}	drout	L dependence coefficient of the *DIBL* correction parameter in R_{out}	0.56	none
P_{SCBE1}	pscbe1	First substrate current induced body-effect parameter	4.24×10^{8}	V/m

P_{SCBE2}	pscbe2	Second substrate current induced body-effect parameter	$1.0\text{x}10^{-5}$	m/V
P_{VAG}	pvag	Gate dependence of Early voltage	0.0	none
δ	delta	Effective V_{ds} parameter	0.01	V
N_{GATE}	ngate	Poly gate doping concentration	0	cm^{-3}
D_{SUB}	dsub	*DIBL* coefficient exponent in subthreshold region	D_{ROUT}	none
C_{IT}	cit	Interface trap capacitance	0.0	F/m^2
C_{DSC}	cdsc	Drain/Source to channel coupling capacitance	$2.4\text{x}10^{-4}$	F/m^2
C_{DSCD}	cdscd	Drain-bias sensitivity of C_{DSC}	0.0	F/Vm^2
C_{DSCB}	cdscb	Body-bias sensitivity of C_{DSC}	0.0	F/Vm^2
W_{LN}	wln	Power of length dependence of width offset	1.0	none
W_L	wl	Coefficient of length dependence for width offset	0.0	$m^{W\ln}$
W_{WN}	wwn	Power of width dependence of width offset	1.0	none
W_W	ww	Coefficient of width dependence for width offset	0.0	m^{Wwn}
W_{WL}	wwl	Coefficient of length and width cross term for width offset	0.0	$m^{Wwn+W\ln}$
L_{LN}	lln	Power of length dependence for length offset	1.0	none
L_L	ll	Coefficient of length dependence for length offset	0.0	$m^{L\ln}$
L_{WN}	lwn	Power of width dependence for length offset	1.0	none
L_W	lw	Coefficient of width dependence for length offset	0.0	m^{Lwn}
L_{WL}	lwl	Coefficient of length and width cross term for length offset	0.0	$m^{Lwn+L\ln}$

References

[4.1] N. Arora, *MOSFET Models for VLSI Circuit Simulation*, Springer-Verlag, Wien New York, 1994.

[4.2] E. H. Nicollian and J. R. Brews, *MOS Physics and Technology*, Wiely-Interscience, New York, 1982.

[4.3] Y. Cheng et al., "ICM--An analytical Inversion charge model for accurate modeling of thin gate oxide MOSFETs," *1997 International Conf. on Simulation of Semiconductor Processes and Devices*, Sept. 1997, Boston.

[4.4] R. Rios et al., "A physical compact MOSFET model, including quantum mechanical effects, for statistical circuit design applications," *IEDM Tech. Dig*., pp. 937-940, 1995.

[4.5] Y. Cheng et al., "A unified MOSFET channel charge model for device modeling in circuit simulation," *IEEE Trans. Computer-aided Design of Integrated Circuits and Systems*, vol.17, pp.641-644, 1998.

[4.6] M. C. Jeng, *Design and modeling of deep-submicrometer MOSFETs,* ERL memorandum ERL M90/90, University of California, Berkeley, 1990.

[4.7] Y. Cheng et al., *BSIM3 version 3.0 User's Manual*, University of California, Berkeley, 1995.

[4.8] Y. Cheng et al., *BSIM3 version 3.1 User's Manual*, University of California, Berkeley, Memorandum No. UCB/ERL M97/2, 1997.

[4.9] Y. Cheng et al., "An Investigation on the robustness, accuracy and simulation performance of a physics-based deep-submicrometer BSIM model for analog/digital circuit simulation", *CICC'96*, pp. 321-324, May 1996.

[4.10] M. S. Liang et al., "Inversion layer capacitance and mobility of very thin gate oxide MOSFETs," *IEEE Trans. Electron Devices*, ED-33, p. 409, 1986.

[4.11] K. Lee et al., "Physical understanding of low field carrier mobility in silicon inversion layer," *IEEE Trans. Electron Devices*, ED-38, p. 1905, 1991.

[4.12] A.G. Sabnis and J.T. Clemens, "Characterization of electron velocity in the inverted <100> Si surface," *IEDM Tech. Dig.,* pp. 18-21, 1979.

[4.13] K. Chen et al., "MOSFET carrier mobility model based on gate oxide thickness, threshold and gate voltages", *Solid-State Electronics*, pp.1515-1518, Vol. 39, No. 10, October 1996.

[4.14] J. H. Huang et al., *BSIM3 Manual (Version 2.0),* University of California, Berkeley, March 1994.

[4.15] S. M. Sze, *Semiconductor Devices: Physics and Technology*, John Wiley & Sons, New York, 1985.

[4.16] Y. Cheng et al., "A physical and scalable BSIM3v3 I-V model for analog/ digital circuit simulation", *IEEE Trans. Electron Devices*, Vol. 44, pp.277-287, Feb. 1997.

[4.17] J. H. Huang et al., "A physical model for MOSFET output resistance", *IEDM, Technical Digest*, Dec. of 1992.

[4.18] Y. A. El-Mansy and A. R. Boothroyd, "A simple two dimensional model for IGFET." *IEEE Trans. Electron Devices*, ED-24, pp. 254-262, 1977.

[4.19] M. E. Banna and M. E. Nokali, "A pseudo-two-dimensional analysis of short channel MOSFETs," *Solid-state Electronics*, Vol. 31 pp.269-274, 1988.

[4.20] R. R. Troutman, "VLSI Limitations from Drain-Induced Barrier Lowering," *IEEE Trans. on Electron Devices*, vol. ED-26, p.461, 1979.

[4.21] T. A. Fjeldly and M. Shur, "Threshold voltage modeling and the subthreshold regime of operation of short channel MOSFETs," *IEEE Trans. on Electron Devices*, vol. ED-40, pp. 137-145, 1993.

[4.22] Z. H. Liu et al., "Threshold voltage model for deep-submicron MOSFET's," *IEEE Trans. on Electron Devices*, vol. ED-40, pp.86-98, 1993.

[4.23] P. K. Ko, *Hot carrier Effects in MOSFETs*, Ph. D dissertation, Dept. of Electrical Engineering and Computer Sciences, University of California, Berkeley, 1982.

[4.24] K. F. Schuegraf et al., "Impact of polysilicon depletion in thin oxide MOS technology," *Proc. Int. Symp. VLSI Tech., Sys. and Appl.*, pp. 86-90, 1993.

[4.25] R. Rios et al., "An analytical polysilicon depletion effect model for MOSFETs," *IEEE Electron Device Letters*, Vol. 15, pp.129-131, 1994.

[4.26] K. K. Ng, and J. R. Brews, "Measuring the effective channel length of MOSFETs," *IEEE Circuit and Devices*, vol. 6, pp. 33-38, 1990.

[4.27] Y. Cheng et al., "Quarter-micron surface and buried channel P-MOSFET modeling for circuit simulation", *Semiconductor Science and Technology*, pp. 1763-1769, Vol. 11, No. 12, December 1996.

[4.28] B. J. Sheu, D. L. Scharfetter, P. K. Ko, and M. C. Jeng, "BSIM: Berkeley short -channel IGFET model for MOS transistors," *IEEE J. solid-state Circuits*, vol. SC-22, pp.558-565, 1987.

[4.29] Y. P. Tsividis, *Operation and Modeling of the MOS Transistor*, McGraw-Hill, New York, 1987.

[4.30] Y. Cheng et al., "Modeling of small size MOSFETs with reverse short channel and narrow width effects for circuit simulation", *Solid State Electronics*, vol. 41, (9), pp. 1227-1231, 1997.

CHAPTER 5 Capacitance Model

We have discussed the DC models in the previous chapters. In real circuit operation, the device operates under time-varying terminal voltages. Depending on the magnitude of the time-varying voltages, the dynamic operation can be classified as large signal operation or small signal operation. If the variation in voltages is sufficiently small, the device can be modeled with linear resistors, capacitors, current source, etc. Such a model is called a small-signal model. Otherwise, the device must be represented by an analytical, nonlinear "large-signal" model. Both types of dynamic operation are influenced by the device's capacitive effects. Thus, a capacitance model describing the intrinsic and extrinsic components of the device capacitance, is another essential part of a compact MOSFET model for circuit simulation besides the DC model.

In most circuit simulators the same capacitance model is used for both the large-signal transient analysis and the small-signal AC analysis. The capacitance model is almost always based on the quasi-static approximation, which assumes that the charges in the device can follow the varying terminal voltages immediately without any delay or that the voltages do not change much in the span of the "transit time" of the device [5.1]. In this chapter, we will discuss the basic concepts related to the intrinsic charge and capacitance, and then introduce the capacitance models of BSIM3v3. Finally we discuss some significant issues in capacitance modeling.

5.1 Capacitance Components in a MOSFET

Before we discuss the modeling of MOSFET capacitances, let us understand the different capacitances in a MOSFET shown in Fig. 5.1.1. Generally, MOSFET capacitance can be divided into two groups, the intrinsic and the extrinsic capacitances. The intrinsic capacitance is related to the region between the metallurgical source and drain junctions. The extrinsic capacitance, or the parasitic capacitance, is further divided into five components: 1) the outer fringing capacitance between the polysilicon gate and the source/drain, C_{FO}; 2) the inner fringing capacitance between the polysilicon gate and the source/drain, C_{FI}; 3) the overlap capacitances between the gate and the heavily doped S/D regions (and the bulk region), C_{GSO} & C_{GDO} (C_{GBO}), which are relatively insensitive to terminal voltages; 4) the overlap capacitances between the gate and lightly doped S/D region, C_{GSOL} & C_{GDOL}, which changes with bias; and 5) the source/drain junction capacitances, C_{JD} & C_{JS}. The remaining capacitances shown in Fig. 5.1.1, are the intrinsic capacitances. We will discuss the extrinsic capacitances such as C_{FO}, C_{FI}, and the capacitances related to the overlaps of gate to source/drain and gate to bulk in this chapter, and leave the discussion of C_{JS} and C_{JD} to Chapter 8.

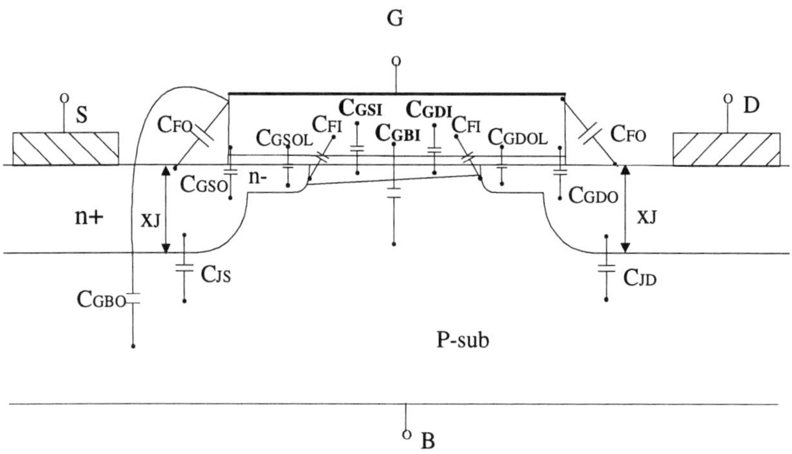

Fig. 5.1.1 An n-channel MOSFET: a few of the intrinsic (bold-faced) and extrinsic capacitances are shown.

The intrinsic capacitance is much more complex than the extrinsic components. In Fig. 5.1.1, C_{GS} is the gate-to-source capacitance, C_{GD} is the gate-to-drain capacitance, and C_{GB} is the gate-to-bulk capacitance. This picture of the intrinsic capacitance is overly simplistic. As we will discuss later in this chapter, the intrinsic capacitance consists of up to 16 nonreciprocal capacitance components.

5.2 Intrinsic Capacitance Model

We first consider the intrinsic part of a MOSFET. The early intrinsic capacitances models, such as the Meyer model [5.2], simply treated the MOSFET capacitance as three separate lumped capacitances, gate-to-source capacitance (C_{gs}[a]), gate-to-drain capacitance (C_{gd}), and gate-to-bulk capacitance (C_{gb}). It is inaccurate for short channel devices and has the now-well-known charge non-conservation problem as we will discuss later. However, this model has been used widely in simulators and continues to be used occasionally as an optional model for its simplicity and efficiency. We will discuss the basic idea and derivation of the Meyer model first, and then present the charge-based capacitance model which guarantees charge conservation.

5.2.1 Meyer model

In the Meyer model, the following assumptions are made to derive the capacitance expressions [5.2]:

a. Capacitances in a MOSFET are reciprocal, that is, $C_{gb} = C_{bg}$, $C_{gd} = C_{dg}$, $C_{gs} = C_{sg}$.
b. The change rate of gate charge Q_g is equal to the change rate of channel charge Q_{inv} when gate, source, and drain bias changes. That is,

$$\left|\frac{\partial Q_g}{\partial V_{gs}}\right| = \left|\frac{\partial Q_{inv}}{\partial V_{gs}}\right| \tag{5.2.1}$$

$$\left|\frac{\partial Q_g}{\partial V_{gd}}\right| = \left|\frac{\partial Q_{inv}}{\partial V_{gd}}\right| \tag{5.2.2}$$

The total charges on both side of the gate oxide are neutral,

a. Symbols with lower-case subscripts are for the capacitances (charges) per unit area or per unit length; Symbols with upper-case subscripts are for the total capacitances (charges). C_{GS},C_{GD}, C_{GB},C_{gd}, C_{gs}, and C_{gb} here include both intrinsic and extrinsic components.

$$Q_g + Q_{inv} + Q_{ox} + Q_b = 0 \tag{5.2.3}$$

where Q_g is the charge on the gate, Q_{inv} is the mobile carrier charge in the inversion channel, Q_{ox} is the charge related to the interface defects, and Q_b is the bulk charge in the depletion layer under the channel.

For simplicity, we ignore the Q_{ox} component in the following derivation because Q_{ox} is small compared with other charge components. Thus, we have

$$Q_g = -(Q_{inv} + Q_b) \tag{5.2.4}$$

In strong inversion, the channel charge density along the channel length direction can be written as:

$$Q_{inv} = -C_{ox}(V_{gs} - V_{th} - V(y)) \tag{5.2.5}$$

where C_{ox} is the gate oxide capacitance per unit area, and $V(y)$ is the channel potential at any point y along the channel length direction, referenced to the source junction.

As discussed in Chapter 4,

$$I_{ds} = W\mu_s Q_{inv}\frac{dV}{dy} \tag{5.2.6}$$

where μ_s is the carrier mobility, and W is the channel width.

The I_{ds} expression in the linear operation region can be obtained easily by integrating Eq.(5.2.6) from source to drain and remembering that $V_{gd} = V_{gs} - V_{ds}$:

$$I_{ds} = \frac{W\mu_s C_{ox}}{2L}[(V_{gs} - V_{th})^2 - (V_{gd} - V_{th})^2] \tag{5.2.7}$$

Also Eq. (5.2.6) can be rewritten as:

$$dy = \frac{W\mu_s Q_{inv}}{I_{ds}}dV \tag{5.2.8}$$

The total charge in the channel is

$$Q_G = -W\int_0^L Q_{inv}dy - Q_B \tag{5.2.9}$$

Combining Eqs. (5.2.5), (5.2.8), (5.2.9), and performing the integration, we have

$$Q_G = \frac{2}{3}WLC_{ox}\left[\frac{(V_{gd}-V_{th})^3-(V_{gs}-V_{th})^3}{(V_{gd}-V_{th})^2-(V_{gs}-V_{th})^2}\right]-Q_B \tag{5.2.10}$$

The capacitances C_{GS}, C_{GD}, and C_{GB} in the linear region can be obtained from the following definitions:

$$C_{GS} \equiv \frac{\partial Q_G}{\partial V_{gs}}\bigg|_{V_{gd},V_{gb}} \tag{5.2.11a}$$

$$C_{GD} \equiv \frac{\partial Q_G}{\partial V_{gd}}\bigg|_{V_{gs},V_{gb}} \tag{5.2.11b}$$

$$C_{GB} \equiv \frac{\partial Q_G}{\partial V_{gb}}\bigg|_{V_{gs},V_{gd}} \tag{5.2.11c}$$

By differentiating Eq. (5.2.10) according to Eq. (5.2.11), we can calculate the gate capacitances C_{GS}, C_{GD}, and C_{GB} in the linear region:

$$C_{GS} = \frac{2}{3}WLC_{ox}\left[1-\frac{(V_{gd}-V_{th})^2}{(V_{gs}-V_{th}+V_{gd}-V_{th})^2}\right] \tag{5.2.12a}$$

$$C_{GD} = \frac{2}{3}WLC_{ox}\left[1-\frac{(V_{gs}-V_{th})^2}{(V_{gs}-V_{th}+V_{gd}-V_{th})^2}\right] \tag{5.2.12b}$$

$$C_{GB} = 0 \tag{5.2.12c}$$

It is to be expected that C_{GB} is zero in strong inversion since the inversion layer in the channel from the drain to the source shields the gate from the bulk and prevents any response of the gate charge to a change in the substrate bias,

V_{bs}. This is approximately true in the strong inversion case. However, C_{GB} can not be considered zero in the weak inversion and accumulation regions.

For $V_{ds}>V_{dsat}$ the gate charge can be obtained by replacing V_{ds} in Eq. (5.2.10) with V_{dsat}. Assuming a long channel device, $V_{dsat} = V_{gs}-V_{th}$ and the gate charge can be given as

$$Q_G = \frac{2}{3}WLC_{ox}(V_{gs} - V_{th}) - Q_B \tag{5.2.13}$$

It is easy to obtain the C_{GS}, C_{GD}, and C_{GB} in the saturation region,

$$C_{GS} = \frac{2}{3}WLC_{ox} \tag{5.2.14a}$$

$$C_{GD} = 0 \tag{5.2.14b}$$

$$C_{GB} = 0 \tag{5.2.14c}$$

There is a physical explanation for Eq. (5.2.14b). In the saturation region, the channel is pinched off at the drain end of the channel. This electrically isolates the channel from the drain so that the charge on the gate is not influenced by a change in the drain voltage, and the capacitance C_{GD} vanishes.

In weak inversion, the charge in the inversion layer can be ignored compared with the depletion charge so that Eq. (5.2.4) becomes,

$$Q_g = -Q_b \tag{5.2.15}$$

The depletion charge density in the bulk for a long channel device can be written as (see Eq. (2.3.12))

$$Q_b = -C_{ox}\gamma\sqrt{\phi_s} \tag{5.2.16}$$

where γ is the body effect coefficient given in Eq. (2.2.3). ϕ_s is the surface potential in weak inversion which is given by

$$\phi_s = \left[-\frac{\gamma}{2}+\sqrt{\frac{\gamma^2}{4}+V_{gb}-V_{FB}}\right]^2 \tag{5.2.17}$$

The total depletion charge can be obtained by performing the following integration:

$$Q_B = -W\int_0^L Q_b dy \tag{5.2.18}$$

Therefore we can calculate the total gate charge in the weak inversion region,

$$Q_G = -Q_B = W\int_0^L Q_b dy = -\frac{1}{2}WLC_{ox}\gamma^2\left[1-\sqrt{1+\frac{4}{\gamma^2}(V_{gb}-V_{FB})}\right] \tag{5.2.19}$$

By differentiating Eq. (5.2.19) according to Eq. (5.2.11), the capacitances C_{GS}, C_{GD}, and C_{GB} in the weak inversion region can be given as

$$C_{GS} = 0 \tag{5.2.20a}$$

$$C_{GD} = 0 \tag{5.2.20b}$$

$$C_{GB} = \frac{WLC_{ox}}{\sqrt{1+\frac{4}{\gamma^2}(V_{gb}-V_{FB})}} \tag{5.2.20c}$$

According to Eq. (5.2.14a), C_{GS}=2/3Cox when $V_{gs}=V_{th}$ in the saturation region. However, when $V_{gs}<V_{th}$, C_{GS} =0 according to (5.2.20a). To avoid this large discontinuity at $V_{gs}=V_{th}$, it was proposed that C_{GS} decreases linearly from 2/3 C_{ox} at $V_{gs}=V_{th}$ to zero at $V_{gs}=V_{th}-\phi_B$ [5.3]. This is reasonable because the channel charge decreases gradually as V_{gs} drops below V_{th}. C_{GS} should not be zero until the inversion layer vanishes totally (at the intrinsic condition $V_{gs}-V_{th}=-\phi_B$).

In the accumulation region, $C_{GB}=C_{ox}$, and $C_{GS}=C_{GD}=0$ [5.3].

Fig. 5.2.1 illustrates the capacitances in the Meyer model. Fig. 5.2.2 shows C_{GS}, C_{GD}, and C_{GB} vs. V_{gs} at several V_{ds} for a long channel MOSFET (L=5μm) using the above expressions (Eqs. (5.2.12), (5.2.14) and (5.2.20)).

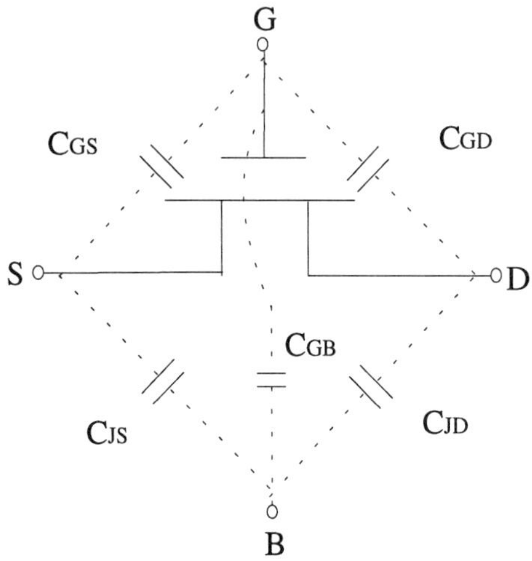

Fig. 5.2.1 An illustration of the capacitances presented by the Meyer capacitance model. C_{JS} and C_{JD} are the capacitances of S/B and D/B junctions.

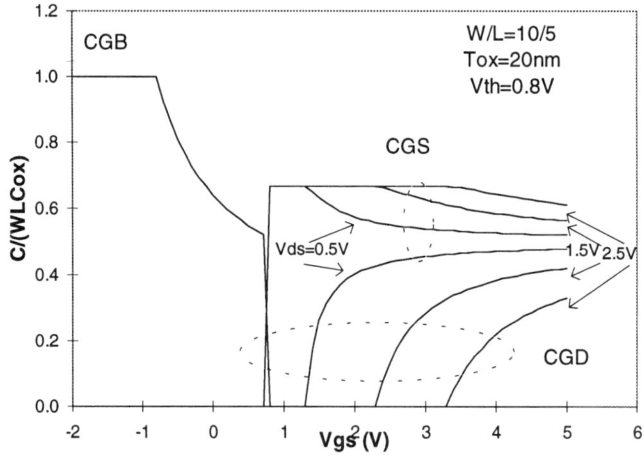

Fig. 5.2.2 The capacitance characteristics calculated according to the Meyer model.

One advantage of the Meyer model is that it can be described by a simple equivalent circuit as shown in Fig. 5.2.3 [5.4]. The Meyer model is still used widely by circuit designers for this reason and its efficiency although it has the charge non-conservation problem discussed in the next section.

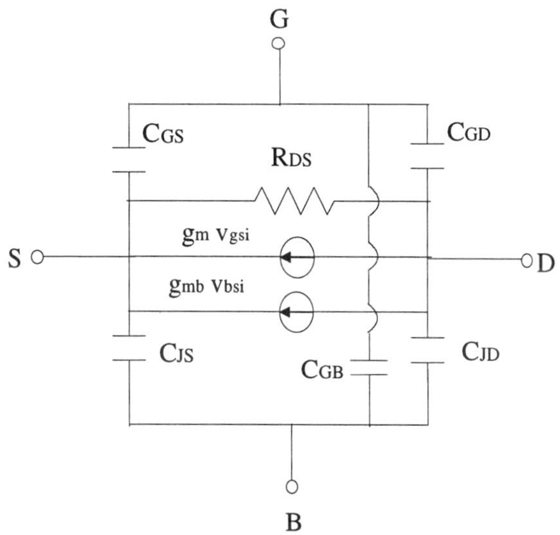

Fig. 5.2.3 An equivalent circuit with the capacitances represented by the Meyer model for the intrinsic MOSFET. g_m and g_{mb} are the gate and substrate transconductances. v_{gsi} and v_{bsi} are the gate and substrate biases (reference to the source) in the intrinsic MOSFET.

5.2.2 Shortcomings of the Meyer model

The Meyer model is simple and sufficiently accurate for many circuit applications and has been used over many years since it was implemented in SPICE [5.5]. However, it has been found to yield non-physical results when used to simulate circuits that have charge storage nodes. Charge built-up on these nodes are incorrectly predicted by the simulation. This problem shows up in MOS charge pumps [5.6], silicon-on-sapphire(SOS) circuits [5.7], static RAM and switched-capacitor circuits [5.8]. It is termed the charge non-conservation problem [5.9].

The charge non-conservation problem has been investigated in detail [5.7, 5.8, 5.9]. It is known that the proper way to model MOSFET capacitances is to assign charges to each of the terminals. With the quasi-static assumption, these charges at each time point t only depend on the values of the terminal voltages at the same time. So generally we have,

$$Q_G = Q_G(V_{gs}, V_{gd}, V_{gb}) \tag{5.2.21}$$

$$Q_S = Q_S(V_{gs}, V_{gd}, V_{gb}) \tag{5.2.22}$$

$$Q_D = Q_D(V_{gs}, V_{gd}, V_{gb}) \tag{5.2.23}$$

$$Q_B = Q_B(V_{gs}, V_{gd}, V_{gb}) \tag{5.2.24}$$

It is now understood that the capacitances in a MOSFET cannot be arbitrary functions. For example C_{GG}, C_{DG}, $C_{SG,}$ and C_{BG} must satisfy

$$C_{GG}(V_{gs}, V_{ds}, V_{bs}) \equiv \frac{\partial Q_G}{\partial V_g} \tag{5.2.25}$$

$$C_{DG}(V_{gs}, V_{ds}, V_{bs}) \equiv \frac{\partial Q_D}{\partial V_g} \tag{5.2.26}$$

$$C_{SG}(V_{gs}, V_{ds}, V_{bs}) \equiv \frac{\partial Q_S}{\partial V_g} \tag{5.2.27}$$

$$C_{BG}(V_{gs}, V_{ds}, V_{bs}) \equiv \frac{\partial Q_B}{\partial V_g} \tag{5.2.28}$$

and that the sum of charges in the devices must meet the charge neutrality relationship given in the following:

$$Q_G + Q_D + Q_S + Q_B = 0 \tag{5.2.29}$$

Otherwise, charge would not be conserved.

Alternative capacitance models have been developed to solve the charge non-conservation problem [5.10, 5.11]. Some Meyer-like models, carefully formulated and implemented to preserve the necessary relationship among the capacitances have been reported to be satisfactory in conserving charge [5.12]. Another approach is the charge-based model [5.13, 5.15].

The failure of the reciprocity assumption in Meyer model has been discussed. It can be shown that the reciprocity of the Meyer model requires Q_S to be independent of V_{ds} and V_{bs}, and Q_D to be independent of V_{gs} and V_{bs} if charge conservation is to be ensured [5.11]. The following proof between the double lines is included only for the most interested readers.

==

According to the reciprocity assumption in the Meyer model,

$$C_{gs} \equiv \frac{\partial Q_g}{\partial V_{gs}} = C_{sg} \equiv \frac{\partial Q_s}{\partial V_{sg}} = -\frac{\partial Q_s}{\partial V_{gs}} \qquad (5.2.30a)$$

$$C_{gd} \equiv \frac{\partial Q_g}{\partial V_{gd}} = C_{dg} \equiv \frac{\partial Q_d}{\partial V_{dg}} = -\frac{\partial Q_d}{\partial V_{gd}} \qquad (5.2.30b)$$

$$C_{gb} \equiv \frac{\partial Q_g}{\partial V_{gb}} = C_{bg} \equiv \frac{\partial Q_b}{\partial V_{bg}} = -\frac{\partial Q_b}{\partial V_{gb}} \qquad (5.2.30c)$$

Therefore, the gate-source reciprocity implies that a given change in $V_{gs}(\Delta V_{gs})$ causes equal and opposite changes in Q_g and Q_s (ΔQ_g=-ΔQ_s). Similar arguments hold for gate-drain and gate-bulk reciprocity.

Differentiating the charge conservation equation (5.2.3) with respect to V_{gs} and substituting for $\partial Q_g/\partial V_{gs}$ from Eq. (5.2.30a), we have,

$$-\frac{\partial Q_g}{\partial V_{gs}} + \frac{\partial Q_s}{\partial V_{gs}} + \frac{\partial Q_d}{\partial V_{gs}} + \frac{\partial Q_b}{\partial V_{gs}} = 0 \qquad (5.2.31)$$

According to Eq. (5.2.30a), $-\frac{\partial Q_g}{\partial V_{gs}} = \frac{\partial Q_s}{\partial V_{gs}}$ so that Eq. (5.2.31) becomes

$$\frac{\partial Q_d}{\partial V_{gs}} + \frac{\partial Q_b}{\partial V_{gs}} = 0 \qquad (5.2.32)$$

In the derivation of the Meyer model, it has been assumed that the depletion charge Q_b is a constant respect to the V_{gs}, that is $\frac{\partial Q_b}{\partial V_{gs}} = 0$, so that we finally have $\frac{\partial Q_d}{\partial V_{gs}} = 0$ according to Eq. (5.2.32).

Similar analysis can be made with respect to V_{gd} and we have

$$\frac{\partial Q_s}{\partial V_{gd}} + \frac{\partial Q_b}{\partial V_{gd}} = 0 \tag{5.2.33}$$

So we get the result of $\frac{\partial Q_s}{\partial V_{gd}} = 0$ because $\frac{\partial Q_b}{\partial V_{gd}} = 0$ according to the assumption in the model derivation.

==

According to the above analysis, it is clear that the reciprocity of the Meyer model requires Q_s to depend only on V_{gs} and Q_d to depend only on V_{gd} if charge conservation is to be ensured. In other words, C_{gs}=C_{sg}≡dQ_s/dV_g cannot be a function of V_{ds} or V_{bs}, etc. This is non-physical because the channel charge can be modulated by both V_{ds} and V_{bs} [5.11]. Non-reciprocal effects arise because the channel charge in a MOSFET is controlled by three or more bias voltages. Reciprocal capacitors simply cannot be used to model the capacitive effects in a MOSFET.

5.2.3 Charge-based capacitance model

In a charge-based approach, the emphasis is put on the charge, rather than the capacitance, from derivation through model implementation. The approach is to determine the charges in the drain, gate, source, and bulk of a MOSFET, and use them as state variables in the circuit simulation. The transient currents and the capacitances are obtained through mathematical differentiation of the charge with respect to time or voltage, respectively. The charge-based capacitance model automatically ensures the charge conservation, as long as the following equation is satisfied,

$$Q_G + Q_D + Q_S + Q_B = 0 \tag{5.2.34}$$

The capacitive currents can be rewritten as

$$i_G = \frac{dQ_G}{dt} = \frac{\partial Q_G}{dV_{gb}}\frac{dV_{gb}}{dt} + \frac{\partial Q_G}{dV_{gd}}\frac{dV_{gd}}{dt} + \frac{\partial Q_G}{dV_{gs}}\frac{dV_{gs}}{dt} \tag{5.2.35a}$$

$$i_D = \frac{dQ_D}{dt} = \frac{\partial Q_D}{dV_{db}}\frac{dV_{db}}{dt} + \frac{\partial Q_D}{dV_{dg}}\frac{dV_{dg}}{dt} + \frac{\partial Q_D}{dV_{ds}}\frac{dV_{ds}}{dt} \tag{5.2.35b}$$

$$i_S = \frac{dQ_S}{dt} = \frac{\partial Q_S}{dV_{sb}}\frac{dV_{sb}}{dt} + \frac{\partial Q_S}{dV_{sd}}\frac{dV_{sd}}{dt} + \frac{\partial Q_S}{dV_{sg}}\frac{dV_{sg}}{dt} \tag{5.2.35c}$$

$$i_B = \frac{dQ_B}{dt} = \frac{\partial Q_B}{dV_{bg}}\frac{dV_{bg}}{dt} + \frac{\partial Q_B}{dV_{bd}}\frac{dV_{bd}}{dt} + \frac{\partial Q_B}{dV_{bs}}\frac{dV_{bs}}{dt} \tag{5.2.35d}$$

By defining the following,

$$\begin{cases} C_{ij} = \dfrac{\partial Q_i}{\partial V_{ij}} & i \neq j, i, j = G, D, S, B \\ C_{ij} = -\dfrac{\partial Q_i}{\partial V_{ij}} & i = j \end{cases} \tag{5.2.36a}$$

or

$$\begin{cases} C_{ij} = -\dfrac{\partial Q_i}{\partial V_j} & i \neq j, i, j = G, D, S, B \\ C_{ij} = \dfrac{\partial Q_i}{\partial V_j} & i = j \end{cases} \tag{5.2.36b}$$

and substituting Eq. (5.2.36) into Eq. (5.2.35), we can derive (see, for example, [5.1])

$$\sum_{i \neq j} C_{ij} = \sum_{i \neq j} C_{ji} \tag{5.2.37}$$

Only 9 of the 16 capacitances are independent according to Eq. (5.2.37). For example, if we select C_{gb}, C_{gd}, C_{gs}, C_{bg}, C_{bd}, C_{bs}, C_{dg}, C_{db}, and C_{ds} as independent capacitances, then C_{sg}, C_{sb}, and C_{sd} can be obtained by

$$C_{sg} = C_{gg} + C_{gd} + C_{gs} - C_{bg} - C_{dg} \tag{5.2.38a}$$

$$C_{sb} = C_{bg} + C_{bd} + C_{bs} - C_{gb} - C_{db} \tag{5.2.38b}$$

$$C_{sd} = C_{dg} + C_{db} + C_{ds} - C_{gd} - C_{bd} \tag{5.2.38c}$$

and C_{gg}, C_{dd}, C_{ss}, and C_{bb} can be calculated by

$$C_{gg} = C_{gs} + C_{gd} + C_{gb} \tag{5.2.39a}$$

$$C_{dd} = C_{ds} + C_{dg} + C_{db} \tag{5.2.39b}$$

$$C_{ss} = C_{sg} + C_{sd} + C_{sb} \tag{5.2.39c}$$

$$C_{bb} = C_{bs} + C_{bd} + C_{bg} \tag{5.2.39d}$$

The charge-based capacitance model needs the charge equations for all four terminals, that is Q_G, Q_S, Q_D, and Q_B. Q_G and Q_B can be obtained directly by integrating the corresponding charge density over the channel [5.13],

$$Q_G = W\int_0^L Q_g dy \tag{5.2.40a}$$

$$Q_B = W\int_0^L Q_b dy \tag{5.2.40b}$$

$$Q_{INV} = -(Q_G + Q_B) \tag{5.2.40c}$$

Q_g and Q_b are given in the following

$$Q_{g(y)} = C_{ox}(V_{gs} - V_{FB} - \phi_s - V_y) \tag{5.2.41}$$

$$Q_{b(y)} = -C_{ox}[V_{th} - V_{FB} - \phi_s - (1 - A_{bulk})V_y] \tag{5.2.42}$$

where A_{bulk} is the bulk charge coefficient discussed in Chapter 2.

The derivation of Q_G and Q_B is straightforward according to Eq. (5.2.41):

$$Q_G = W\,L\,C_{ox}[V_{gs} - V_{FB} - \phi_s - \frac{1}{2}V_{ds} + \frac{A_{bulk}V_{ds}^{\ 2}}{12(V_{gs} - V_{th} - \frac{A_{bulk}}{2}V_{ds})}] \qquad (5.2.43)$$

$$Q_B = W\,LC_{ox}[V_{FB} - V_{th} + \phi_s - \frac{1 - A_{bulk}}{2}V_{ds} - \frac{(1 - A_{bulk})A_{bulk}V_{ds}^{\ 2}}{12(V_{gs} - V_{th} - \frac{A_{bulk}}{2}V_{ds})}] \qquad (5.2.44)$$

It is easy to calculate the total inversion charge in the channel. However, it is difficult to model the charges on the source and drain terminals because only the total mobile channel charge $Q_{INV}=Q_D+Q_S$ is known, and a partition of Q_{INV} into Q_D and Q_S is needed. At $V_{ds}=0$, the partition should be $Q_S = Q_D = Q_{INV}/2$ due to symmetry. Several charge partition approaches have been suggested for the saturation region ($V_{ds}>V_{dsat}$) [5.7, 5.8]. They are 50/50, 40/60 and 0/100, and are usually distinguished in compact model with a model parameter called X_{PART} [5.14]. When $X_{PART}>0.5$, the 0/100 charge partition is chosen, which assumes that $Q_S=Q_{INV}$ and $Q_D=0$ in the saturation region. When $X_{PART}=0.5$, the 50/50 charge partition is used, which assumes that the ratio of the drain charge to source charge is 50/50. When $X_{PART}<0.5$, the 40/60 charge partition is used, which assumes that ratio of the drain charge to source charge is 40/60 in saturation region.

The 40/60 partition is physically correct under the quasi-static condition as proven by 2-D device simulation and experiments [5.15]. One derivation of the 40/60 model is given in [5.7]

$$Q_S = W\int_0^L (1 - \frac{y}{L})Q_{inv}(y)dy \qquad (5.2.45)$$

$$Q_D = W\int_0^L \frac{y}{L}Q_{inv}(y)dy \qquad (5.2.46)$$

where L is the channel length of the device.

By performing the integration in Eq. (5.2.45) and Eq. (5.2.46), the following expressions for Q_S and Q_D in a long channel device at linear operation regime can be obtained

$$Q_S = -WLC_{ox}\left(\frac{V_{gs}-V_{th}}{2} - \frac{A_{bulk}V_{ds}}{6} + \frac{A_{bulk}^2 V_{ds}^2 [5(V_{gs}-V_{th}) - 3A_{bulk}V_{ds}]}{120\left(V_{gs}-V_{th}-\frac{A_{bulk}V_{ds}}{2}\right)^2}\right) \quad (5.2.47)$$

$$Q_D = -WLC_{ox}\left(\frac{V_{gs}-V_{th}}{2} - \frac{A_{bulk}V_{ds}}{3} + \frac{A_{bulk}^2 V_{ds}^2 [5(V_{gs}-V_{th}) - 2A_{bulk}V_{ds}]}{120\left(V_{gs}-V_{th}-\frac{A_{bulk}V_{ds}}{2}\right)^2}\right) \quad (5.2.48)$$

The corresponding charges in the saturation region can be obtained by replacing V_{ds} in the above equations with V_{dsat}, which is equal to $(V_{gs}\text{-}V_{th})/A_{bulk}$ for long channel devices [5.10],

$$Q_D = -WLC_{ox}\frac{4(V_{gs}-V_{th})}{15} \quad (5.2.49)$$

$$Q_S = -WLC_{ox}\frac{2(V_{gs}-V_{th})}{5} \quad (5.2.50)$$

$$Q_G = W\,L\,C_{ox}\left(V_{gs}-V_{FB}-2\phi_B-\frac{1}{3}(V_{gs}-V_{th})\right) \quad (5.2.51)$$

$$Q_B = W\,LC_{ox}\left(V_{FB}-V_{th}+2\phi_B-\frac{1-A_{bulk}}{3}(V_{gs}-V_{th})\right) \quad (5.2.52)$$

Fig. 5.2.4 (a) and (b) show the charge and capacitance versus gate bias for the BSIM3v3 ***capMod***=0 model [5.17], which will be discussed in detail in section 5.4. Fig. 5.2.5 shows the simulated capacitance versus drain bias.

Based on the formula of the charges and the charge partition for drain and source charges, an admittance matrix for the device can be created. All of the capacitance terms in the matrix are non-zero and non-reciprocal. Some capacitances, such as C_{sd} and C_{ds}, are negative as they should be [5.16].

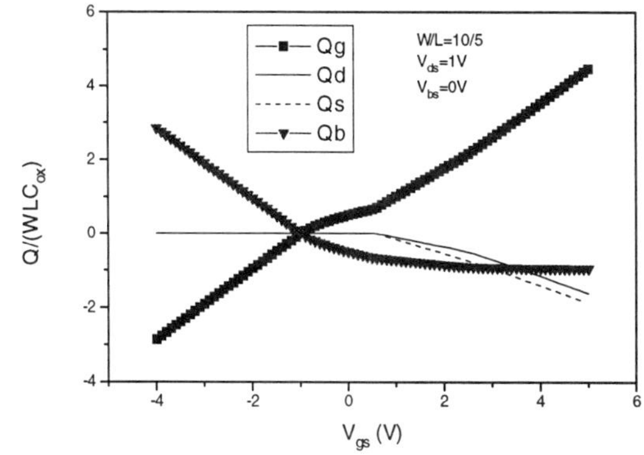

Fig. 5.2.4 (a) Charges associated with gate, bulk, source, and drain terminals simulated with the ***capMod***=0 model in BSIM3v3.

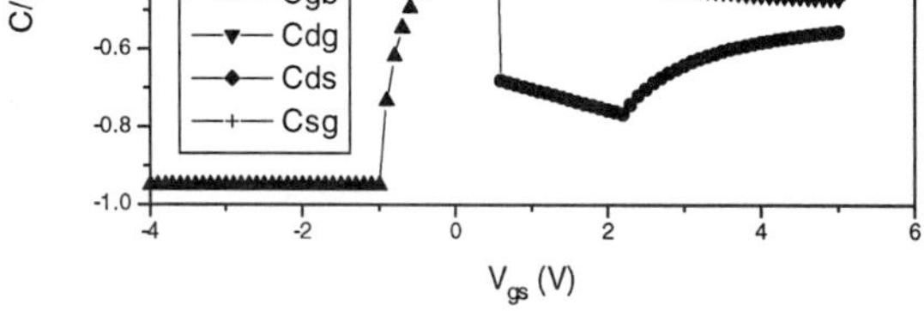

Fig. 5.2.4 (b) Capacitance vs. V_{gs} simulated with the ***capMod***=0 model.

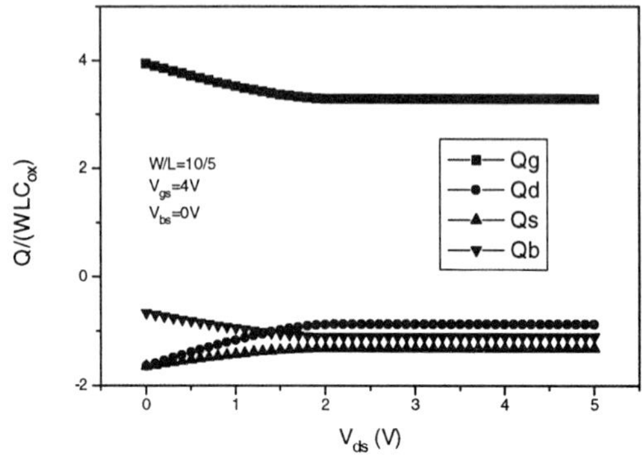

Fig. 5.2.5 (a) Charges associated with gate, bulk, source, and drain terminals vs. V_{ds} simulated with the ***capMod***=0 model in BSIM3v3.

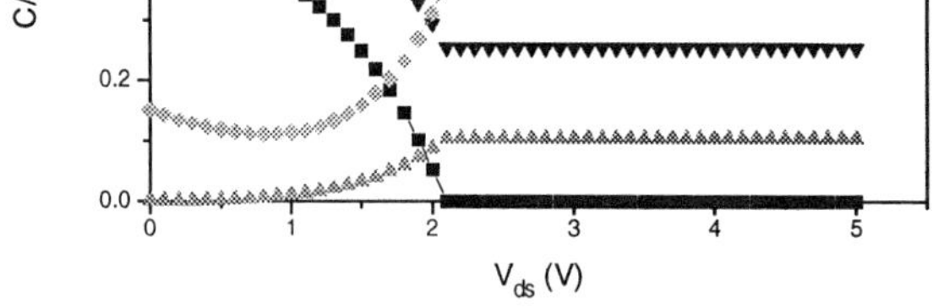

Fig. 5.2.5 (b) Capacitance versus drain voltage simulated with the ***capMod***=0 model in BSIM3v3.

5.3 Extrinsic Capacitance Model

We will now analyze the extrinsic capacitances. As shown in Fig. 5.1.1, the extrinsic capacitance consists of four components, gate overlap capacitance C_{ov} (C_{GSO}/C_{GDO}, C_{GSOL}/C_{GDOL} and C_{GBO}) in source/drain region, outer fringing field capacitance C_{FO}, inner fringing field capacitance C_{FI}, and source/drain junction capacitance C_j (C_{JS} and C_{JD}). We discuss C_{ov}, C_{FI}, and C_{FO} in this section, and will discuss the parasitic capacitance of source/drain junctions in Chapter 8.

With the shrinking device sizes in VLSI circuit chips accurate modeling of the gate-to-drain and gate-to-source overlap capacitance becomes increasingly important. The overlap capacitance is an important parameter in determining device and circuit performance (speed). There used to be a large gate-source/drain overlap in MOSFETs years ago. In that case, the overlap capacitance can be modeled simply as a parallel plate capacitance

$$C_{ov} = \frac{\varepsilon_{ox} d}{T_{ox}} \tag{5.3.1}$$

where T_{ox} is the oxide thickness, and d is the width of the gate-drain/source overlap.

Besides C_{ov} at the source ($C_{ov,GS}$) and drain ($C_{ov,GD}$), there is an additional parasitic capacitance between the gate and bulk caused by the over-layer of the poly-silicon gate required at one or both ends. The width of the polysilicon over-layer is in fact the channel length of the device. Thus, the gate-bulk overlap capacitance can be given as

$$C_{ov,GB} = C_{gbo} L \tag{5.3.2}$$

where C_{gbo} is the gate-bulk overlap capacitance with the unit length. Usually, the $C_{ov,GB}$ is much smaller than $C_{ov,GD}$ and $C_{ov,GS}$ so that it can be ignored.

The overlap capacitance models discussed above were used in SPICE MOSFET models many years ago when the model accuracy did not have to meet today's high standards. However, it has been found that the measured overlap capacitances of today's devices are significantly different from the above equations. As the device dimension shrinks the fringing capacitance associated with the gate perimeter and the finite poly-silicon gate thickness becomes a significant contributor to the capacitance.

The device structure shown in Fig. 5.3.1 may be used to analyze the fringing capacitance [5.18]. In Fig. 5.3.1, T_{ox} is the oxide thickness, X_p is the poly-silicon gate thickness, and the junction depth is X_j. α is the slope angle of the (poly-silicon) gate. The overlap capacitance can be approximated by the sum of the following three components:

a. fringing capacitance C_{fo} on the outer side between the gate and the source/drain;
b. direct overlap capacitance C_{ov} between gate-source/drain, which is described by parallel plate capacitance formula;
c. fringing capacitance C_{fi} on the channel side (inner side) between the gate and side wall of the source/drain junction.

The three capacitance components for unit width of the device are given by these expressions [5.18]:

$$C_{fo} = \frac{\varepsilon_{ox}}{\alpha}\ln(1+\frac{X_p}{T_{ox}}) \tag{5.3.3}$$

$$C_{ov} = \frac{\varepsilon_{ox}(d+\Delta)}{T_{ox}} \tag{5.3.4}$$

$$C_{fi} = \frac{\varepsilon_{ox}}{\beta}\ln(1+\frac{X_j \sin\beta}{T_{ox}}) \tag{5.3.5}$$

where Δ is a correction of the length of the overlap region to account for some higher order effects, and is given in the following:

$$\Delta = \frac{T_{ox}}{2}(\frac{1-\cos\alpha}{\sin\alpha}+\frac{1-\cos\beta}{\sin\beta}) \tag{5.3.6}$$

β is given by

$$\beta = \frac{\pi\varepsilon_{ox}}{2\varepsilon_{si}} \tag{5.3.7}$$

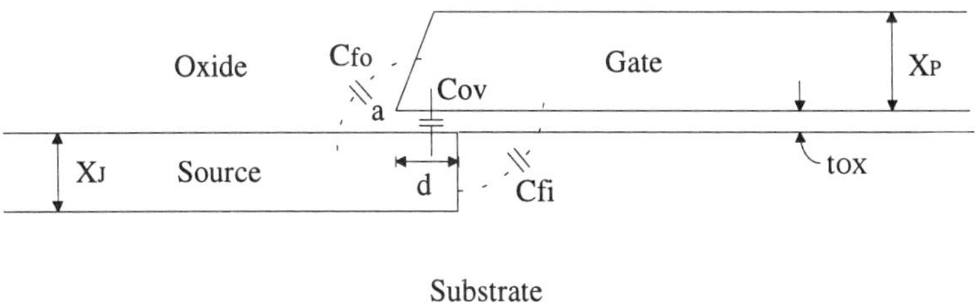

Fig. 5.3.1 Approximate structure to model the overlap capacitance. After Shrivastava and Fitzpatrick [5.18].

The total overlap capacitance per unit width of the MOS device is given by

$$C_{ovt} = C_{fo} + C_{ov} + C_{fi} \tag{5.3.8}$$

It should be noted that the above discussion did not account for the bias dependence of the overlap capacitances. In fact, especially for *LDD* devices, the overlap capacitances are functions of the terminal voltages. C_{fi} given in Eq. (5.3.5) may be taken as the maximum value of the inner fringing capacitance, and the inner fringing capacitance becomes smaller when gate bias increases from subthreshold to strong inversion, and vanishes when the device operates in strong inversion. The bias dependence of the overlap capacitance may not be ignored in *LDD* devices with thin oxide thickness. The bias dependence of the overlap capacitance, mainly C_{ov}, has been modeled recently [5.17, 5.20]. Modeling of the bias dependence of the inner overlap capacitance is still an issue to be dealt with.

5.4 Capacitance Model of BSIM3v3

The MOSFET capacitance models discussed above are piece-wise models. In these models, different sets of charge-voltage equations are employed for different regions of the device operation, i.e. accumulation, depletion and inversion regions. The model equations are derived in each specific region. However, the piece-wise models usually contain discontinuities in the capacitance-voltage (*C-V*) characteristics in transition regions such as near the threshold and flat-band voltages. These discontinuities are believed to be a potential cause for non-convergence in transient circuit simulations and also reduce the model accuracy.

Unified models to improve the continuity have been reported [5.21, 5.22]. However, both these piece-wise models and unified model did not consider the short channel effects, poly-silicon gate depletion, and channel quantization effects that have become important recently. Therefore, these models are not suitable for the simulation of short channel devices.

In low power and analog applications, designers are interested in device operation near the threshold voltage. Thus, the model must also be accurate in the transition region from the subthreshold to the strong inversion region as well. To ensure proper behavior, both the *I-V* and *C-V* model equations should be developed from an identical set of charge equations so that C_{ij}/I_{ds} is well behaved.

In BSIM3v3 [5.17, 5.23], there are several different model options for users to select through a model parameter called ***capMod***. ***capMod*** can be 0, 1, 2, or 3. When ***capMod***=0, a long channel charge-based capacitance model is used. The capacitance model with ***capMod***=0 is a modified version of the BSIM1 capacitance model [5.13]. When ***capMod***=1 and 2, two capacitance models with short channel effects are used [5.17]. The difference between ***capMod***=1 and ***capMod***=2 is that the capacitance model for ***capMod***=2 introduces smoothing functions for both V_{gs} and V_{ds}, and hence has better continuity and smoothness than for ***capMod***=1 [5.17]. With ***capMod***=3, which is the default capacitance model in BSIM3v3.2 [5.23], the quantization effects are modeled. We next give the details of both the intrinsic and extrinsic capacitance models.

5.4.1 Long channel capacitance model (*capMod=0*)

1. Intrinsic charge and capacitance

As discussed previously, the charge model is the basis of the capacitance model. The space charge of the MOS structure is made up of three fundamental components: the charge on the gate electrode, Q_G, the charge in the bulk depletion layer, Q_B, and the mobile charge in the channel region, Q_{INV}. The following relationship holds,

$$Q_G + Q_{INV} + Q_B = 0 \tag{5.4.1}$$

$$Q_{INV} = Q_S + Q_D \tag{5.4.2}$$

The capacitance characteristics are often divided into accumulation, depletion, linear and saturation regions as shown in Fig. 5.4.1, and use piece-wise equations to describe the charge/capacitance characteristics in each regime.

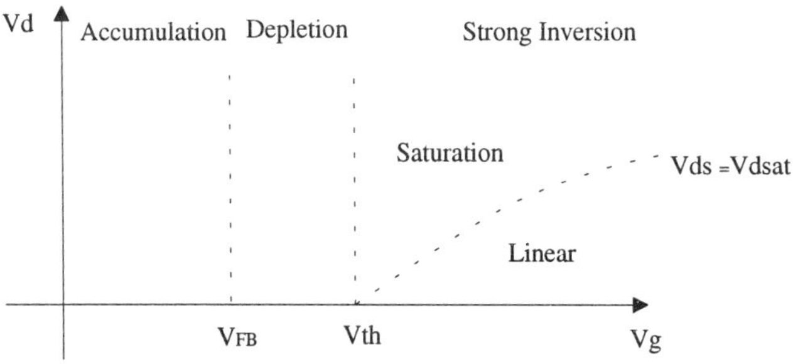

Fig. 5.4.1 Different operation regimes of capacitance modeling.

If $V_{gs} < \boldsymbol{V_{FBCV}} + V_{bs}$, the device operates in the accumulation region

$$Q_B = -W_{active} L_{active} C_{ox} \left(V_{gs} - V_{bseff} - \boldsymbol{V_{FBCV}} \right) \tag{5.4.3}$$

$$Q_G = -Q_B \tag{5.4.4}$$

where W_{active} and L_{active} are the effective channel width and channel length of the device in the capacitance models; C_{ox} is the oxide capacitance in per unit area, and $\boldsymbol{V_{FBCV}}$ is the flat-band voltage and is a user defined model parameter (***capMod***=0 only). V_{bseff} is the effective body bias introduced in Chapter 3.

If $\boldsymbol{V_{FBCV}}+V_{bs}<V_{gs}<V_{th}$, the device is in the subthreshold region and the charge expression becomes

$$Q_B = -W_{active} L_{active} C_{ox} \frac{K_{1OX}{}^2}{2} \left(-1 + \sqrt{1 + \frac{4\left(V_{gs} - \boldsymbol{V_{FBCV}} - V_{bseff} \right)}{K_{1OX}{}^2}} \right) \tag{5.4.5}$$

$$Q_G = -Q_B \tag{5.4.6}$$

where K_{1OX} is the parameter for the body effect coefficient defined in Eq. (3.4.25d).

If $V_{gs}>V_{th}$, the device is in strong inversion. Similar to the DC case, the device operates either in the linear or saturation regime depending on the drain voltage. The saturation voltage expression has been changed from that in the BSIM1 capacitance model to be consistent with that in other capacitance model options in BSIM3v3 and is [5.17]

$$V_{dsat,cv} = \frac{V_{gs} - V_{th}}{A_{bulk}'} \tag{5.4.7}$$

$$A_{bulk}' = A_{bulk0}\left(1 + \left(\frac{\boldsymbol{C_{LC}}}{L_{active}}\right)^{\boldsymbol{C_{LE}}}\right) \tag{5.4.8}$$

$$A_{bulk0} = \left(1 + \frac{K_{1OX}}{2\sqrt{\phi_s - V_{bseff}}}\{\frac{\boldsymbol{A_0} L_{eff}}{L_{eff} + 2\sqrt{\boldsymbol{X_J} X_{dep}}} + \frac{\boldsymbol{B_0}}{W_{eff}' + \boldsymbol{B_1}}\}\right)\frac{1}{1 + \boldsymbol{K_{ETA}} V_{bseff}} \tag{5.4.9}$$

$$V_{th} = \boldsymbol{V_{FBCV}} + \phi_s + K_{1OX}\sqrt{\phi_s - V_{bseff}} \tag{5.4.10}$$

Here the long channel device threshold voltage expression is used. A_{bulk0} is borrowed from the A_{bulk} in the *I-V* model but the gate bias dependence is ignored. $\boldsymbol{C_{LC}}$ and $\boldsymbol{C_{LE}}$ are fitting parameters introduced to improve the model accuracy which we will discuss again later. The other parameters are all from the *I-V* model.

The different charge partitions, controlled by the model parameter $\boldsymbol{X_{PART}}$, are described in the following. The capacitances can be obtained according to the definition given in Eq. (5.2.36) from the charge expressions.

(1) 50/50 charge partition (X_{PART}=0.5)

When $V_{ds} < V_{dsat}$, the device is biased in the linear region. The gate and bulk charge density expressions can be given as

$$Q_g(y) = C_{ox}(V_{gs} - \boldsymbol{V_{FBCV}} - \phi_s - V_y(y)) \tag{5.4.11}$$

$$Q_b(y) = -C_{ox}[V_{th} - \boldsymbol{V_{FBCV}} - \phi_s - (1 - A_{bulk}')V_y(y)] \tag{5.4.12}$$

The total charge in the gate, channel, and bulk regions can be obtained by integrating the distributed charge densities, Q_g and Q_b, over the gate area:

$$Q_G = W_{active} \int_0^{L_{active}} Q_g(y)dy \tag{5.4.13}$$

$$Q_B = W_{active} \int_0^{L_{active}} Q_b(y)dy \tag{5.4.14}$$

$$Q_{INV} = -Q_G - Q_B \tag{5.4.15}$$

Therefore, we have

$$Q_G = W_{active} L_{active} C_{ox} \left(V_{gs} - V_{FBCV} - \phi_s - \frac{1}{2} V_{ds} + \frac{A_{bulk}' V_{ds}^2}{12\left(V_{gs} - V_{th} - \frac{A_{bulk}'}{2} V_{ds} \right)} \right) \tag{5.4.16}$$

$$Q_S = Q_D = -\frac{W_{active} L_{active} C_{ox}}{2} \left(V_{gs} - V_{th} - \frac{A_{bulk}'}{2} V_{ds} + \frac{A_{bulk}'^2 V_{ds}^2}{12\left(V_{gs} - V_{th} - \frac{A_{bulk}'}{2} V_{ds} \right)} \right) \tag{5.4.17}$$

$$Q_B = W_{active} L_{active} C_{ox} \left(V_{FBCV} - V_{th} + \phi_s - \frac{1 - A_{bulk}'}{2} V_{ds} - \frac{(1 - A_{bulk}') A_{bulk}' V_{ds}^2}{12\left(V_{gs} - V_{th} - \frac{A_{bulk}'}{2} V_{ds} \right)} \right) \tag{5.4.18}$$

When $V_{ds} > V_{dsat,cv}$, the device works in the saturation region. The total gate, channel (source and drain), and bulk charge can be expressed as:

$$Q_G = W_{active} L_{active} C_{ox} \left(V_{gs} - V_{FBCV} - \phi_s - \frac{1}{3} V_{dsat,cv} \right) \tag{5.4.19}$$

$$Q_S = Q_D = -\frac{W_{active} L_{active} C_{ox}}{3}\left(V_{gs} - V_{th}\right) \tag{5.4.20}$$

$$Q_B = W_{active} L_{active} C_{ox}\left(\boldsymbol{V_{FBCV}} - V_{th} + \phi_s + \frac{1 - A_{bulk}'}{3} V_{dsat,cv}\right) \tag{5.4.21}$$

(2) 40/60 charge partition (X_{PART}<0.5)

Similar to the case for 50/50 charge partition, we can derive the following charges for the 40/60 charge partition.

When $V_{ds}<V_{dsat,cv}$

$$Q_G = W_{active} L_{active} C_{ox}\left(V_{gs} - \boldsymbol{V_{FBCV}} - \phi_s - \frac{1}{2} V_{ds} + \frac{A_{bulk}' V_{ds}^2}{12\left(V_{gs} - V_{th} - \frac{A_{bulk}'}{2} V_{ds}\right)}\right) \tag{5.4.22}$$

$$Q_B = W_{active} L_{active} C_{ox}\left(V_{FBCV} - V_{th} + \phi_s - \frac{1 - A_{bulk}'}{2} V_{ds} - \frac{(1 - A_{bulk}') A_{bulk}' V_{ds}^2}{12\left(V_{gs} - V_{th} - \frac{A_{bulk}'}{2} V_{ds}\right)}\right) \tag{5.4.23}$$

$$Q_D = -W_{active} L_{active} C_{ox}\left[\frac{V_{gs} - V_{th}}{2} - \frac{A_{bulk}'}{2} V_{ds} + \frac{A_{bulk}' V_{ds}\left[\frac{(V_{gs} - V_{th})^2}{6} - \frac{A_{bulk}'(V_{gs} - V_{th}) V_{ds}}{8} + \frac{(A_{bulk}' V_{ds})^2}{40}\right]}{\left(V_{gs} - V_{th} - \frac{A_{bulk}'}{2} V_{ds}\right)^2}\right] \tag{5.4.24}$$

$$Q_S = -(Q_G + Q_B + Q_D) \tag{5.4.25}$$

When $V_{ds} \geq V_{dsat,cv}$

$$Q_G = W_{active} L_{active} C_{ox}\left(V_{gs} - \boldsymbol{V_{FBCV}} - \phi_s - \frac{1}{3} V_{dsat,cv}\right) \tag{5.4.26}$$

$$Q_D = -\frac{4W_{active}L_{active}C_{ox}}{15}\left(V_{gs} - V_{th}\right) \tag{5.4.27}$$

$$Q_B = W_{active}L_{active}C_{ox}\left(\boldsymbol{V_{FBCV}} - V_{th} + \phi_s + \frac{1 - A_{bulk}'}{3}V_{dsat,cv}\right) \tag{5.4.28}$$

$$Q_S = -(Q_G + Q_B + Q_D) \tag{5.4.29}$$

(3) 0/100 charge partition (X_{PART} >0.5)

When V_{ds}<$V_{dsat,cv}$

$$Q_G = W_{active}L_{active}C_{ox}\left(V_{gs} - \boldsymbol{V_{FBCV}} - \phi_s - \frac{1}{2}V_{ds} + \frac{A_{bulk}'V_{ds}^2}{12\left(V_{gs} - V_{th} - \frac{A_{bulk}'}{2}V_{ds}\right)}\right) \tag{5.4.30}$$

$$Q_B = W_{active}L_{active}C_{ox}\left(\boldsymbol{V_{FBCV}} - V_{th} + \phi_s - \frac{1 - A_{bulk}'}{2}V_{ds} - \frac{(1 - A_{bulk}')A_{bulk}'V_{ds}^2}{12\left(V_{gs} - V_{th} - \frac{A_{bulk}'}{2}V_{ds}\right)}\right) \tag{5.4.31}$$

$$Q_D = -W_{active}L_{active}C_{ox}\left(\frac{V_{gs} - V_{th}}{2} + \frac{A_{bulk}'}{4}V_{ds} - \frac{(A_{bulk}'V_{ds})^2}{24\left(V_{gs} - V_{th} - \frac{A_{bulk}'}{2}V_{ds}\right)}\right) \tag{5.4.32}$$

$$Q_S = -(Q_G + Q_B + Q_D) \tag{5.4.33}$$

When V_{ds}≥$V_{dsat,cv}$

$$Q_G = W_{active}L_{active}C_{ox}\left(V_{gs} - \boldsymbol{V_{FBCV}} - \phi_s - \frac{1}{3}V_{dsat,cv}\right) \tag{5.4.34}$$

$$Q_D = 0 \tag{5.4.35}$$

$$Q_B = W_{active} L_{active} C_{ox} \left(V_{FBCV} - V_{th} + \phi_s + \frac{1 - A_{bulk}'}{3} V_{dsat,cv} \right) \tag{5.4.36}$$

$$Q_S = -(Q_G + Q_B + Q_D) \tag{5.4.37}$$

2. Extrinsic charge and capacitance

The overlap capacitance for ***capMod***=0 does not include any bias dependence, and the charge is a linear function of the gate bias.

a. Source overlap capacitance

The source overlap charge is given as follows:

$$\frac{Q_{overlap,s}}{W_{active}} = C_{GSO} V_{gs} \tag{5.4.38}$$

where C_{GSO} is a model parameter for the source overlap capacitance.

b. Drain overlap capacitance

The drain overlap charge is given as follows

$$\frac{Q_{overlap,d}}{W_{active}} = C_{GDO} V_{gd} \tag{5.4.39}$$

where C_{GDO} is a model parameter for the drain overlap capacitance.

c. Gate overlap capacitance

The gate charge contributed to the overlap capacitances is simply the sum of the source and drain charges given above

$$Q_{overlap,g} = -\left(Q_{overlap,s} + Q_{overlap,d}\right) \tag{5.4.40}$$

5.4.2 Short channel capacitance (*capMod*=1)

1. Intrinsic charge and capacitance

Similar to the case for ***capMod***=0, to ensure charge conservation, terminal charges are used as the state variables instead of the terminal voltages. The terminal charges Q_G, Q_B, Q_S, and Q_D are the charges associated with the gate, bulk, source, and drain. The gate charge is comprised of mirror charges from 3 components - the channel inversion charge (Q_{INV}), the accumulation charge (Q_{ACC}) and the depletion charge (Q_{DEP}). The accumulation and depletion charges are associated with the bulk node. The channel charge comes from the source and drain nodes. The ratio of Q_S and Q_D is the charge partition ratio. As in ***capMod***=0, the charge partition schemes in ***capMod***=1 are 0/100, 40/60, and 50/50. However, the depletion charge is divided into two components - the depletion charge at zero source-drain bias (Q_{DEP0}), which is a function of the gate to bulk bias, and the additional non-uniform depletion charge in the presence of a drain bias (δQ_{DEP}). Thus we have the following equations:

$$Q_G = -(Q_B + Q_{INV}) \tag{5.4.41}$$

$$Q_B = Q_{DEP} + Q_{ACC} \tag{5.4.42}$$

$$Q_{DEP} = Q_{DEP0} + \delta Q_{DEP} \tag{5.4.43}$$

Q_{ACC} and Q_{DEP0} can be divided into three regions:

a. Accumulation region ($V_{gs} < v_{fb} + V_{bs}$):

$$Q_{DEP0} = 0 \tag{5.4.44}$$

$$Q_{ACC} = -W_{active} L_{active} C_{ox} \left(V_{gs} - v_{fb} - V_{bseff} \right) \tag{5.4.45}$$

where v_{fb} is a variable corresponding to the flat-band voltage and calculated according to V_{th}. We will provide some discussion on this parameter in section 5.6.

b. Subthreshold region ($v_{fb} + V_{bs} < V_{gs} < V_{th}$):

$$Q_{DEP0} = -W_{active} L_{active} C_{ox} \frac{K_{1OX}^2}{2} \left(-1 + \sqrt{1 + \frac{4\left(V_{gs} - v_{fb} - V_{bseff}\right)}{K_{1OX}^2}} \right) \tag{5.4.46}$$

$$Q_{ACC} = 0 \tag{5.4.47}$$

c. Inversion region: $V_{gs} > V_{th}$

$$Q_{DEP0} = -C_{ox} W_{active} L_{active} K_{1OX} \sqrt{\left(2\phi_B - V_{bseff}\right)} \tag{5.4.48}$$

$$Q_{ACC} = 0 \tag{5.4.49}$$

In ***capMod***=0 model, the inversion capacitance changes abruptly from 0 to C_{ox} at the threshold voltage, which can cause oscillations during circuit simulation or result in large simulation error for circuits operating near V_{th}.

To avoid this problem, a smooth inversion channel capacitance model was developed based on the unified charge model used in the *I-V* model. It uses a continuous equation to formulate the channel charge in the subthreshold, transition, and inversion regions. Because of the fact that the inversion charge is much less than the depletion charge in the subthreshold region and the contribution from inversion charge in the C-V model is not as important as that in the subthreshold *I-V* model, the inversion charge model expression in the C-V model is simplified in the subthreshold region from that used in the *I-V* model. A new function called $V_{gsteff,cv}$ is introduced as follows

$$V_{gsteff,cv} = n v_t \boldsymbol{N_{OFF}} \ln\left(1 + \exp\left(\frac{V_{gs} - V_{th} - \boldsymbol{V_{OFFCV}}}{n v_t \boldsymbol{N_{OFF}}}\right)\right) \tag{5.4.50}$$

where parameters $\boldsymbol{N_{OFF}}$ and $\boldsymbol{V_{OFFCV}}$ are model parameters introduced to improve the model accuracy in the transition region from the subthreshold to the strong inversion regions. n is the geometry and bias dependent subthreshold swing parameter given in Eq. (4.5.12). v_t is the thermal voltage, and V_{th} is the threshold voltage.

Notice that when $(V_{gs}-V_{th})>3\boldsymbol{N_{OFF}}\cdot nv_t$, $V_{gsteff,cv}$ becomes $V_{gs}-V_{th}$, and for $V_{gs} <V_{th}$, $V_{gsteff,cv}$ decreases exponentially with $V_{gs}-V_{th}$ with a subthreshold slope of $\boldsymbol{N_{OFF}}\cdot nv_t$ and rapidly drops to zero when $(V_{gs} - V_{th})<-3\boldsymbol{N_{OFF}}\cdot nv_t$.

In Eq. (5.4.50), the "inversion" (minority) charge is always non-zero, even in the accumulation region. However it decreases exponentially with the gate bias in the subthreshold region, and become negligible compared with the depletion charge component in the depletion region. In the model implementation, a lower bound is used for the exponential term in Eq. (5.4.50) to avoid under-flow problems.

For backward compatibility, the charge model (***capMod***=1) is based on the BSIM1 charge model. In deriving the BSIM1 long channel charge model, mobility is assumed to be constant with no velocity saturation. Therefore in the saturation region (V_{ds}>V_{dsat}) the carrier density at the drain end is zero. Since no channel length modulation is assumed, the channel charge will remain a constant throughout the saturation region. In essence, the channel charge in the saturation region is assumed to be zero. This is a good approximation for long channel devices but fails when the channel length becomes short. If we define a drain bias, $V_{dsat,cv}$, at which the channel charge becomes a constant and does not continue to vary as the drain bias increases, we will find that $V_{dsat,cv}$ in general is larger than $V_{dsat,iv}$, the saturation voltage determined from the *I-V* characteristics, but smaller than the long channel saturation voltage, given by V_{gt}/A_{bulk} (V_{gt}= V_{gs}-V_{th} and A_{bulk} is the parameter modeling the body charge effect) [5.17]. However, in the long channel charge model (such as BSIM1 [5.16]) $V_{dsat,cv}$ is set to V_{gt}/A_{bulk} which is independent of channel length. Consequently, C_{ij}/L_{eff} also has no channel length dependence, which is not accurate because some physical effects such as velocity saturation can become stronger as device channel length becomes shorter and hence influence the saturation voltage and capacitance characteristics. Therefore, the *L* dependence of saturation voltage needs to be included in a short channel capacitance model.

The difficulty in developing a short channel model lies in calculating the charge in the saturation region. Although current continuity stipulates that the charge density in the saturation region is almost constant, it is difficult to calculate accurately the length of the saturation region. Moreover, due to the exponentially increasing lateral electric field, most of the charge in the saturation region is not controlled by the gate electrode. One would expect that the total charge in the channel will exponentially decrease with drain bias. However, it will result in very complex model equations. In BSIM3v3, a simple model is adopted to empirically fit $V_{dsat,cv}$ to channel length based on the following experimental observation,

$$V_{dsat,iv} < V_{dsat,cv} < V_{dsat,iv}\Big|_{L_{active}\to\infty} = \frac{V_{gsteff}}{A_{bulk}} \tag{5.4.51}$$

where V_{gsteff} is given by Eq. (4.2.16),

and $V_{dsat,cv}$ is modeled empirically by the following [5.17]:

$$V_{dsat,cv} = \frac{V_{gsteff,cv}}{A_{bulk}'} \tag{5.4.52}$$

$$A_{bulk}' = A_{bulk0}\left(1+\left(\frac{CLC}{L_{active}}\right)^{CLE}\right) \tag{5.4.53}$$

The expression of A_{bulk0} is given in Eq. (5.4.9).

The effects of body bias and *DIBL* are included in the capacitance model by using the same threshold voltage as in the *I-V* model.

With the $V_{gsteff,cv}$ function and the V_{bseff} function introduced in Chapter 3, the charge expressions can be given in the following forms

1. Gate and substrate charges without the influence of drain bias

When $V_{gs}<v_{fb}+V_{bs}+V_{gsteff,cv}$, without considering the influence of V_{ds}, we have

$$Q_{G0} = -W_{active}L_{active}C_{ox}\left(V_{gs} - v_{fb} - V_{bseff} - V_{gsteff,cv}\right) \tag{5.4.54}$$

$$Q_B = Q_{ACC} = -Q_{G0} \tag{5.4.55}$$

When $V_{gs}\geq v_{fb}+V_{bs}+V_{gsteff,cv}$, we have

$$Q_{G0} = W_{active}L_{active}C_{ox}\frac{K_{1OX}^2}{2}\left(-1+\sqrt{1+\frac{4\left(V_{gs} - v_{fb} - V_{gsteffcv} - V_{bseff}\right)}{K_{1OX}^2}}\right) \tag{5.4.56}$$

$$Q_B = Q_{DEP0} = -Q_{G0} \tag{5.4.57}$$

Please note that V_{bs} has been replaced by V_{bseff} in the above equations.

2. Drain bias effect on the terminal charges

a. Linear region

When $0<V_{ds} \le V_{dsat,cv}$, the gate charge can be given as

$$Q_G = Q_{G0} + W_{active}L_{active}C_{ox}\left(\left(V_{gsteff,cv} - \frac{V_{ds}}{2}\right) + \frac{A_{bulk}'^2 V_{ds}^2}{12\left(V_{gsteff,cv} - \frac{A_{bulk}'}{2}V_{ds}\right)}\right) \quad (5.4.58)$$

The increment of depletion charge δQ_{DEP} caused by V_{ds} can be derived:

$$\delta Q_{DEP} = W_{active}L_{active}C_{ox}\left(\frac{1-A_{bulk}'}{2}V_{ds} - \frac{(1-A_{bulk}')A_{bulk}'V_{ds}^2}{12\left(V_{gsteff,cv} - \frac{A_{bulk}'}{2}V_{ds}\right)}\right) \quad (5.4.59)$$

Thus, the total depletion charge is

$$Q_{DEP} = Q_{DEP0} + \delta Q_{DEP} \quad (5.4.60)$$

Depending on the charge partition, the source and drain charges have the following model expressions.

i. 50/50 charge partition

$$Q_S = Q_D = -\frac{W_{active}L_{active}C_{ox}}{2}\left(V_{gsteff,cv} - \frac{A_{bulk}'}{2}V_{ds} + \frac{A_{bulk}'^2 V_{ds}^2}{12\left(V_{gsteff,cv} - \frac{A_{bulk}'}{2}V_{ds}\right)}\right)$$

(5.4.61)

ii. 40/60 charge partition

$$Q_S = -\frac{W_{active}L_{active}C_{ox}}{2\left(V_{gsteff,cv} - \frac{A_{bulk}'}{2}V_{ds}\right)^2}[V_{gsteff,cv}^3 - \frac{4}{3}V_{gsteff,cv}^2(A_{bulk}'V_{ds})$$

$$+\frac{2}{3}V_{gsteff,cv}(A_{bulk}'V_{ds})^2 - \frac{2}{15}(A_{bulk}'V_{ds})^3] \quad (5.4.62)$$

$$Q_D = -(Q_G + Q_B + Q_S) \quad (5.4.63)$$

iii. 0/100 charge partition

$$Q_S = -W_{active}L_{active}C_{ox}\left(\frac{V_{gsteff,cv}}{2} + \frac{A_{bulk}'V_{ds}}{4} - \frac{(A_{bulk}'V_{ds})^2}{24\left(V_{gsteff,cv} - \frac{A_{bulk}'}{2}V_{ds}\right)}\right) \quad (5.4.64)$$

$$Q_D = -(Q_G + Q_B + Q_S) \quad (5.4.65)$$

b. Saturation region

When $V_{ds} > V_{dsat,cv}$, the gate charge can be given by

$$Q_G = Q_{G0} + W_{active}L_{active}C_{ox}\left(V_{gsteff,cv} - \frac{V_{dsat,cv}}{3}\right) \quad (5.4.66)$$

The increment of depletion charge δQ_{DEP} caused by V_{ds} can be derived in the following:

$$\delta Q_{DEP} = -W_{active}L_{active}C_{ox}\frac{V_{gsteff,cv} - V_{dsat,cv}}{3} \quad (5.4.67)$$

The total bulk charge is

$$Q_B = Q_{DEP} = Q_{DEP0} + \delta Q_{DEP} \quad (5.4.68)$$

Depending on the charge partition methodology, the source and drain charges have different model expressions.

i. 50/50 charge partition

$$Q_S = Q_D = -\frac{W_{active}L_{active}C_{ox}}{3}V_{gsteff,cv} \quad (5.4.69)$$

ii. 40/60 charge partition

$$Q_S = -\frac{2W_{active}L_{active}C_{ox}}{5}V_{gsteff,cv} \quad (5.4.70)$$

$$Q_D = -(Q_G + Q_B + Q_S) \tag{5.4.71}$$

iii. 0/100 charge partition

$$Q_S = -W_{active} L_{active} C_{ox} \frac{2V_{gsteff,cv}}{3} \tag{5.4.72}$$

$$Q_D = -(Q_G + Q_B + Q_S) \tag{5.4.73}$$

2. Extrinsic charge and capacitance

1). Source overlap capacitance

In the capacitance model for ***capMod***=1, the bias dependence of overlap capacitance is considered.

When V_{gs}<0, the charge contributed from the source overlap can be written as

$$\frac{Q_{overlap,s}}{W_{active}} = C_{GSO} V_{gs} - \frac{C_{KAPPA}\,C_{GSL}}{2}\left(-1 + \sqrt{1 - \frac{4V_{gs}}{C_{KAPPA}}}\right) \tag{5.4.74}$$

where $\boldsymbol{C_{GSL}}$ and $\boldsymbol{C_{KAPPA}}$ are model parameters that account for the gate bias dependence of the gate charge due to the source/bulk overlap.

When V_{gs}≥0, the charge contributed from the source overlap can be written as

$$\frac{Q_{overlap,s}}{W_{active}} = (C_{GSO} + C_{KAPPA}\,C_{GSL})\,V_{gs} \tag{5.4.75}$$

2). Drain overlap capacitance

When V_{gd} <0, the charge contributed from the drain overlap can be written as

$$\frac{Q_{overlap,d}}{W_{active}} = C_{GDO} V_{gd} - \frac{C_{KAPPA}\,C_{GDL}}{2}\left(-1 + \sqrt{1 - \frac{4V_{gd}}{C_{KAPPA}}}\right) \tag{5.4.76}$$

where $\boldsymbol{C_{GDL}}$ is a model parameter to account for the gate bias dependence of the gate charge due to the drain/bulk overlap. The same $\boldsymbol{C_{KAPPA}}$ parameter is used in both the gate/drain and gate/source overlap capacitance models.

When $V_{gd} \geq 0$, the charge contributed from the drain overlap can be written as

$$\frac{Q_{overlap,d}}{W_{active}} = (\mathbf{C_{GDO}} + \mathbf{C_{KAPPA}}\ \mathbf{C_{GDL}})\, \mathrm{V_{gd}} \tag{5.4.77}$$

3. Gate overlap capacitance

The total charge for the gate overlap over the source and drain region is the sum of $Q_{overlap,s}$ and $Q_{overlap,d}$:

$$Q_{overlap,g} = -\left(Q_{overlap,s} + Q_{overlap,d}\right) \tag{5.4.78}$$

5.4.3 Single-equation short channel capacitance model (*capMod=2)*

1. Intrinsic charge and capacitance model

1). Basic formulation

Both ***capMod***=0 and ***capMod***=1 use piece-wise expressions in the linear and saturation regimes and from accumulation to depletion regions. With improved model continuity, the capacitance model (***capMod***=2) is developed from the ***capMod***=1 model. The derivation of the capacitance model for ***capMod***=2 is somewhat the same as that for ***capMod***=1.

The terminal charges Q_G, Q_B, Q_S, and Q_D are the charges associated with the gate, bulk, source, and drain:

$$Q_G = -(Q_B + Q_{INV}) \tag{5.4.79}$$

$$Q_B = Q_{DEP} + Q_{ACC} \tag{5.4.80}$$

$$Q_{INV} = Q_S + Q_D \tag{5.4.81}$$

$$Q_{DEP} = Q_{DEP0} + \delta Q_{DEP} \tag{5.4.82}$$

The total charge is computed by integrating the charge along the channel. Therefore,

$$Q_{INV} = W_{active} \int_0^{L_{active}} Q_{inv} dy = W_{active} C_{ox} \int_0^{L_{active}} \left(V_{gt} - A_{bulk} V_y\right) dy \tag{5.4.83}$$

$$Q_G = W_{active} \int_0^{L_{active}} Q_g dy = W_{active} C_{ox} \int_0^{L_{active}} \left(V_{gt} + V_{th} - v_{fb} - \phi_s - V_y\right) dy \tag{5.4.84}$$

$$Q_{DEP} = W_{active} \int_0^{L_{active}} Q_{dep} dy = -W_{active} C_{ox} \int_0^{L_{active}} \left(V_{th} - v_{fb} - \phi_s - (1 - A_{bulk}) V_y\right) dy \tag{5.4.85}$$

where A_{bulk} is the parameter for the body charge effect.

Substituting $dy = dV_y/dE_y$, where V_y and E_y are the potential and electric field (refer to the source) along the channel caused by the applied drain bias respectively, and using the following from the I-V model,

$$I_{ds} = W_{eff} \mu_{eff} C_{ox} \left(V_{gt} - A_{bulk} V_{ds}\right) E_y \tag{5.4.86}$$

we obtain (assuming $W_{eff} = W_{active}$ and $L_{eff} = L_{active}$)

$$Q_{INV} = -W_{active} L_{active} C_{ox} \left(\left(V_{gt} - \frac{A_{bulk}}{2} V_{ds}\right) + \frac{A_{bulk}^2 V_{ds}^2}{12\left(V_{gt} - \frac{A_{bulk}}{2} V_{ds}\right)} \right) \tag{5.4.87}$$

$$Q_G = -Q_{DEP0} + W_{active} L_{active} C_{ox} \left(\left(V_{gt} - \frac{V_{ds}}{2}\right) + \frac{A_{bulk} V_{ds}^2}{12\left(V_{gt} - \frac{A_{bulk}}{2} V_{ds}\right)} \right) \tag{5.4.88}$$

$$Q_{DEP0} = -C_{ox} W_{active} L_{active} \gamma \sqrt{\left(2\phi_B - V_{bseff}\right)} \tag{5.4.89a}$$

$$\delta Q_{DEP} = W_{active} L_{active} C_{ox} \left(\frac{1 - A_{bulk}}{2} V_{ds} - \frac{(1 - A_{bulk}) A_{bulk} V_{ds}^2}{12\left(V_{gt} - \frac{A_{bulk}}{2} V_{ds}\right)} \right) \tag{5.4.89b}$$

2). Velocity saturation

Like ***capMod***=1, a simple model is adopted which empirically fits the saturation voltage $V_{dsat,cv}$ to channel length based on the following observed experimentally,

$$V_{dsat,iv} < V_{dsat,cv} < V_{dsat,iv}\Big|_{L_{active}\to\infty} = \frac{V_{gsteff}}{A_{bulk}} \tag{5.4.90}$$

$V_{dsat,cv}$ is fitted empirically using the following equation:

$$V_{dsat,cv} = \frac{V_{gsteff,cv}}{A_{bulk}'} \tag{5.4.91}$$

$$A_{bulk}' = A_{bulk0}\left(1+\left(\frac{CLC}{L_{active}}\right)^{CLE}\right) \tag{5.4.92}$$

where A_{bulk0} is given in Eq. (5.4.9).

Because of the introduction of A_{bulk}', the A_{bulk} terms in the above equations from Eq. (5.4.85) to Eq. (5.4.88) are replaced with A_{bulk}' given in Eq. (5.4.92).

3). Implementation of the polysilicon gate depletion effect

As in the *I-V* model, the implementation of the polysilicon depletion effect in the capacitance model is realized by using an effective gate voltage V_{gs_eff} given in Eq. (4.4.57) to replace the gate voltage V_{gs} in the model equations. As in the *I-V* model, the poly-gate depletion effect is turned on in the *C-V* model in the operation regime where V_{gs} is larger than $V_{FB}+\phi_s$ when N_{GATE} is given in the model card with a value larger than $1\text{x}10^{18}\text{cm}^{-3}$ but less than $1\text{x}10^{25}\text{cm}^{-3}$.

4). Continuous equation formulation

In the piece-wise capacitance models, the capacitance is divided into different operation regions, such as accumulation, depletion, weak inversion, and strong inversion or triode and saturation regimes. There are separate equations modeling the nodal charges in each region. From one region to another region the charges are continuous, but not the slopes. Therefore the capacitances at the transitions are discontinuous. To solve this problem, a single equation is used in the capacitance model for ***capMod***=2 to model each charge for all regions.

a. Transition from the accumulation region to the depletion region

A function called V_{FBeff} is used to smooth out the transition between the accumulation and depletion regions. It affects the accumulation and depletion charges. V_{FBeff} becomes v_{fb} when $V_{gb} > v_{fb}$ and approaches to V_{gb} when $V_{gb} < v_{fb}$ as shown in Fig. 5.4.4.

$$V_{FBeff} = v_{fb} - 0.5\left\{V_3 + \sqrt{V_3^2 + 4\delta_3 |v_{fb}|}\right\} \quad (5.4.93)$$

$$v_{fb} = V_{th} - \phi_s - K_{1OX}\sqrt{\phi_s - V_{bseff}} \quad (5.4.94)$$

$$V_3 = v_{fb} - V_{gb} - \delta_3 \quad \text{where } \delta_3 = 0.02 \quad (5.4.95)$$

$$Q_{ACC} = -W_{active} L_{active} C_{ox}\left(V_{FBeff} - v_{fb}\right) \quad (5.4.96)$$

$$Q_{DEP0} = -W_{active} L_{active} C_{ox} \frac{K_{1OX}^2}{2}\left(-1 + \sqrt{1 + \frac{4\left(V_{gs} - V_{FBeff} - V_{gsteff,cv} - V_{bseff}\right)}{K_{1OX}^2}}\right) \quad (5.4.97)$$

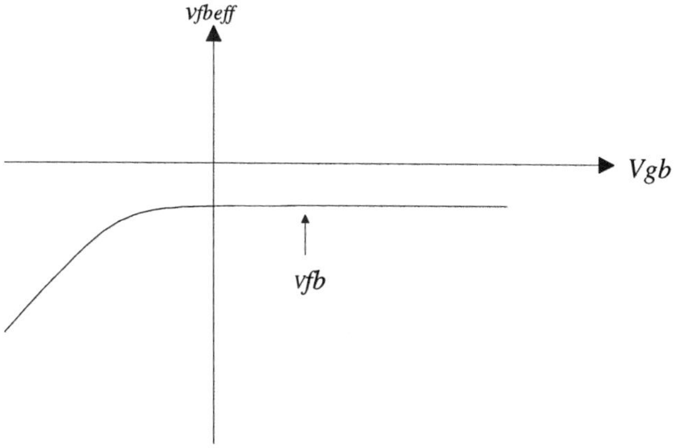

Fig. 5.4.4 V_{FBeff} function becomes v_{fb} when $V_{gb} > v_{fb}$, and V_{gb} when $V_{gb} < v_{fb}$.

b. Transition from the depletion region to the inversion region

The earlier compact models used a step function for the inversion capacitance that changes abruptly from 0 to C_{ox}. On the other hand, the substrate charge becomes a constant in strong inversion while it is a function of the gate and body biases in subthreshold region, and therefore the substrate capacitance drop abruptly to 0 at threshold voltage. Both can cause oscillations during circuit simulation. For analog and low power circuits an accurate capacitance model around the threshold voltage is very important.

The smooth inversion channel capacitance and depletion capacitance model is developed in ***capMod***=1. It uses a single equation to formulate the weak inversion, moderate inversion, and strong inversion regions by introducing the $V_{gseff,cv}$ function, which is copied into the capacitance model for ***capMod***=2.

c. Transition from triode to saturation region

A function V_{cveff} is used to smooth out the transition between triode and saturation regions. It affects the inversion charge. V_{cveff} tends to $V_{dsat,cv}$ when $V_{ds} > V_{dsat,cv}$ and becomes V_{ds} when $V_{ds} < V_{dsat,cv}$.

$$V_{cveff} = V_{dsat,cv} - 0.5\left\{V_4 + \sqrt{V_4^2 + 4\delta_4 V_{dsat,cv}}\right\} \tag{5.4.98}$$

$$V_4 = V_{dsat,cv} - V_{ds} - \delta_4 \tag{5.4.99}$$

where δ_4=0.02V. With the introduction of V_{cveff}, we have the following,

$$Q_{INV} = -W_{active}L_{active}C_{ox}\left(\left(V_{gsteff,cv} - \frac{A_{bulk}'}{2}V_{cveff}\right) + \frac{A_{bulk}'^2 V_{cveff}^2}{12\left(V_{gsteff,cv} - \frac{A_{bulk}'}{2}V_{cveff}\right)}\right) \tag{5.4.100}$$

$$\delta Q_{DEP} = W_{active}L_{active}C_{ox}\left(\frac{1 - A_{bulk}'}{2}V_{cveff} - \frac{(1 - A_{bulk}')A_{bulk}'V_{cveff}^2}{12\left(V_{gsteff,cv} - \frac{A_{bulk}'}{2}V_{cveff}\right)}\right) \tag{5.4.101}$$

i. 50/50 charge partition

$$Q_S = Q_D = -\frac{W_{active}L_{active}C_{ox}}{2}\left(V_{gsteff,cv} - \frac{A_{bulk}'}{2}V_{cveff} + \frac{A_{bulk}'^2 V_{cveff}^2}{12\left(V_{gsteff,cv} - \frac{A_{bulk}'}{2}V_{cveff}\right)}\right) \quad (5.4.102)$$

ii. 40/60 charge partition

$$Q_S = -\frac{W_{active}L_{active}C_{ox}}{2\left(V_{gsteff,cv} - \frac{A_{bulk}'}{2}V_{cveff}\right)^2}$$
$$\left(V_{gsteff,cv}^3 - \frac{4}{3}V_{gsteff,cv}^2\left(A_{bulk}'V_{cveff}\right) + \frac{2}{3}V_{gsteff,cv}\left(A_{bulk}'V_{cveff}\right)^2 - \frac{2}{15}\left(A_{bulk}'V_{cveff}\right)^3\right)$$

(5.4.103)

$$Q_D = -\frac{W_{active}L_{active}C_{ox}}{2\left(V_{gsteff,cv} - \frac{A_{bulk}'}{2}V_{cveff}\right)^2}$$
$$\left(V_{gsteff,cv}^3 - \frac{5}{3}V_{gsteff,cv}^2\left(A_{bulk}'V_{cveff}\right) + V_{gsteff,cv}\left(A_{bulk}'V_{cveff}\right)^2 - \frac{1}{5}\left(A_{bulk}'V_{cveff}\right)^3\right)$$

(5.4.104)

iii. 0/100 charge partition

$$Q_S = -W_{active}L_{active}C_{ox}\left(\frac{V_{gsteff,cv}}{2} + \frac{A_{bulk}'V_{cveff}}{4} - \frac{\left(A_{bulk}'V_{cveff}\right)^2}{24\left(V_{gsteff,cv} - \frac{A_{bulk}'}{2}V_{cveff}\right)}\right) \quad (5.4.105)$$

$$Q_D = -W_{active}L_{active}C_{ox}\left(\frac{V_{gsteff,cv}}{2} - \frac{3A_{bulk}'V_{cveff}}{4} + \frac{\left(A_{bulk}'V_{cveff}\right)^2}{8\left(V_{gsteff,cv} - \frac{A_{bulk}'}{2}V_{cveff}\right)}\right)$$

(5.4.106)

5). Bias dependent threshold voltage effects on capacitance

As in the ***capMod***=1 model, the effects of body bias and DIBL are included in the capacitance model by using the same threshold voltage as in the *I-V* model. The intrinsic capacitances can be derived based on the charge equations according to Eq. (5.2.36).

2. Extrinsic charge and capacitance models

An accurate model for the overlap capacitance is essential. This is especially true for the drain side where the effect of the capacitance is amplified by the transistor gain. In the earlier SPICE models this capacitance is assumed to be bias independent. However experimental data shows that the overlap capacitance changes with gate to source and gate to drain biases and the bias dependence is the result of surface depletion of the source and drain regions. In single drain structures (or the gate to the heavily doped S/D overlap region in a LDD structure) the modulation is expected to be very small so we can model this region with a constant capacitance. In LDD MOSFETs a substantial portion of the LDD region can be depleted, leading to a large reduction of the overlap capacitance. This LDD region can also be in accumulation. A single equation for the overlap capacitance in both the accumulation and depletion regions is found through the smoothing functions $V_{gs,overlap}$ and $V_{gd,overlap}$ for the source and drain side respectively. Unlike the case with the intrinsic capacitance, the overlap capacitances are reciprocal, i.e. $C_{gs,overlap} = C_{sg,overlap}$ and $C_{gd,overlap} = C_{dg,overlap}$.

1). Charge in the gate/source overlap region

With the introduction of the $V_{gs,overlap}$ function the charge in the gate/source overlap region can be given by

$$\frac{Q_{overlap,s}}{W_{active}} = CGSOV_{gs} + CGSL[V_{gs} - V_{gs,overlap} - \frac{CKAPPA}{2}(-1+\sqrt{1-\frac{4V_{gs,overlap}}{CKAPPA}})]$$

(5.4.107)

$$V_{gs,overlap} = \frac{1}{2}[(V_{gs}+\delta_1) - \sqrt{(V_{gs}+\delta_1)^2 + 4\delta_1}] \qquad (5.4.108)$$

where $\delta_1 = 0.02$V.

C_{KAPPA} can be calculated by $2\varepsilon_{si}qN_{LDD}/C_{ox}^2$ if the average doping in the LDD region is known. A typical value for N_{LDD} is $5\text{x}10^{17}$ cm^{-3}. In BSIM3v3, C_{KAPPA} is a user input parameter.

2). Charge in the gate/drain overlap region

With the introduction of the $V_{gd,\text{overlap}}$ function the charge in the gate/drain overlap region can be modeled with

$$\frac{Q_{overlap,d}}{W_{active}} = CGD0V_{gd} + CGDL[V_{gd} - V_{gd,overlap} - \frac{CKAPPA}{2}(-1+\sqrt{1-\frac{4V_{gd,overlap}}{CKAPPA}})]$$

(5.4.109)

$$V_{gd,overlap} = \frac{1}{2}[(V_{gd}+\delta_2) - \sqrt{(V_{gd}+\delta_2)^2 + 4\delta_2}] \quad (5.4.110)$$

where δ_2=0.02V.

Fig. 5.4.5 shows the simulated overlap capacitance C_{gd} as a function of V_{gd} for V_{bs} =0.

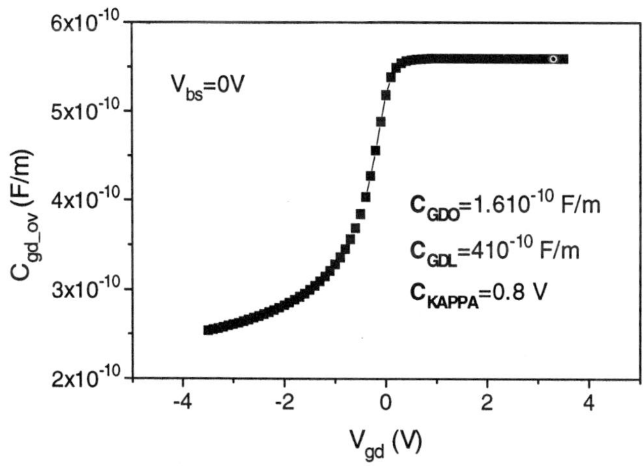

Fig. 5.4.5 Simulated overlap capacitance C_{gd} versus V_{gd} for an n-MOSFET.

3). Gate overlap charge

$$Q_{overlap,g} = -\left(Q_{overlap,s} + Q_{overlap,d}\right) \quad (5.4.111)$$

In the above expressions, if C_{GS0} and C_{GD0} (the heavily doped S/D region to gate overlap capacitance) are not given, they are calculated according to the following:

$$C_{GS0} = D_{LC}\, C_{ox} - C_{GSL} \quad \text{(If } D_{LC} \text{ is given and } D_{LC} > 0\text{)} \tag{5.4.112}$$

$$C_{GS0} = 0.6\, X_J C_{ox} \quad \text{(Otherwise)} \tag{5.4.113}$$

$$C_{GD0} = D_{LC}\, C_{ox} - C_{GDL} \quad \text{(If } D_{LC} \text{ is given and } D_{LC} > 0\text{)} \tag{5.4.114}$$

$$C_{GD0} = 0.6\, X_J C_{ox} \quad \text{(Otherwise)} \tag{5.4.115}$$

The fringing capacitance consists of a bias independent outer fringing capacitance and a bias dependent inner fringing capacitance. In the present release of BSIM3v3, only the bias independent outer fringing capacitance is implemented. Experimentally it is impossible to separate this capacitance from the overlap capacitance but it can be calculated theoretically by:

$$C_F = \frac{\varepsilon_{ox}}{\alpha} \ln\left(1 + \frac{T_{poly}}{T_{ox}}\right);\ \alpha = \frac{\pi}{2} \tag{5.4.116}$$

where T_{poly} is the thickness of the polysilicon gate. In the model implementation, T_{poly} is set to 400nm.

5.4.4 Short channel capacitance model with quantization effect (*capMod=3)*

As CMOS technologies rapidly advance into the deep sub-micron regime with extremely thin gate oxides ($T_{ox} \leq$ 7nm), physical effects such as the polysilicon gate depletion and inversion layer charge quantization effects will significantly affect the C-V characteristics of CMOS devices [5.24, 5.25, 5.26, 5.27]. To account for these effects, a new capacitance model with ***capMod***=3 has been developed and released in BSIM3v3.2 [5.23]. This model introduces computationally efficient and accurate compact equations for thin-oxide MOSFET intrinsic capacitance, and includes the finite charge thickness from the accumulation through depletion to inversion regions as well as the polysilicon depletion effects, and shows good accuracy and continuity in all regions of operation. This new model preserves the high scaleability and accurate modeling of the non-uniform doping, mobility degradation, and *DIBL* effects that are characteristic of BSIM3 [5.17, 5.28].

In this section formulations for the finite charge thickness from the accumulation through depletion to inversion regions, and bias-dependent surface potential due to the bulk charge in the inversion region are discussed first, followed by the charge equations and channel charge partitioning.

1. The implementation of poly-silicon gate depletion effects

As discussed in the capacitance model for ***capMod***=2, the implementation of poly-silicon gate depletion effects uses an effective gate voltage V_{gs_eff} given in Eq. (5.4.57) to replace the gate voltage V_{gs} in all the model equations [5.29]. As in the *I-V* model and other capacitance model options, the poly-gate depletion effect is turned on in the operation regime where V_{gs} is larger than $V_{FB}+\phi_s$ when N_{GATE} is given in the model card with a value larger than $1\text{x}10^{18}\text{cm}^{-3}$ but less than $1\text{x}10^{25}\text{cm}^{-3}$ [5.17].

2. Finite inversion charge layer thickness and its formulation

As the gate oxide thickness T_{ox} continues to scale down (<10nm), the finite inversion charge thickness, X_{DC}, cannot be ignored. It may be represented as a capacitance, C_{cen}, which is in series with the gate oxide capacitance C_{ox}. This results in a reduced effective gate oxide capacitance C_{oxeff}, which can be expressed as [5.30, 5.31]

$$C_{oxeff} = \frac{C_{ox}C_{cen}}{C_{ox}+C_{cen}} \tag{5.4.117}$$

$$C_{cen} = \varepsilon_{si} / X_{DC} \tag{5.4.118}$$

Based on 1-D quantum mechanical simulation results, a universal expression for the finite charge thickness from the accumulation to depletion region has been developed [5.31]:

$$X_{DC} = \frac{1}{3} L_{debye} \exp[\boldsymbol{A_{CDE}}(\frac{\boldsymbol{N_{CH}}}{2x10^{16}})^{-0.25} \frac{V_{gs}-V_{bs}-v_{fb}}{\boldsymbol{T_{OX}}}] \text{ [cm]} \tag{5.4.119}$$

$$L_{debye} = \sqrt{\frac{\varepsilon_{si} v_{t0}}{q \boldsymbol{N_{CH}}}} \tag{5.4.120}$$

where v_{t0} is the thermal voltage at T_{NOM}, $(V_{gs}-V_{bs}-v_{fb})/T_{OX}$ has the units of MV/cm, V_{bs} is the body bias, and A_{CDE} is a fitting parameter with the default value of 1. However, Eq. (5.4.119) is not implemented directly in the simulator. Instead, the following expression is proposed to increase simulation robustness,

$$X_{DCeff} = X_{DC\max} - \frac{1}{2}\left(X_0 + \sqrt{X_0^2 + 4\delta_x X_{DC\max}}\right) \tag{5.4.121}$$

$$X_0 = X_{DCmax} - X_{DC} - \delta_x \tag{5.4.122}$$

where $\delta_x = 10^{-3} \cdot T_{ox}$ and X_{DCmax} is an upper bound for the accumulation and depletion charge layer thickness for the simulation stability:

$$X_{DC\max} = X_{DC}\big|_{V_{gs}=V_{bs}+v_{fb}} = \frac{1}{3} L_{debye} \tag{5.4.123}$$

Eq. (5.4.121) reduces to Eq. (5.4.119) for $V_{gs} < (V_{bs} + v_{fb})$ and Eq. (5.4.123) for $V_{gs} > (V_{bs} + v_{fb})$.

In the inversion region, the inversion charge layer thickness proposed in [5.31] is reformulated as

$$X_{DCeff} = \frac{1.9x10^{-7}}{1 + \left[\frac{V_{gsteff,cv} + 4(V_{th} - v_{fb} - 2\phi_B)}{2T_{OX}}\right]^{0.7}} \text{ [cm]} \tag{5.4.124}$$

where the second term in the denominator has the units of MV/cm.

Through v_{fb} given in Eq. (5.4.94), Eq. (5.4.124) is found to be applicable to both N^+ and P^+ polysilicon gates [5.31].

3. Bias dependent surface potential in the inversion region

The classical condition for strong inversion is defined by the surface potential being equal to $2\phi_B$ [5.4] even when V_{gs} exceeds the threshold voltage V_{th}. In reality, the surface potential varies with the gate bias even in strong inversion. This approximation of constant surface potential in strong inversion is one cause of the sharp turn in C-V around V_{th} in the modeled capacitance which can give rise to inaccuracies for analog and low voltage/power circuit designs.

Considering both the inversion charge (Q_{inv}) and bulk charge (Q_b) layer thickness in the inversion region, the surface potential can be written as

$$\phi_s = -\frac{X_{DCeff} \cdot Q_{inv}}{\varepsilon_{si}} - \frac{X_{dep} \cdot Q_b}{\varepsilon_{si}} \tag{5.4.125}$$

Q_{inv} can now be formulated as

$$Q_{inv} = -C_{oxeff}(V_{gsteff,cv} - \Phi_\delta) \tag{5.4.126}$$

where $\Phi_\delta = -X_{dep} \cdot Q_b / \varepsilon_{si} - \phi_s$. By solving the Poisson equation and assuming zero inversion charge layer thickness, an analytical formulation for Φ_δ is proposed as

$$\Phi_\delta = \phi_s - 2\phi_B = v_t \ln[\frac{V_{gseff,cv}(V_{gsteff,cv} + 2K_{1OX}\sqrt{2\phi_B})}{M_{OIN} K_{1OX} v_t^2} + 1] \tag{5.4.127}$$

where M_{OIN} is a fitting parameter with a typical value of 15. Note that Eq. (5.4.127) rapidly drops to zero for $(V_{gs} - V_{th}) < -3N_{OFF} \cdot nv_t$ as the inversion layer disappears.

4. Charge model

a. Charge equations for the accumulation region

In the accumulation region, the inversion charge Q_{INV} is close to zero, and the gate charge Q_G is mirrored in the bulk as the accumulation charge Q_{ACC} near the silicon surface. Q_{ACC} is computed by

$$Q_{ACC} = W_{active} L_{ative} C_{oxeff} V_{gbacc} \tag{5.4.128}$$

where V_{gbacc} is the effective gate-to-body voltage and is given by

$$V_{gbacc} = \frac{1}{2}\left(V_0 + \sqrt{V_0^2 + 4\delta_v |v_{fb}|}\right) \tag{5.4.129}$$

where $V_0 = v_{fb} + V_{bseff} - V_{gs} - \delta_v$ and $\delta_v = 0.02\text{V}$ so that V_{gbacc} reduces to $v_{fb} + V_{bseff} - V_{gs}$ in the accumulation region $(V_{gs} - v_{fb} - V_{bseff} << -\delta_v)$ and zero in other operating regions $(V_{gs} - v_{fb} - V_{bseff} >> \delta_v)$.

b. Charge equations for the depletion region

Under the depletion approximation, the bulk charge in the depletion region without the influence of drain bias can be obtained [5.1, 5.3]

$$Q_{DEP0} = -W_{active}L_{active}C_{oxeff}\frac{K_{1OX}^2}{2}[-1+\sqrt{1+\frac{4(V_{gs}-v_{fbx}-V_{bseff}-V_{gsteff,cv})}{K_{1OX}^2}}] \quad (5.4.130)$$

$$v_{fbx} = v_{fb} - V_{gbacc} \quad (5.4.131)$$

Note that Eq. (5.4.130) becomes close to zero in the accumulation region.

c. Charge equations for the inversion region

All previous analytic MOSFET models use the assumption of zero thickness for the inversion layer in the inversion regime. In BSIM3v3.2, the finite thickness of the inversion layer is considered, and the inversion charge in the linear region is modeled as

$$Q_{INV} = -W_{active}C_{oxeff}\int_0^{L_{active}}(V_{gsteff,cv} - \Phi\delta - A_{bulk}'V_y)dy \quad (5.4.132)$$

Performing integration for Eq. (5.4.132) by replacing dy with dV_y/E_y, as discussed in previous capacitance models, the expression for Q_{INV} in the linear region is given as

$$Q_{INV} = -W_{active}L_{active}C_{oxeff}[V_{gsteff,cv} - \Phi\delta - \frac{1}{2}A_{bulk}'V_{ds} + \frac{A_{bulk}'^2 V_{ds}^2}{12(V_{gsteff,cv} - \Phi\delta - A_{bulk}'V_{ds}/2)}] \quad (5.4.133)$$

As discussed earlier in ***capMod*** =2, Q_{DEP} can be divided into two parts. One is V_{ds} independent denoted by Q_{DEP0} while the other component is a function of V_{ds} called δQ_{DEP}. Q_{DEP0} is given by Eq. (5.4.132) where $V_{gsteff,cv}$ reduces to $(V_{gs}$-$V_{th})$ in the strong inversion region, while δQ_{DEP} is formulated as

$$\delta Q_{DEP} = W_{active}L_{active}C_{oxeff}[\frac{1-A_{bulk}'}{2}V_{ds} - \frac{(1-A_{bulk}')A_{bulk}'V_{ds}^2}{12(V_{gsteff,cv} - \Phi\delta - A_{bulk}'V_{ds}/2)}] \quad (5.4.134)$$

In order for Eq. (5.4.133) and Eq. (5.4.134) to be applicable in the saturation region, all V_{ds} terms in Eqs. (5.4.133) and (5.4.134) are replaced by V_{cveff} given in Eq. (5.4.98).

The saturation voltage $V_{dsat,cv}$ is proposed as

$$V_{dsat,cv} = \frac{V_{gsteff,cv} - \Phi\delta}{A_{bulk}'} \tag{5.4.135}$$

d. Channel charge partitioning

The channel charge Q_{INV} given by Eq. (5.4.133) can be separated into drain and source charge components Q_S and Q_D by following the same partition schemes as discussed in the other capacitance models.

i. 50/50 charge partition

$$Q_S = Q_D = -\frac{W_{active}L_{active}C_{oxeff}}{2}[V_{gsteff,cv} - \Phi\delta - \frac{A_{bulk}'}{2}V_{cveff} + \frac{A_{bulk}'^2 V_{cveff}^2}{12(V_{gsteff,cv} - \Phi\delta - \frac{A_{bulk}'}{2}V_{cveff})}] \tag{5.4.136}$$

ii. 40/60 charge partition

$$Q_S = -\frac{W_{active}L_{active}C_{oxeff}}{2\left(V_{gsteff,cv} - \Phi\delta - \frac{A_{bulk}'}{2}V_{cveff}\right)^2}\{(V_{gsteff,cv} - \Phi\delta)^3 - \frac{4}{3}(V_{gsteff,cv} - \Phi\delta)^2 (A_{bulk}'V_{cveff}) + \frac{2}{3}(V_{gsteff,cv} - \Phi\delta)(A_{bulk}'V_{cveff})^2 - \frac{2}{15}(A_{bulk}'V_{cveff})^3\} \tag{5.4.137}$$

$$Q_D = -\frac{W_{active}L_{active}C_{oxeff}}{2\left(V_{gsteff,cv} - \Phi\delta - \frac{A_{bulk}'}{2}V_{cveff}\right)^2} \{(V_{gsteff,cv} - \Phi\delta)^3 - \frac{5}{3}(V_{gsteff,cv} - \Phi\delta)^2 (A_{bulk}'V_{cveff}) + (V_{gsteff,cv} - \Phi\delta)(A_{bulk}'V_{cveff})^2 - \frac{1}{5}(A_{bulk}'V_{cveff})^3\} \tag{5.4.138}$$

iii. 0/100 charge partition

$$Q_S = -W_{active}L_{active}C_{oxeff}[\frac{V_{gsteff,cv} - \Phi\delta}{2} + \frac{A_{bulk}'V_{cveff}}{4} - \frac{(A_{bulk}'V_{cveff})^2}{24(V_{gsteff,cv} - \Phi\delta - \frac{A_{bulk}'}{2}V_{cveff})}] \quad (5.4.139)$$

$$Q_D = -W_{active}L_{active}C_{oxeff}[\frac{V_{gsteff,cv} - \Phi\delta}{2} - \frac{3(A_{bulk}'V_{cveff})}{4} + \frac{(A_{bulk}'V_{cveff})^2}{8(V_{gsteff,cv} - \Phi\delta - \frac{A_{bulk}'}{2}V_{cveff})}] \quad (5.4.140)$$

Q_G can be obtained directly from the charge conservation principle.

The MOSFET intrinsic capacitance can be obtained by differentiating the terminal charges described above with respect to the terminal voltages, as given in Eq. (5.2.36). Fig. 5.4.5 to 5.4.9 show the charge and capacitance versus gate bias for the ***capMod***=3 model. Fig. 5.4.10 to 5.4.14 show the charge and capacitance characteristics versus drain bias.

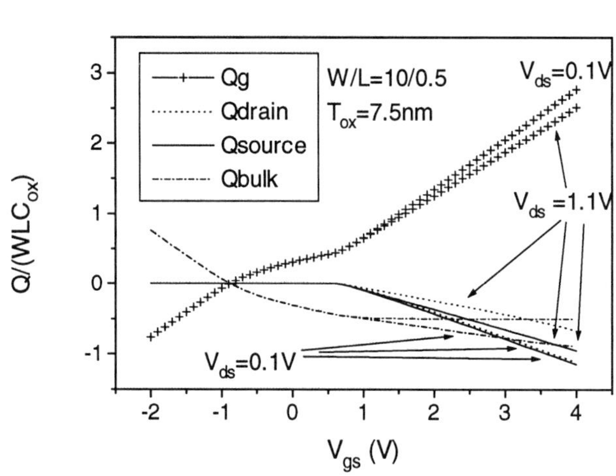

Fig. 5.4.5 Simulated Q_G, Q_D, Q_S, Q_B as a function of V_{gs} for several V_{ds} (V_{bs}=0).

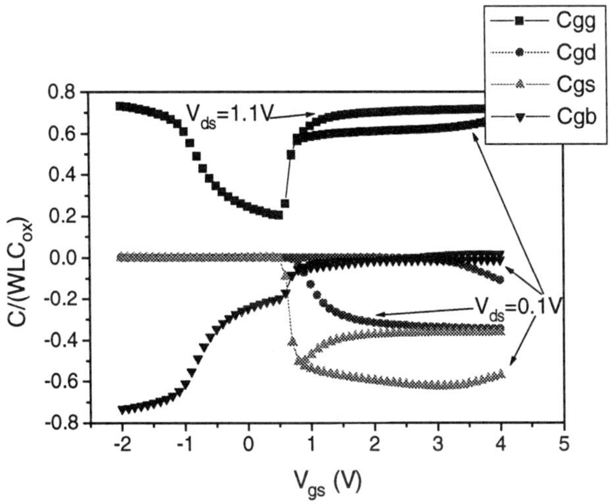

Fig. 5.4.6 Simulated capacitance $C_{GG,}$ $C_{GD,}$ $C_{GS,}$ C_{GB} as a function of V_{gs} for several V_{ds} (V_{bs}=0).

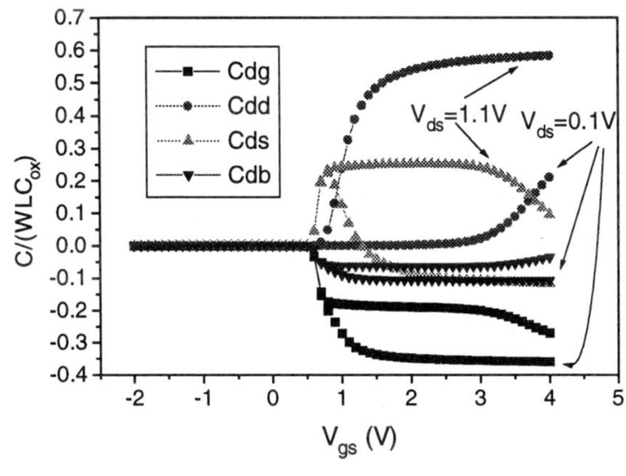

Fig. 5.4.7 Simulated capacitance $C_{DG},$ $C_{DD,}$ $C_{DS,}$ C_{DB} as a function of V_{gs} for several V_{ds} (V_{bs}=0).

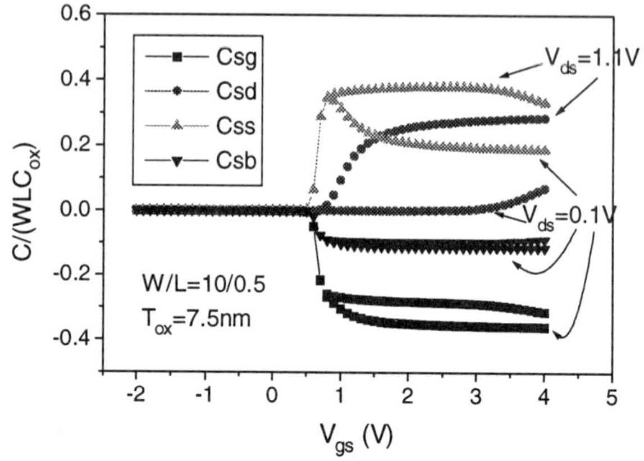

Fig. 5.4.8 Simulated capacitance C_{SG}, C_{SD}, C_{SS}, C_{SB} as a function of V_{gs} for several V_{ds} (V_{bs}=0).

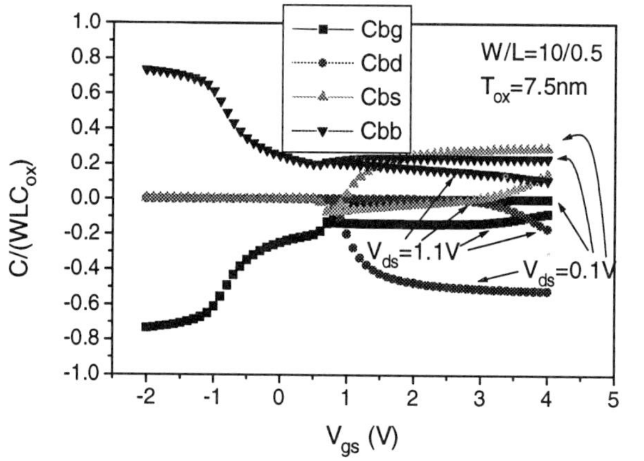

Fig. 5.4.9 Simulated capacitance C_{BG}, C_{BD}, C_{BS}, C_{BB} as a function of V_{gs} for several V_{ds} (V_{bs}=0).

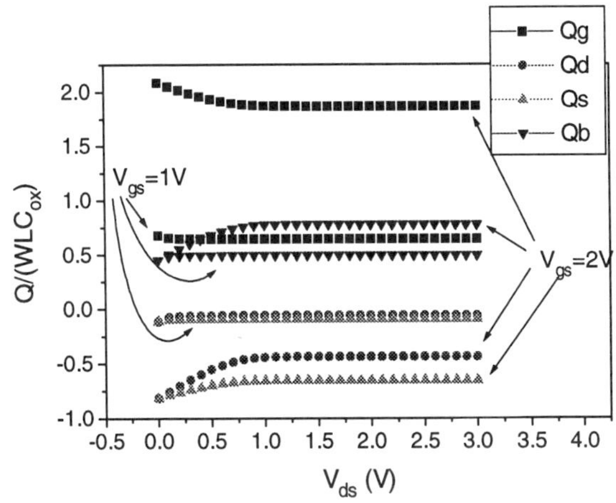

Fig. 5.4.10 Simulated Q_G, Q_D, Q_S, Q_B as a function of V_{ds} for several V_{gs} (V_{bs}=0).

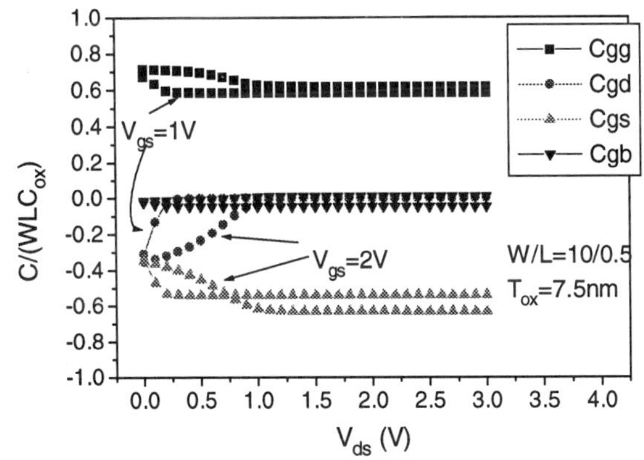

Fig. 5.4.11 Simulated capacitance C_{GG}, C_{GD}, C_{GS}, C_{GB} as a function of V_{ds} for several V_{gs} (V_{bs}=0).

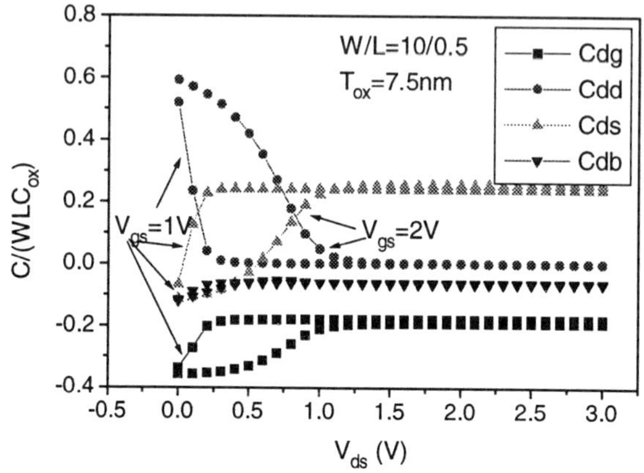

Fig. 5.4.12 Simulated capacitance C_{DG}, C_{DD}, C_{DS}, C_{DB} as a function of V_{ds} for several V_{gs} (V_{bs}=0).

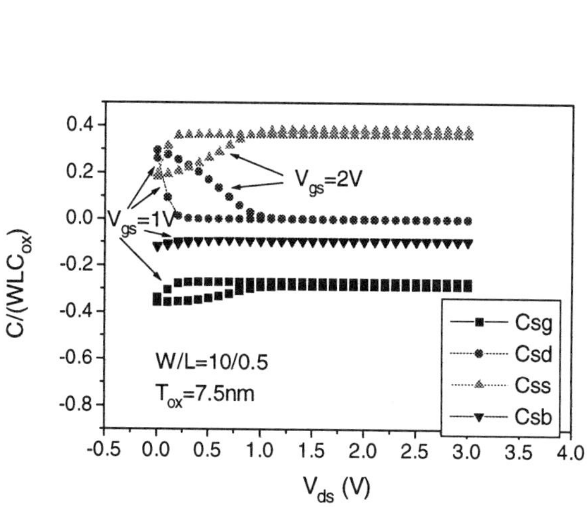

Fig. 5.4.13 Simulated capacitance C_{SG}, C_{SD}, C_{SS}, C_{SB} as a function of V_{ds} for several V_{gs} (V_{bs}=0).

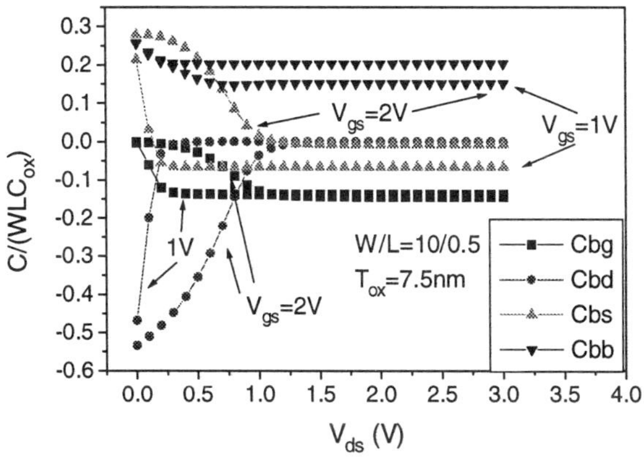

Fig. 5.4.14 Simulated capacitance C_{BG}, C_{BD}, C_{BS}, C_{BB} as a function of V_{ds} for different V_{gs} (V_{bs}=0).

5.5 Channel Length/Width in Capacitance Model

The inconsistency of the effective channel length extracted from DC and capacitance measurements has been long observed [5.33]. It is not surprising and can be explained. The DC current depends on the movement of the carriers from source to drain and this distance between the source and drain junctions is characterized as the effective channel length. Charge and capacitance behaviors depend on the electric field flux distribution between the poly-silicon gate and bulk silicon substrate material. Thus, the corresponding effective channel length obtained from *C-V* characteristics is not necessarily equal to that from DC current measurements.

In most previous compact models, both the *I-V* and *C-V* models use the same effective channel length and width. This is not the case in BSIM3v3. The channel length and width used in charge and capacitance models are given as the following and are independent of the "channel length" used in the DC model:

$$L_{active} = L_{drawn} - 2\delta L_{eff} \tag{5.5.1}$$

$$W_{active} = W_{drawn} - 2\delta W_{eff} \tag{5.5.2}$$

$$\delta W_{eff} = \boldsymbol{Dwc} + \frac{\boldsymbol{Wlc}}{L^{\boldsymbol{Wln}}} + \frac{\boldsymbol{Wwc}}{W^{\boldsymbol{Wwn}}} + \frac{\boldsymbol{Wwlc}}{L^{\boldsymbol{Wln}} W^{\boldsymbol{Wwn}}} \tag{5.5.3}$$

$$\delta L_{eff} = \boldsymbol{Dlc} + \frac{\boldsymbol{Llc}}{L^{\boldsymbol{Lln}}} + \frac{\boldsymbol{Lwc}}{W^{\boldsymbol{Lwn}}} + \frac{\boldsymbol{Lwlc}}{L^{\boldsymbol{Lln}} W^{\boldsymbol{Lwn}}} \tag{5.5.4}$$

$\boldsymbol{D_{WC}}$ and $\boldsymbol{D_{LC}}$ are separate and different from $\boldsymbol{W_{INT}}$ and $\boldsymbol{L_{INT}}$ in the *I-V* model. L_{active} and W_{active} are the effective length and width of the intrinsic device for capacitance calculations. Unlike the case with *I-V*, it is assumed that these dimensions have no voltage bias dependence. The parameter δL_{eff} is equal to the source/drain to gate overlap length plus the difference between drawn and actual polysilicon gate due to processing (gate printing, etching and oxidation) on one side. Overall, a distinction should be made between the effective channel length extracted from the capacitance measurement and from the *I-V* measurement.

The L_{active} parameter extracted from the capacitance method is found to be closer to the metallurgical junction length (physical length) than L_{eff}. If neither $\boldsymbol{D_{WC}}$ and $\boldsymbol{D_{LC}}$ are specified in the model card, BSIM3v3's capacitance model will assume that the device has the same effective dimensions for *I-V* and *C-V* models (i.e. $\boldsymbol{D_{WC}} = \boldsymbol{W_{INT}}$ and $\boldsymbol{D_{LC}}=\boldsymbol{L_{INT}}$).

5.6 Helpful Hints

1. The reciprocity of the charge-based capacitances

Many people think of the capacitances in a MOSFET as a group of conventional two terminal capacitors. For a two terminal capacitor, the capacitance is reciprocal, that is $C_{12}=C_{21}$, where 1 and 2 are node numbers of the terminals. In other words,

$$\frac{\partial Q_1}{\partial V_2} = \frac{\partial Q_2}{\partial V_1} \tag{5.6.1}$$

Furthermore, C_{12} and C_{21} can only be either a constant or a function of V_{12} or V_{21} (since there are no other voltages). These are the all too familiar properties of a capacitor, but they only apply to two-terminal capacitors. For a MOSFET with three or four terminals, the capacitances are in general non-reciprocal, that is $C_{xy} \neq C_{yx}$. Consider the capacitances C_{gd} and C_{dg} of a long channel MOSFET in the saturation regime [5.1]. According to the definition of capacitance, C_{gd} is the gate charge variation caused by the voltage change at the drain, and C_{dg} is the drain charge variation caused by the voltage change at the gate. Assuming the device operates in saturation, the pinch-off condition near the drain isolates the gate from the drain so that the variation of drain voltage does not have any influence on the charge at the gate terminal. In other words, the gate charge will not change as the drain voltage changes and C_{gd} will be zero. However, the inversion charge in the channel changes as the gate voltage changes so that the drain charge (a portion of the channel charge) will change, that is, C_{dg} will not be zero. The non-reciprocity property of MOSFET capacitance has been confirmed by simulations and measurements. If we artificially make $C_{gd} = C_{dg}$ in the compact model and allow C_{gd} to be a function of V_{gs} or V_{bs}, the model would not only be inaccurate but also incorrectly predict charge build-up at floating circuit nodes.

The confusion and misunderstanding of the non-reciprocity discussed above may be caused by the term "capacitance". One may want to call the capacitances in 3 or 4 terminal devices by a different name, transcapacitances, which are not reciprocal in general.

2. The quasi-static (QS) assumption and non-quasi-static (NQS) effects

It should be emphasized that all of the capacitance models discussed in this chapter are based on the quasi-static assumption, that is, the charges can follow the change in voltages immediately without any delay. In other words, the signals vary slowly (relative to the device transit time) so that the channel charge is in steady-state at all times. It has been found that the QS assumption is acceptable for short channel devices in digital applications according to a rule of thumb given in [5.1],

$$T_R > 20\tau_d \tag{5.6.2}$$

where T_R is the rise time of the input signal and τ_d is the transit time of the carriers leaving the source and arriving at the drain.

According to [5.1], τ_d depends on the channel length and V_g. For long channel devices, τ_d is

$$\tau_d = \frac{0.4L^2}{\mu(V_{gs} - V_{th})} \tag{5.6.3}$$

where L is the channel length of the device, μ is the carrier mobility, V_{gs} is the gate bias, and V_{th} is the threshold voltage.

For the limiting case short channel device in which the velocity saturation effect is overwhelming, τ_d is given by

$$\tau_d = \frac{L}{v_{sat}} \tag{5.6.4}$$

where v_{sat} is the carrier saturation velocity.

We can estimate the limitation of the *QS* models according to the above expressions. Taking an n-channel MOSFET as an example and assuming v_{sat}=1x10^5 m/s, the transit time τ_d is 2.5ps for a device of 0.25μm channel length. Thus, the QS assumption should be valid if T_R is larger than 50ps. In a digital circuit the clock frequency is typically 1/20T_R. So the *QS* assumption is acceptable for a 1GHz clock rate in 0.25μm technology. As the channel length decreases, the transit time decreases so that the QS assumption is acceptable for most digital applications.

However, the case may be different in high frequency analog applications at radio frequencies (RF). Some problems have been found for using the capacitance models based on QS assumption for an n-channel MOSFET of channel length larger than 1μm and a p-channel MOSFET of channel length larger than 0.7μm in RF applications when the operation frequency is higher than 5GHz [5.34]. An NQS (non-quasi-static) model is needed to ensure the accurate simulation of circuits at such high operation frequencies. The difficulty of developing an NQS C-V model is that the charges on the terminals will be functions of the past history of terminal voltages, not just the present voltages. An NQS model has been implemented in BSIM3v3 and will be discussed in Chapter 10.

3. Charge partition

Charge-based capacitance models ensure charge conservation. However, they need to partition the inversion charge into the drain and source charge.

How to partition the channel charge into source and drain accurately is a difficult issue that has not been solved since the appearance of the charge based capacitance model. As mentioned before there are three different charge partition schemes existing in today's charge-based models used in circuit simulators, such as 50/50, 40/60, and 0/100, which is distinguished in circuit simulation using a model parameter called X_{PART}.

50/50 (X_{PART}=0.5) is the simplest of all partition scheme in that the inversion charge is divided equally between the source and drain nodes. Despite it's simplicity it is found to approximate the 2-D simulation data well.

40/60 (X_{PART}<0.5) is the most physical model of the three partition schemes. However both the 40/60 and 50/50 models predict a nonphysical negative I_d pulse when the V_g of an n-MOSFET is ramped rapidly up and crosses V_{th} even if the drain terminal is at a high voltage, e.g. V_{dd}. This is due to the strong NQS effect at V_{gs} around V_{th}.

The 0/100 (X_{PART}>0.5) partition scheme is developed to artificially suppress the negative drain current spike by assigning all inversion charge in the saturation region to the source electrode. Notice that this charge partition scheme will give a drain current spike in the linear region and aggravate the source current spike problem.

These constant charge partition schemes are sufficient to meet the accuracy requirements of the simulation of logic gate delays as discussed in the previous subsection. However, at V_{gs}=V_{th} the transient time becomes very large and nonphysical artifacts can easily be observed with all the QS capacitance models.

4. The overlap capacitances

As mentioned earlier, the overlap capacitances can be divided into several different components. It is extremely difficult (if not impossible) to measure the different overlap components separately. 2-D or 3-D device simulation may be necessary for extracting the parameters of a complex capacitance model. This is a good reason to keep the capacitance model simple and use the fewest parameters possible. Table 5.5.1 summarizes the bias dependence of different parasitic capacitance components. It can be seen that both the direct overlap capacitance component and the inner fringing capacitance are bias-dependent. However, the inner fringing capacitance no doubt exhibits its maximum value in the depletion region and vanishes in the strong inversion region because the inversion layer screens out the coupling between the source/drain and the gate. Because the modeling of the bias dependence of inner fringing capacitance is very difficult and not available in compact MOSFET models at the present time, it may cause some problem for circuit simulation, especially for analog and high frequency applications where the accuracy of the capacitance model is critical.

Table 5.5.1 Bias dependence of parasitic capacitance components in a NMOSFET

Capacitance components	Bias dependence	Modeling work	Model status
n+ overlap capacitance	Small	Easy	Exist
n- overlap capacitance	Strong	Moderate	Exist
Outer fringing capacitance	No	Easy	Exist
Inner fringing capacitance	Strong	Difficult	No
S/D Junction capacitance	Strong	Easy	Exist
Outer sidewall capacitance	Strong	Easy	Exist
Inner sidewall capacitance	Strong	Moderate	Exist

BSIM3v3 has not included the bias dependence of C_{fi}, However, it does account for the bias dependence of the overlap source/drain capacitance. $\boldsymbol{C_{GDO}}/\boldsymbol{C_{GSO}}$ can be considered as an overlap capacitance when V_{gd} and V_{gs} are zero. The measured capacitance characteristics at zero bias includes both outer and inner fringing capacitances and overlap capacitances. Users should be aware that these parameters, extracted from the measured data at zero bias, may not be suitable for use in strong inversion due to the disappearance of the inner fringing capacitance in that region. Therefore, depending on the circuit application, users can adopt different approaches to use the overlap capacitance in BSIM3v3. For example, in digital circuit applications the speed delay of a circuit, say a ring oscillator, is often used as the figure of merit to judge the accuracy of the model. It has been found that the change in the circuit delay time is not very significant (less than 2%) when the bias dependent overlap capacitance is included in the simulation. So one can ignore the bias dependence of the overlap capacitance (by setting $\boldsymbol{C_{GDL}}$ and $\boldsymbol{C_{GSL}}$ to 0) and find constant overlap capacitance values for $\boldsymbol{C_{GDO}}$ and $\boldsymbol{C_{GSO}}$ to match the circuit delay for use in a digital application. However, for analog applications the bias dependence of the overlap capacitances becomes more important. In that case, a practical approach is to extract the values of $\boldsymbol{C_{GDO}}$, $\boldsymbol{C_{GDL}}$, and $\boldsymbol{C_{KAPPA}}$ together by using optimization to match the measured overlap capacitance characteristics as well as possible in all operation regions from accumulation through depletion to strong inversion.

5. The flat-band voltage parameter v_{fb} in the capacitance models

Flat-band voltage is an important parameter in MOSFET capacitance modeling. In the ***capMod***=0 model the flat-band voltage is treated as a model parameter and different V_{th} models are used in the DC and capacitance models. However, in the ***capMod***>0 capacitance models in BSIM3v3, the same threshold voltage is used in both *I-V* and *C-V* models, and the threshold voltage is characterized using the measured *I-V* data. Thus, the flat-band voltage parameter v_{fb} can be calculated from V_{th} according to the following expression

$$v_{fb} = V_{th} - \phi_s - K_{1OX}\sqrt{\phi_s - V_{bseff}} \tag{5.6.5}$$

where V_{th} is the threshold voltage and K_{1OX} is the model parameter defined in Chapter 3.

Since V_{th} includes non-uniform doping, short channel and narrow width effects, the calculated v_{fb} parameter also contains these effects. Discussions about the dependence of flat band voltage on non-uniform doping and short channel effects have been reported [5.36].

The v_{fb} parameter calculated with Eq. (5.6.5) is used to determine the boundary between the accumulation and depletion regimes. Thus, it will influence the capacitance model accuracy around the transition region from accumulation to depletion regions. A suitable characterization methodology of the threshold voltage, which considers the case for both *I-V* and *C-V*, may be necessary to ensure *C-V* model accuracy. It has been found that the recommended methodology to extract V_{th} from the measured *I-V* data can match the V_{th} obtained with the *C-V* measurement [5.37].

It should be noted that the threshold voltage in the above discussion should be measured at low drain bias and zero body bias in order to use Eq. (5.6.5) in calculating v_{fb}. In BSIM3v3.0 and BSIM3v3.1 the v_{fb} parameter is implemented inappropriately in which a bias dependent V_{th} expression given in Eq. (3.4.25) is used in Eq. (5.6.5). This brings about too strong a bias dependence of v_{fb} causing some continuity problems in the accumulation region. In the implementation of BSIM3v3.2, the bias dependence in short channel and narrow width effects of V_{th} is removed in Eq. (3.4.25) to correct this problem and improve the model continuity

$$
\begin{aligned}
V_{th} = {} & V_{th0OX} + K_{1OX}\left(\sqrt{\phi_s - V_{bseff}} - \sqrt{\phi_s}\right) \\
& + K_{1OX}\left(\sqrt{1+\frac{N_{LX}}{L_{eff}}} - 1\right)\sqrt{\phi_s} + K_3 \frac{T_{OX}}{W_{eff}' + W_0}\phi_s \\
& - D_{VT0}\left(\exp(-D_{VT1}\frac{L_{eff}}{2l_{t0}}) + 2\exp(-D_{VT1}\frac{L_{eff}}{l_{t0}})\right)(V_{bi} - \phi_s) \\
& - D_{VT0w}\left(\exp(-D_{VT1w}\frac{W_{eff}' L_{eff}}{2l_{tw0}}) + 2\exp(-D_{VT1w}\frac{W_{eff}' L_{eff}}{l_{tw0}})\right)(V_{bi} - \phi_s)
\end{aligned}
\tag{5.6.6}
$$

6. The C_{LC} and C_{LE} parameters

According to the understanding of the *C-V* characteristics in short channel devices, which shows that the saturation voltage in *C-V* characteristics is larger than the saturation voltage in *I-V* characteristics, a different saturation voltage expression from that in the *I-V* model is used in the *C-V* models of BSIM3v3. C_{LC} and C_{LE} are two empirical fitting parameters introduced to accurately describe the saturation voltage with the influence of velocity saturation effects in short channel devices, that is, as the channel length decreases the saturation voltage decreases at a given gate voltage. C_{LC} and C_{LE} are important parameters in determining the accuracy of the *C-V* model, especially in analog and RF applications because the saturation voltage will influence the cut-off frequency (f_T) characteristics and determine the point at which f_T begins to drop from the maximum value as the gate voltage increases at a fixed drain voltage [5.38]. An accurate determination of C_{LC} and C_{LE} is necessary when using the capacitance models in BSIM3v3.

The C_{LC} and C_{LE} parameters can be extracted from the measured characteristics of C_{gd} vs. V_{ds} at different V_{gs} for devices with different L in the strong inversion region.

7. The non-symmetry issue at V_{ds} =0

Model symmetry is a desirable feature of MOSFET capacitance model because a real MOSFET is symmetric and it may also help the convergence. From symmetry considerations, some capacitances should be equal at V_{ds} =0, such as C_{gd} and C_{gs}, C_{dd} and C_{ss}, and C_{bd} and C_{bs}. However, such symmetry has not been achieved in the capacitance models of BSIM3v3 as some trans-capacitances such as C_{bd} and C_{bs}, C_{dd} and C_{ss} show asymmetry at V_{ds} =0, as shown in Fig. 5.6.1. It is apparently non-physical and may result in simulation errors when the devices are biased at or near V_{ds} =0.

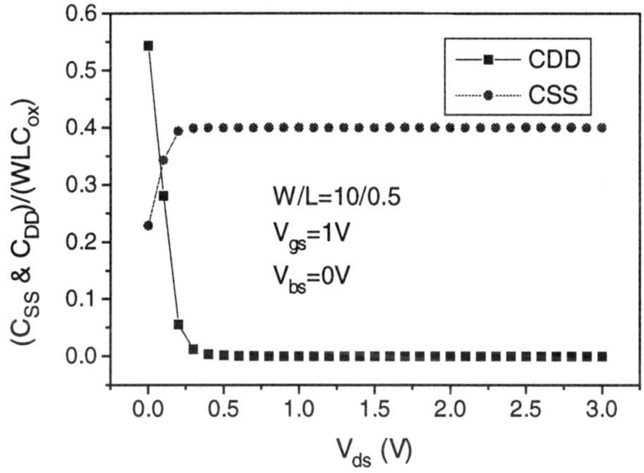

Fig. 5.6.1 (a) Simulated C_{SS} and C_{DD} as a function of V_{ds}. $C_{SS} \neq C_{DD}$ when V_{ds}=0 indicating the existence of asymmetry.

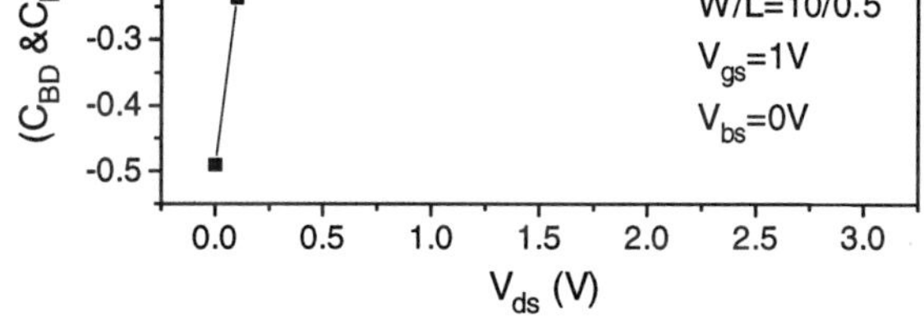

Fig. 5.6.1 (b) Simulated capacitance C_{BD} and C_{BS} as a function of V_{ds}. $C_{BD} \neq C_{BS}$ when V_{ds}=0.

The problem is caused by the fact that all of the body bias dependence is modeled with V_{bs} (without any V_{bd}) terms in the equations. This issue is brought out so the users are aware of this limitation of the model.

8 C-V model parameters

Parameters in the BSIM3v3 C-V models are listed in Table 5.6.2.

Table 5.6.2 C-V model parameters

Symbols in equation	Symbols in source code	Description	Default	Unit
capMod	capmod	Capacitance model selector	3	none
X_{PART}	xpart	Charge partitioning parameter	0	none
C_{GSO}	cgso	Non LDD region source-gate overlap capacitance per channel length	calculated	F/m
C_{GDO}	cgdo	Non LDD region drain-gate overlap capacitance per channel length	calculated	F/m
C_{GBO}	cgbo	Gate bulk overlap capacitance per unit channel length	0.0	F/m
C_{GSL}	cgsl	Lightly doped source-gate region overlap capacitance	0.0	F/m
C_{GDL}	cgdl	Lightly doped drain-gate region overlap capacitance	0.0	F/m
C_{KAPPA}	ckappa	Coefficient for lightly doped region overlap capacitance	0.6	F/m
C_F	cf	Fringing field capacitance	calculated	F/m
C_{LC}	clc	Constant term for the short channel model	1.0×10^{-7}	m
C_{LE}	cle	Exponential term for the short channel model	0.6	none
D_{LC}	dlc	Length offset fitting parameter	lint	m
D_{WC}	dwc	Width offset fitting parameter	wint	m
V_{FBCV}	vfbcv	Flat-band voltage parameter (for ***capMod***=0 only)	-1	V
N_{OFF}	noff	*C-V* parameter for $V_{gsteff,cv}$	1.0	none
V_{OFFCV}	voffcv	Offset voltage parameter of V_{th} from weak to strong inversion in *C-V* model	0.0	V
A_{CDE}	acde	Exponential coefficient for the charge thickness in accumulation and depletion regions	1.0	m/V

M_{OIN}	moin	Coefficient for the gate- bias dependent surface potential	15.0	$V^{1/2}$
L_{LC}	llc	Coefficient of length dependence for channel length offset in *C-V* models	L_L	m^{LLN}
L_{WC}	lwc	Coefficient of width dependence for channel length offset in *C-V* models	L_W	m^{LWN}
L_{WLC}	lwlc	Coefficient of length and width dependence for channel length offset in *C-V* models	L_{WL}	$m^{LWN+LLN}$
W_{LC}	wlc	Coefficient of length dependence for channel width offset in *C-V* models	W_L	m^{WLN}
W_{WC}	wwc	Coefficient of width dependence for channel width offset in *C-V* models	W_W	m^{WWN}
W_{WLC}	wwlc	Coefficient of length and width dependence for channel width offset in *C-V* models	W_{WL}	$m^{WLN+WWN}$

References

[5.1] Y. P. Tsividis, *Operation and Modeling of the MOS Transistor*, McGraw-Hill, New York, 1987.

[5.2] J. Meyer, "MOS models and circuit simulation", *RVA Review*, vol. 32, pp. 42-63 (1971).

[5.3] N. Arora, *MOSFET Models for VLSI Circuit Simulation*, Springer-Verlag, Wien New York, 1994

[5.4] S. Liu and L. W. Nagel, "Small-signal MOSFET models for analog circuit design," *IEEE J. Solid-state Circuits*, vol. SC-17, pp. 983-998, 1982.

[5.5] L. W. Nagel, *SPICE2: A computer program to simulate semiconductor circuits*, ERL-M520, Electronics Research Laboratory, University of California, Berkeley, 1975.

[5.6] M. A. Cirit, "The Meyer model revisited: Why is charge not conserved," *IEEE Trans. on Computer-aided Design*, vol. 8, pp. 1033-1037, 1989.

[5.7] D. E. Ward and R. W. Dutton, "A charge oriented model for MOS transistor capacitances," *IEEE J. Solid-state Circuits*, vol. SC-13, pp. 703-707, 1978.

[5.8] P. Yang et al., "An investigation of the charge conversation problem for MOSFET circuit simulation," *IEEE J. Solid-state Circuits*, vol. SC-18, pp. 128-138, 1983.

[5.9] K. A. Sakallah et al., "The Meyer model revisited: Explaining and correcting the charge non-conservation problem," in *ICCAD-87*, Dig. Tech. 1987.

[5.10] B. J. Sheu et al., "An MOS transistor charge model for VLSI design," *IEEE Trans. on Computer-aided Design*, vol. 7, pp. 520-527, 1988.

[5.11] K. A. Sakalllah et al., "A first-order charge conserving MOS capacitance model," *IEEE Trans. on Computer-aided Design*, vol. 9, pp. 99-108, 1990.

[5.12] C. Turchetti et al., "A Meyer-like approach for the transient analysis of digital MOS IC's," *IEEE Trans. on Computer-aided Design*, vol. 5, pp.490-507, 1986.

[5.13] B. J. Sheu et al., "A compact IGFET charge model," *IEEE Trans. on Circuits and Systems*, vol. CAS-31, pp. 745-748, 1984.

[5.14] B. J. Sheu, et al., *Compact short channel IGFET model (CSIM)*, Electronics Res. Lab, M84/20, University of California, Berkeley, 1984.

[5.15] D. E. Ward, *Charge-based modeling of capacitance in MOS transistors*, Stanford Electronics Laboratory, Tech. G201-11, Stanford University, CA., 1981.

[5.16] B. J. Sheu et al., "BSIM – Berkeley short channel IGFET model for MOS transistors," *IEEE J. Solid-state Circuits*, vol. SC-22, pp. 558-565, 1987.

[5.17] Y. Cheng et al., *BSIM3 version 3.1 User's Manual*, University of California, Berkeley, Memorandum No. UCB/ERL M97/2, 1997.

[5.18] R. Shrivastava and K. Fitzpatrick, "A simple model for the overlap capacitance of a VLSI MOS device," *IEEE Trans. on Electron Devices*, vol. ED-29, pp. 1870-1875, 1982.

[5.19] Y. Cheng et al., "A unified MOSFET channel charge model for device modeling in circuit simulation," *IEEE Trans. Computer-aided Design of Integrated Circuits and Systems*, vol. 17, pp. 641-644, 1998.

[5.20] P. Klein et al, "Short channel charge LDD-MOSFET model for analog and digital circuits with low overdrive voltage," *IEEE 1995 Custom Integrated Circuit Conference*, pp. 229-232, 1995.

[5.21] M. Miura-Mattausch, U. Feldmann, A. Rahm, M. Bollu, and D. Savignac, "Unified complete MOSFET model for analysis of digital and analog circuits," *IEEE Trans. CAD of Integrated Circuis. and Systems* vol. 15, pp. 1-7, 1996.

[5.22] K. M. Rho, K. Lee, M. Shur, and T. A. Fjeldly, "Unified quasi-static MOSFET capacitance model," *IEEE Trans. Electron Devices*, vol. 40, pp. 131-136, 1990.

[5.23] W. Liu et al., *BSIM3 version 3.2 User's Manual,* University of California, Berkeley, 1998.

[5.24] Y. Cheng et al., "ICM--An analytical Inversion charge model for accurate modeling of thin gate oxide MOSFETs," *1997 International Conference on Simulation of Semiconductor Processes and Devices*, Sept. 1997, Boston.

[5.25] R. Rios, N. D. Arora, C.-L. Huang, N. Khalil, J. Faricelli, and L. Gruber, "A physical compact MOSFET model, including quantum mechanical effects, for statistical circuit design applications," *IEDM Tech. Dig.*, pp. 937-940, 1995.

[5.26] S. A. Hareland, S. Krishnamurthy, S. Jallepalli, C.-F. Yeap, K. Hasnat, A. F. Tasch, Jr., and C. M. Maziar, "A computationally efficient model for inversion layer quantization effects in deep submicron n-channel MOSFET's," *IEDM Tech. Dig.*, pp. 933.936, 1995.

[5.27] S. A. Hareland, S. Krishnamurthy, S. Jallepalli, C.-F. Yeap, K. Hasnat, Al F. Tasch, Jr., and C. M. Maziar, "A computationally efficient model for inversion layer quantization effects in deep submicron n-channel MOSFET's," *IEEE Trans. Electron Devices*, vol. 43, pp. 90-96, 1996.

[5.28] J. H. Huang et al., *BSIM3 Manual (Version 2.0),* University of California, Berkeley, March 1994.

[5.29] K. F. Schuegraf, C. C. King, and C. Hu, "Impact of polysilicon depletion in thin oxide MOS technology," *Proc. 1993 Int. Symp. VLSI Tech., Sys. and Appl.* (VLSI-TSA), Taiwan, pp. 86-90, 1993.

[5.30] Y. King et al., "AC charge centroid model for quantization of inversion layer in NMOSFET," *Int. Symp. VLSI Technology, Systems and Applications, Proc. of Tech. Papers,* Taipei, Taiwan, pp. 245-249, June 1997.

[5.31] W. Liu et al., *An accurate MOSFET intrinsic capacitance model considering quantum mechanic effect for BSIM3v3.2,* Memorandum no. UCB/ERL M98/47, University of California, Berkeley, 1998.

[5.32] P. Yang, "Capacitance modeling for MOSFETs," in *Advances in CAD for VLSI*, vol. 3 pt.I, A. E. Ruehli, Ed. Amsterdam, The Netherlands: North Holland, pp. 107-130, 1986.

[5.33] K. K. Ng and J. R. Brews, "Measuring the effective channel length of MOSFETs," *IEEE Circuit and Devices*, vol. 6, pp. 33-38, 1990

[5.34] C. Enz and Y. Cheng, "MOS transistor modeling issues for rf ic design", Workshop of Advances in Analog Circuit Design, France, March 1999.

[5.35] N. D. Arora, "Modeling submicron MOSFET transistor capacitances," *Meta-Software Journal*, pp.11-13, Dec. 1994.

[5.36] J. S. T. Huang, J. W. Schrankler, et al., "Flat-band voltage dependence on channel length in short channel threshold model," *IEEE Trans. on Electron Devices*, vol. ED-32, pp.1001-1002, 1985.

[5.37] B. J. Sheu and P. K. Ko, "A capacitance method to determine channel length for conventional and LDD MOSFETs," *IEEE Electron Device Letters,* vol. EDL-5, p. 491, 1984.

[5.38] Y. Cheng et al., "RF modeling issues of deep-submicron MOSFETs for circuit design," *1998 International Conference on Solid-state and Integrated Circuit Technology*, pp. 416-419, 1998.

CHAPTER 6

Substrate Current Model

For submicrometer MOSFETs, the modeling of substrate current is important, especially for the analog circuit design. In this chapter, we will briefly describe the substrate current generation and introduce the substrate current model of BSIM3v3.

6.1 Substrate Current Generation

As discussed in Chapter 2, impact ionization is the physical mechanism for the generation of substrate current. As the channel length of MOSFETs is reduced to the submicrometer regime, the electric field near the drain region causes impact ionization at a significant rate. As shown in Fig. 2.7.2, the generated hole current (taking an n-channel FET as an example) flows into the substrate as the substrate current. The substrate current can cause problem such as latchup, shift of threshold voltage, and the degradation of transconductance in short channel devices [6.1, 6.2, 6.3]. The substrate current also contributes to the output conductance in the saturation region and the breakdown characteristics [6.4, 6.5]. The substrate current is determined by the drain current, I_{ds}, and the peak lateral electric field in the channel, E_m, according to the lucky electron model [6.6, 6.7]:

$$I_{sub} \propto I_{ds} \exp(-\frac{\phi_i}{\lambda E_m}) \qquad (6.1.1)$$

where the exponential term describes the probability of a carrier, while traversing a mean free path λ in an electric field E_m, to gain a critical energy of $q\phi_i$ for impact ionization.

6.2 Substrate Current Model in BSIM3v3

As discussed in section 2.7, the substrate current expression can be derived by integrating the carrier impact ionization coefficient over the velocity-saturated region of the channel [6.8]:

$$I_{sub} = I_{ds} A_i \int_{y=0}^{y=ld} \exp(-\frac{B_i}{E_s(y)}) dy \qquad (6.2.1)$$

where A_i and B_i are the impact ionization coefficients. E_s(y) is the electric field along the channel direction, y=0 is at the edge of the velocity-saturation region in the channel and l_d is the length of the velocity-saturation region. I_{ds} is the drain current without consideration of the impact ionization effect.

A pseudo-two-dimensional analysis can find E_s(y) in the velocity saturated region [6.9, 6.10]. There is an exponential relationship of E_s(y) versus distance,

$$E_s(y) = E_{sat} \cosh(y / l_t) \qquad (6.2.2)$$

where E_{sat} is the critical field for velocity saturation and l_t is the characteristic length of the exponentially rising electric field and is given by [6.11, 6.12]

$$l_t = \sqrt{\frac{\varepsilon_{si}}{\varepsilon_{ox}} T_{OX} X_J} \qquad (6.2.3)$$

where T_{OX} is the gate oxide thickness and X_J is the drain/source junction depth. ε_{ox} and ε_{si} are the dielectric permittivity of silicon dioxide and silicon, respectively.

The above expression can also be expressed in terms of voltage within the saturation region [6.10],

$$E_s(y) = \sqrt{\left(\frac{V_s(y) - V_{dsat}}{l_t}\right)^2 + E_{sat}^2} \tag{6.2.4}$$

By changing the variable, the substrate current in Eq. (6.2.1) can be rewritten as

$$I_{sub} = I_{ds} l_t A_i \int_{E_{sat}}^{E_d} \frac{\exp(-\frac{B_i}{E_s})}{\sqrt{E_s^2 - E_{sat}^2}} dE_s \approx \frac{I_d A_i l_t E_d}{B_i} \exp(-\frac{B_i}{E_d}) \tag{6.2.5}$$

where E_d is the electric field at the drain end and can be found from Eq. (6.2.4) to be

$$E_d = \sqrt{\left(\frac{V_{ds} - V_{dsat}}{l_t}\right)^2 + E_{sat}^2} \tag{6.2.6}$$

In the saturation region, generally $E_d >> E_{sat}$ so that E_d can be expressed approximately as

$$E_d \approx \frac{V_{ds} - V_{dsat}}{l_t} \tag{6.2.7}$$

Combining Eq. (6.2.5) and Eq. (6.2.7), we can obtain the expression for the substrate current

$$I_{sub} = \frac{A_i}{B_i}(V_{ds} - V_{dsat}) \exp(-\frac{B_i l_t}{V_{ds} - V_{dsat}}) I_{ds} \tag{6.2.8}$$

Eq. (6.2.8) has been used widely to calculate the substrate current in MOS devices.

Since V_{dsat} depends on L_{eff}, I_{sub} is a strong function of L_{eff}. Based on Eq. (6.2.8) and the unified *I-V* equations discussed in Chapter 4, the substrate current model in BSIM3v3 is obtained. It is a single equation for all operation regions,

$$I_{sub} = (\alpha_1 + \frac{\alpha_0}{L_{eff}})(V_{ds} - V_{dseff}) \exp(-\frac{\beta_0}{V_{ds} - V_{dseff}}) I_{dsa} \tag{6.2.9a}$$

$$I_{dsa} = \frac{I_{dso}}{1 + \frac{R_{ds} I_{dso}}{V_{dseff}}} \left(1 + \frac{V_{ds} - V_{dseff}}{V_A}\right) \tag{6.2.9b}$$

where I_{dsa} is the drain current without consideration of impact ionization [6.8]. The expressions of I_{dso}, R_{ds}, V_{dseff}, and V_A can be found in Chapter 4. $\boldsymbol{\alpha_0}$, $\boldsymbol{\alpha_1}$, and $\boldsymbol{\beta_0}$ are model parameters extracted from the measured I_{sub} data.

By recalling the I_{ds}-V_{ds} and V_{dseff} expressions presented in Chapter 4, we can understand Eq. (6.2.9) easily. V_{dseff} becomes V_{dsat} when $V_{ds}>V_{dsat}$ and becomes V_{ds} when $V_{ds}<V_{dsat}$. Thus, the substrate current vanishes in the linear region. $\boldsymbol{\beta_0}$ represents the product of B_i and l_t in Eq. (6.2.8). $\boldsymbol{\alpha_1}$ represents the A_i/B_i term in Eq. (6.2.8). Thus, Eq. (6.2.9) can be considered as another form of Eq. (6.2.8). $\boldsymbol{\alpha_0}/L_{eff}$ is an empirical term that improves the accuracy of the I_{sub} dependence on L_{eff}. BSIM3 does not employ a separate model equation to model the substrate current in the subthreshold regime [6.20]. Eq. (6.2.9) can also model the substrate current in the subthreshold regime. It is valid in both the strong inversion and the subthreshold regions.

As an example, Fig. 6.2.1 shows the simulated characteristics of the substrate current versus V_{gs} (at different V_{ds} and fixed V_{bs}) for an n-channel device of W/L=10μm/0.6μm. Fig. 6.2.2 shows the simulated characteristics of the substrate current versus V_{ds} (at several V_{gs} and a fixed V_{bs}) for the same device used in Fig. 6.2.1.

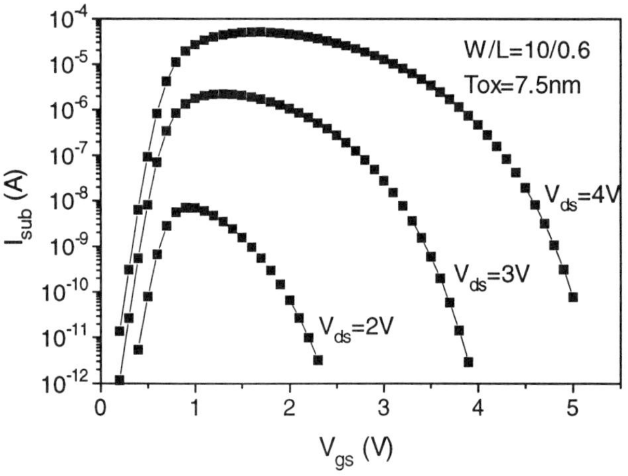

Fig. 6.2.1 Simulated substrate current characteristics versus V_{gs} at several V_{ds} and V_{bs}=0V for a device with W/L=10/0.6.

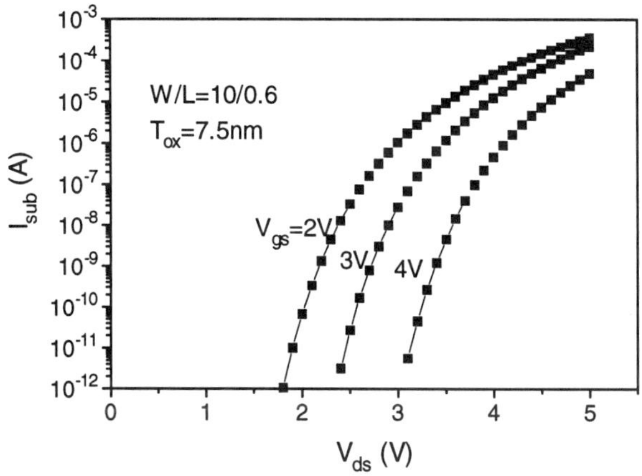

Fig. 6.2.2 Simulated substrate current characteristics versus V_{ds} at several V_{gs} and V_{bs}=0V for a device with W/L=10/0.6.

6.3 Helpful Hints

1. The current I_{dsa} in the substrate current model

Given in Eq. (6.2.9), the substrate current is a function of the channel current. It should be pointed out again that the I_{dsa} used in Eq. (6.2.9) is the drain current without including the contribution of the substrate current induced body effect. In other words, it is the channel current in the absence of impact ionization.

2. The parameters in the substrate current model and in the substrate-current-induced-body-effect (SCBE) of the I-V model

It may appear that we can use the substrate current expression in the current model directly, that is:

$$I_{ds} = Idsa + I_{sub} \tag{6.3.1}$$

where I_{dsa} is the drain current without including the influence of the impact ionization, and I_{sub} is the substrate current given by Eq. (6.2.9a). Eq. (6.3.1) is wrong because I_{sub} contributes significantly more than itself to I_{ds} as discussed in Chapter 4. I_{sub} induces a body bias and a decease in V_{th}, which eventually determines the total contribution of impact ionization to I_{ds}.

Furthermore, separate fitting parameters are introduced while the same I_{sub} equation is used in the substrate current model and *SCBE* model so that accuracy can be achieved in both output resistance and substrate current characteristics. Specifically $\boldsymbol{P_{SCBE1}}$ and $\boldsymbol{P_{SCBE2}}$ are used in the *I-V* model for *SCBE*, and $\boldsymbol{\alpha_0}$, $\boldsymbol{\alpha_1}$, and $\boldsymbol{\beta_0}$ are used in the substrate current model.

3. The α_0 and α_1 parameters

The $\boldsymbol{\alpha_0}$, $\boldsymbol{\alpha_1}$, and $\boldsymbol{\beta_0}$ parameters are introduced in the substrate current model to predict the substrate current accurately. In Eq. (6.2.9a), $\boldsymbol{\alpha_1}$ is similar to the A_i/B_i term in Eq. (6.2.8), and $\boldsymbol{\alpha_0}/L_{eff}$ term is similar to the $\boldsymbol{P_{SCBE2}}/L_{eff}$ term in Eq. (4.4.29). $\boldsymbol{\alpha_0}$ may be set to zero if the substrate current model gives correct scaleability without this term.

4. Drain-induced breakdown simulation

It should be mentioned that no specific consideration for the drain-induced breakdown mechanisms is given in the model derivation of the substrate current model. The drain current due to impact ionization is modeled with the *SCBE* model in the *I-V* model, as discussed in section 4.4. When the drain voltage bias is large enough, the *I-V* model may or may not reflect the drain-induced breakdown characteristics accurately as shown in Fig. 6.3.1.

5. Gate-induced drain leakage at low gate bias region

The gate-induced-drain-leakage (*GIDL*) or band to band tunneling in low gate bias region has been observed in some MOSFETs [6.10]. The band-to-band tunneling current is generated in the drain region that is overlapped by the gate [6.11, 6.12]. *GIDL* occurs when the gate is grounded and V_{ds} is high as discussed in Chapter 2. Like other forms of leakage current, *GIDL* may contribute to standby power and charge loss from charge storage nodes. Analytical models for the *GIDL* effect have been reported [6.13, 6.14], however, they are rarely included in compact models at the present time. The present BSIM3v3 has not included models for *GIDL*.

6. Substrate current model parameters

The parameters in the BSIM3v3 substrate current model are listed in Table 6.3.1.

Table 6.3.1 Substrate current model parameters

Symbols in equation	Symbols in source code	Description	Default	Unit
α_0	alpha0	The first parameter of substrate current	0	m/V
α_1	alpha1	The length scaling parameter of substrate current model	0	1/V
β_0	beta0	The second parameter of substrate current	30	V

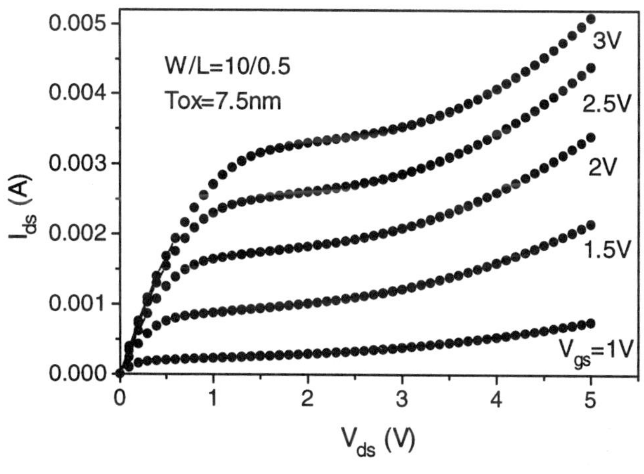

Fig. 6.3.1 *I-V* characteristics with significant impact ionization effects simulated by BSIM3v3.

References

[6.1] J. Matsunaga et al., "Characterization of two step impact ionization and its influence on NMOS and PMOS VLSIs," *IEDM Tech. Dig.*, pp.732-735, 1980.

[6.2] M. S. Liang et al., "Hot-carrier-induced degradation in thin gate oxide MOSFETs," *IEDM Tech. Dig.*, pp.186-189, 1983.

[6.3] E. Takeda, "Hot carrier effects in submicrometer MOS VLSI," *IEE proceedings*, vol. 131, Pt. I, pp153-164, 1984.

[6.4] J. H. Huang et al., "A physical model for MOSFET output resistance", *IEDM, Technical Digest*, Dec. of 1992.

[6.5] J. Chen et al., "Subbreakdown drain leakage current in MOSFETs," *IEEE Electron device Letters*, vol. EDL-8, pp. 515-517, 1987.

[6.6] P. K. Ko, R. S. Muller, and C. Hu, "A unified model for hot-electron currents in MOSFETs," in *IEDM Tech. Dig.*, p. 600, 1980.

[6.7] C. Hu, "Hot carrier effects in MOSFETs," *IEDM Tech. Dig.*, pp. 176-181, 1983.

[6.8] S. M. Sze, *Semiconductor Devices: Physics and Technology*, John Wiley & Sons, New York, 1985.

[6.9] Y. A. El-Mansy and A. R. Boothroyd, "A simple two dimensional model for IGFET." *IEEE Trans. Electron Devices*, ED-24, pp. 254-262, 1977.

[6.10] M. E. Banna and M. E. Nokali, "A pseudo-two-dimensional analysis of short channel MOSFETs," *Solid-state Electronics*, Vol. 31 pp.269-274, 1988.

[6.11] P. K. Ko, "Approaches to scaling," *Chap. 1, in advanced MOS device Physics,* N. G. Einspruch and G. Gildenblatt, Eds., Vol. 18 *VLSI Electronics*. Academic Press 1989.

[6.12] C. Hu et al., "Hot-electron-induced MOSFET degradation - model, monitor, and improvement," *IEEE Trans. Electron Devices*, ED-32, p.375, 1985.

[6.13] Y. Cheng et al., "A physical and scalable BSIM3v3 I-V model for analog/digital circuit simulation", *IEEE Trans. Electron Devices*, Vol. 44, pp.277-287, Feb. 1997.

[6.14] Y. Cheng et al., *BSIM3 version 3.1 User's Manual*, University of California, Berkeley, Memorandum No. UCB/ERL M97/2, 1997.

[6.15] T. Y. Chan, J. Chen, P. K. Ko, and C. Hu, "Impact of gate-induced drain leakage current on device scaling," *IEDM Tech. Dig.*, pp. 718-721, 1987.

[6.16] I. C. Chen et al., "Interface-trap enhanced gate-induced leakage current in MOSFET," *IEEE Electron device Letters*, EDL-10, p.216, 1989.

[6.17] C. Chang and J.Lien, "Corner-field induced drain leakage in thin oxide MOSFETs," *IEDM Tech. Dig.*, pp. 714-717, 1987.

[6.18] R. Shrota et al., "An accurate model of subbreakdown due to band-to-band tunneling and its application," *IEDM Tech. Dig.*, pp. 26-29, 1988.

[6.19] M. Tanizawa et al., "A complete substrate current model including band-to-band tunneling current for circuit simulation," *IEEE trans. on Computer-aided Design in Integrated Circuits*, vol. 12, pp. 1749-1757, 1993.

[6.20] B. Iñiguéz and T. A. Fjeldly, "Unified substrate current model for MOSFETs", *Solid-State Electronics,* vol. 41, No. 1, pp. 87-94, 1997.

CHAPTER 7

Noise Model

The possibility of low-cost integration with logic circuits has made CMOS the technology of choice for many analog and, increasingly, radio frequency (RF) applications. Good noise models in circuit simulators are critical to analog and RF applications. Two types of noise, thermal and flicker, are important in a MOSFET. We will discuss the physical mechanisms of the flicker and thermal noise and present the details of the BSIM3 noise model.

7.1 The Physical Mechanisms of Flicker (1/f) Noise

The basic characteristic of flicker noise is a $1/f$ spectral density. Much effort has been made in understanding the physical origin of flicker noise [7.1-7.6]. Still, the physical mechanism is not very clear. Basically, there are three different theories of flicker noise, (a) carrier density fluctuation models [7.7], (b) mobility fluctuation models [7.8], (c) correlated carrier and mobility fluctuation models [7.9]. In the carrier density fluctuation model, the noise is explained by the fluctuation of channel free carriers due to the random capture and emission of carriers by interface traps at the Si-SiO_2 interface. According to this model, the input noise is independent of the gate bias, and the magnitude of the noise spectrum is proportional to the density of the interface traps. A $1/f$ noise spectrum is predicted if the trap density is uniform in the oxide. The experimental results show a $1/f^{\eta}$ spectrum and η is not always 1, but in the range of 0.7-1.2 [7.5, 7.10]. Recently, some experimental results show that η decreases with increasing gate bias in p-channel MOSFETs [7.11]. Modified charge density fluctuation theories have been proposed to explain these experimental results. The spatial distribution of the active traps in the oxide is assumed to be non-uniform to explain the technology and the gate bias dependence of η [7.7, 7.11].

The mobility fluctuation model considers flicker noise to be the result of fluctuations in carrier mobility based on Hooge's empirical relation for the spectral density of the flicker noise in a homogeneous device [7.12]. It has been proposed that the fluctuation of the bulk mobility in MOSFET's is induced by changes in phonon population [7.13]. The mobility fluctuation models predict a gate bias dependent noise. However, they cannot always account for the magnitude of the noise [7.14].

A unified theory for the origin of the *1/f* noise claims that the capture and emission of carriers by the interface traps cause fluctuation in both the carrier number and the mobility [7.9]. Even though this theory cannot explain all the details of the experimental data, it seems to be the most attractive model available today.

7.2 The Physical Mechanism of Thermal Noise

To understand thermal noise in a MOSFET, we will discuss first the thermal noise model of a resistor.

It is known that the thermal noise of a resistor is directly proportional to temperature *T*. The spectral noise power density $S_i(f)$ (mean-square value of current per frequency bandwidth) of a resistor, *R*, can be given by the following [7.15]:

$$S_i(f) = \frac{\overline{i^2}}{\Delta f} = \frac{4K_BT}{R} \tag{7.2.1}$$

where K_B is the Boltzmann's constant.

The equivalent circuit of the thermal noise can be represented by a shunt current source $\overline{i^2}$, as shown in Fig. 7.2.1.

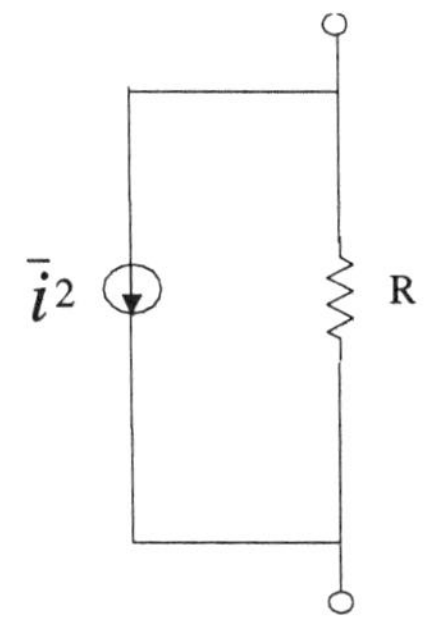

Fig. 7.2.1 Equivalent circuit of the thermal noise of a resistor.

The thermal noise characteristics in a MOSFET operating in the strong inversion region have been studied for over two decades. The origin of thermal noise in a MOSFET has been found to be related to the random thermal motion of carriers in the channel of the device [7.16]. Models have been developed and implemented in circuit simulators [7.17]. Even though using the thermal noise model of a resistor can qualitatively explain the thermal noise in a MOSFET, it is not quantitatively accurate even at low drain bias [7.18, 7.19]. Furthermore, as the moderate inversion region becomes important for low power applications, there is an increasing need for good noise modeling in this region. Therefore, the noise behavior of a transistor should be well modeled from strong inversion through moderate inversion, into weak inversion.

7.3 Flicker Noise Models in BSIM3v3

For users' convenience, two flicker noise models are included in BSIM3v3. One is the SPICE2 flicker noise model [7.20], while the other is the unified flicker noise model [7.21, 7.22]. A model parameter, ***noiMod***, is introduced for the user to select one of these noise models. When ***noiMod*** is 1, the SPICE2 flicker noise model is used, and when ***noiMod*** is 2 the unified flicker noise model is used.

7.3.1 SPICE2 flicker noise model (*noiMod=1*)

The SPICE2 flicker noise model is

$$S_{id}(f) = \frac{K_F I_{ds}^{A_F}}{C_{ox} L_{eff}^2 f^{E_F}} \tag{7.3.1}$$

where S_{id} is the drain current noise power spectral density, I_{ds} is the drain current, $\boldsymbol{A_F}$ is the flicker noise exponent, $\boldsymbol{E_F}$ is the flicker noise frequency coefficient, and $\boldsymbol{K_F}$ is the flicker noise coefficient.

7.3.2 Unified flicker noise model (*noiMod=2)*

1. Model derivation

The coordinate system used in the unified flicker noise model derivation is defined as follows. x is the coordinate along the channel length direction, y is the coordinate along the channel width direction, and z is the coordinate along the direction of oxide thickness perpendicular to both the x and y directions.

For a section of channel width W_{eff} and length Δx in a MOSFET, fluctuations in the amount of trapped interface charge will introduce correlated fluctuations in the channel carrier concentration and mobility. The resulting fractional change in the local drain current can be expressed as [7.22]

$$\frac{\delta I_{ds}}{I_{ds}} = \left[\frac{1}{\Delta N} \frac{\delta \Delta N}{\delta \Delta N_t} \pm \frac{\delta \mu_{eff}}{\delta \Delta N_t} \right] \delta \Delta N_t \tag{7.3.2}$$

where $\Delta N = N W_{eff} \Delta x$, $\Delta N_t = N_t W_{eff} \Delta x$, N is the number of channel carriers per unit area, and N_t is the number of occupied traps per unit area. The sign in front of the mobility term in Eq. (7.3.2) is dependent on whether the trap is neutral or charged when filled [7.22]. The ratio of the fluctuations in the carrier number to the fluctuations in occupied trap number, $R_n = \delta\Delta N / \delta\Delta N_t$, is close to unity at strong inversion but assumes smaller values at other bias conditions [7.23]. A general expression for R_n is

$$R_n = \frac{\delta \Delta N}{\delta \Delta N_t} = -\frac{C_{inv}}{C_{ox} + C_{inv} + C_{dep} + C_{it}} \tag{7.3.3}$$

where C_{inv} is the inversion layer capacitance, C_{dep} is the depletion layer capacitance, and C_{it} is the interface trap capacitance.

The relationship between C_{inv} and N can be given approximately in the following:

$$C_{inv} \approx \frac{q}{v_t} N \tag{7.3.4}$$

where v_t is the thermal voltage.

Thus, Eq. (7.3.3) can be rewritten as,

$$R_n = -\frac{N}{N + N^*} \tag{7.3.5a}$$

where

$$N^* = \frac{v_t}{q}(C_{ox} + C_{dep} + C_{it}) \tag{7.3.5b}$$

To evaluate $\delta\Delta\mu_{eff}/\delta\Delta N_t$, the following model based on Matthiessen's rule is used [7.22],

$$\frac{1}{\mu_{eff}} = \frac{1}{\mu_n} + \frac{1}{\mu_{ox}} = \frac{1}{\mu_n} + \alpha N_t \tag{7.3.6}$$

where μ_{ox} is the mobility limited by oxide charge scattering. α is the scattering coefficient and is a function of the local carrier density due to channel charge screening effect [7.23].

Based on Eq. (7.3.6), the following can be obtained

$$\frac{\delta\mu_{eff}}{\delta\Delta N_t} = -\frac{\alpha\mu_{eff}^{\ 2}}{W_{eff}\Delta x} \tag{7.3.7}$$

Substituting Eq. (7.3.6) and Eq. (7.3.7) into Eq. (7.3.2) yields

$$\frac{\delta I_{ds}}{I_{ds}} = -\left(\frac{R_n}{N} \pm \alpha\mu_{eff}\right)\frac{\delta\Delta N_t}{W_{eff}\Delta x} \tag{7.3.8}$$

The power spectrum density of the local current fluctuations is

$$S_{\Delta Id}(x,f) = \left(\frac{I_{ds}}{W_{eff}\Delta x}\right)^2 \left(\frac{R_n}{N} \pm \alpha\mu_{eff}\right)^2 S_{\Delta Nt}(x,f) \tag{7.3.9}$$

where $S_{\Delta Nt}$ (x,f) is the power spectrum density of the fluctuations in the number of the occupied traps over the area $W_{eff}\Delta$x, and can be given by

$$S_{\Delta Nt}(x,f) = \int_{Ev}^{Ec}\int_{0}^{Wefff}\int_{0}^{Tox} 4N_t(E,x,y,z)\Delta x f_t(1-f_t)\frac{\tau(E,x,y,z)}{1+\omega^2\tau(E,x,y,z)}dzdydE \tag{7.3.10}$$

where N_t (E,x,y,z) is the distribution of the traps in the oxide and over the energy band, τ(E,x,y,z) is the trapping time constant, and $f_t = [1-\exp\frac{(E-E_{fn})}{K_BT}]^{-1}$ is the trap occupancy function. E_{fn} is the electron quasi-Fermi level, $\omega=2\pi f$ is the angular frequency, T_{OX} is the oxide thickness, and E_c-E_v is the silicon energy gap.

To evaluate the integral in Eq. (7.3.10), two assumptions are needed:

a. The oxide traps have a uniform spatial distribution near the interface, that is, N_t(E,x,y,z)=N_t(E).
b. The probability of an electron penetrating into the oxide decreases exponentially with the distance from the interface, and as a result the trapping time constant is given by

$$\tau = \tau_o(E)\exp(\gamma z) \tag{7.3.11}$$

where τ_o(E) is the time constant at the interface and γ is the attenuation coefficient of the electron wave function in the oxide.

Since $f_t(1-f_t)$ in Eq. (7.3.10) behaves like a delta function around the quasi-Fermi level, the major contribution to the integral will be from the trap level around E_{fn}. Thus, N_t (E) can be approximated by N_t (E_{fn}) and taken out of the integral. Replacing $f_t(1-f_t)$ in Eq. (7.3.10) by $-K_BTdf_t/dE$ and carrying out the integration yields

$$S\Delta N_t(x,f) = N_t(E_{fn})\frac{K_BTW_{eff}\Delta x}{\gamma f} \tag{7.3.12}$$

The total drain current noise power spectrum density can be derived as

$$S_i(f)=\frac{1}{L_{eff}^{\,2}}\int_0^{L_{eff}} S_{\Delta I_{ds}}(x,f)\Delta x dx$$

$$=\frac{K_BTI_{ds}^{\,2}}{\gamma f W_{eff}L_{eff}^{\,2}}\int N_t(E_{fn})[\frac{R_n}{N(x)}+\alpha\mu_{eff}]^2dx$$

$$=\frac{qK_BTI_{ds}\ \mu_{eff}}{\gamma f L_{eff}^{\,2}}\int_o^{V_{ds}} N_t(E_{fn})(1\pm\alpha\mu_{eff}NR_n^{\,-1})^2\frac{R_n^{\,2}}{N}dV \qquad (7.3.13)$$

Since α and μ_{eff} are functions of the local carrier density N, Eq. (7.3.13) can be rewritten as

$$S_i(f)=\frac{qK_BT\,I_{ds}\ \mu_{eff}}{\gamma\, f\ L_{eff}^{\,2}}\int_o^{V_{ds}} N_t^*(E_{fn})\frac{R_n^{\,2}}{N}dV \qquad (7.3.14)$$

where $N_t^*(E_{fn})$ is the equivalent oxide-trap density that produces the same noise power if there were no contributions from the mobility fluctuations.

In the present model $N_t^*(E_{fn})$ is approximated as a three parameter function of the channel carrier density

$$N_t^*(E_{fn}) = A + BN + CN^2 \qquad (7.3.15)$$

where A, B and C are technology-dependent model parameters.

Based on the above, the flicker noise power spectrum density in the different operation regions can be found.

1). Linear region in strong inversion (V_{gs}>V_{th} and V_{ds}<V_{dsat})

In the strong inversion region, the charge density of carrier can be given by

$$qN(x) = C_{ox}[V_{gs} - V_{th} - A_{bulk}V(x)] \qquad (7.3.16)$$

Thus, we have

$$qN_o = qN(0) = C_{ox}[V_{gs} - V_{th}] \qquad (7.3.17)$$

$$qN_L = qN(L_{eff}) = C_{ox}[V_{gs} - V_{th} - A_{bulk}V_{ds}] \tag{7.3.18}$$

where N_0 and N_L are carrier densities at the source and drain ends of the channel, respectively.

By using the above equations, Eq. (7.3.13) can be rearranged as

$$S_i(f) = \frac{q^2 K_B T I_{ds}\ \mu_{eff}}{\gamma f L_{eff}^2 C_{ox}} \int_{N_L}^{N_0} N_t^*(E_{fn}) \frac{R_n^2}{N} dN \tag{7.3.19}$$

Substituting Eq. (7.3.3) and Eq. (7.3.15) into Eq. (7.3.16) and performing the integration

$$S_i(f) = \frac{q^2 K_B T I_{ds}\ \mu_{eff}}{A_{bulk}\gamma f L_{eff}^2 C_{ox}}[A \ln(\frac{N_0 + N^*}{N_L + N^*}) + B(N_0 - N_L) + \frac{1}{2}C(N_0^2 - N_L^2)] \tag{7.3.20}$$

2). Saturation region in strong inversion ($V_{gs}>V_{th}$ and $V_{ds} \geq V_{dsat}$)

As discussed in Chapter 2, the channel can be divided into two parts in the saturation region, as shown in Fig. 2.6.1. One part is from the source to L_s, the velocity-saturation (or "pinched-off") point. The other part, $L_{d,}$ is from the velocity-saturation point to the drain. Accordingly, the flicker noise includes two parts as given in the following

$$S_i(f) = \frac{K_B T I_{ds}^2}{\gamma f L_{eff}^2 W_{eff}} \int_0^{L_s} N_t^*(E_{fn}) \frac{R_n^2}{N^2} dx + \frac{K_B T I_{ds}^2}{\gamma f L_{eff}^2 W_{eff}} \int_{L_s}^{L_{eff}} N_t^*(E_{fn}) \frac{R_n^2}{N^2} dx \tag{7.3.21}$$

The solution of the first term in Eq. (7.3.21) can be obtained by replacing V_{ds} with V_{dsat} in the expression of N_L of Eq. (7.3.20).

To evaluate the flicker noise power contributed from the velocity-saturation (or pinch-off) region, assumptions are made that both the electron quasi-Fermi level and carrier density are uniform in the pinch-off region and equal to those at the pinch-off point where the channel potential is equal to V_{dsat}.

Thus the second term in Eq. (7.3.21) is expressed as $\frac{K_B T I_{ds}^2}{\gamma f L_{eff}^2 W_{eff}} \frac{A + BN_L + CN_L^2}{(N_L + N^*)^2} L_d$, where L_d is the length of the pinch-off region.

The total noise power spectrum density in the saturation region is written as

$$S_i(f) = \frac{q^2 K_B T I_{ds} \mu_{eff}}{A_{bulk} \gamma f L_{eff}^2 C_{ox}} [A \ln(\frac{N_0 + N^*}{N_L + N^*}) + B(N_0 - N_L) + \frac{1}{2} C(N_0^2 - N_L^2)] + \frac{K_B T I_{ds} \mu_{eff}}{\gamma f L_{eff}^2 W_{eff}} \frac{A + BN_L + CN_L^2}{(N_L + N^*)^2} L_d \quad (7.3.22)$$

where $N_L = C_{ox}(V_{gs} - V_{th} - A_{bulk} V_{dsat})/q$.

3. Subthreshold region (V_{gs}<V_{th})

In the subthreshold region, the drain current diminishes exponentially with decreasing the gate voltage. The channel charge density can be expressed as

$$qN(V) = \frac{C_{dep}}{v_t} \exp[\frac{(V_{gs} - V_{gc})}{n v_t} - \frac{\phi_F}{2 v_t} - \frac{V}{v_t}] \quad (7.3.23)$$

where n is the subthreshold swing factor, and V_{gc} is the voltage when the surface potential is equal to $1.5\phi_F$.

Substituting Eq. (7.3.23) into Eq. (7.3.14) and rearranging yields the following expression for the spectral flicker noise power density in the subthreshold region:

$$S_i(f) = \frac{q^2 v_t^2 I_{ds} \mu_0}{\gamma f L_{eff}^2} \int_{N_L}^{N_o} \frac{N_t^*(E_{fn})}{(N + N^*)^2} dN \quad (7.3.24)$$

where

$$qN_0 = \frac{K_s}{v_t} \exp(\frac{V_{gs} - V_{th}}{n v_t}) \quad (7.3.25)$$

$$qN_L = qN_0[1 - \exp(-\frac{V_{ds}}{v_t})] \quad (7.3.26)$$

$$K_s = \frac{C_d}{v_t^2}\exp\left(\frac{V_{th} - V_{gc}}{n v_t} - \frac{\phi_F}{2 v_t}\right) \tag{7.3.27}$$

In the subthreshold region, it is reasonable to assume that $N \ll N^*$ and $N_t^*(E_{fn}) = A + BN + CN^2 \approx A$. Thus, the flicker noise power in the subthreshold region can be simplified to

$$S_i(f) = \frac{A q v_t \; I_{ds}^2}{\gamma f W_{eff} L_{eff} N^{*2}} \tag{7.3.28}$$

2. Model equations implemented in BSIM3v3

In the BSIM3v3 implementation of the flicker noise model, some minor changes are made to the above equations to simplify the model calculation. (1) A constant ($1\text{x}10^8$) is used for γ, the attenuation coefficient of the electron wave function in the oxide; (2) N^* is a constant ($2\text{x}10^{14}$); (3) A_{bulk} parameter is assumed to be equal to 1; (4) The parameters *A, B,* and *C* are termed as N_{OIA}, N_{OIB}, and N_{OIC} respectively; and (5) The frequency exponential coefficient $\boldsymbol{E_F}$ is introduced.

With the above modifications, the noise model equations implemented in BSIM3v3 are given below.

1). Strong inversion region (V_{gs}-V_{th}>0.1V):

$$S_{id}(f) = \frac{v_t q^3 I_{ds} \mu_{eff}}{f^{E_F} L_{eff}^2 C_{ox} 10^8}\left[N_{OIA}\log\left(\frac{N_o + 2\text{x}10^{14}}{N_L + 2\text{x}10^{14}}\right) + N_{OIB}(N_o - N_L)\right.$$
$$\left. + 0.5 N_{OIC}(N_o^2 - N_L^2)\right] + \frac{q v_t I_{ds}^2 \Delta L_{clm}}{f^{E_F} L_{eff}^2 W_{eff}' 10^8} \frac{N_{OIA} + N_{OIB} N_L + N_{OIC} N_L^2}{(N_L + 2\text{x}10^{14})^2}$$

(7.3.29)

$$N_o = \frac{C_{ox}(V_{gs} - V_{th})}{q} \tag{7.3.30}$$

$$N_L = \frac{C_{ox}(V_{gs} - V_{th} - V_{ds}')}{q} \tag{7.3.31}$$

$$V_{ds}' = MIN(V_{ds}', V_{dsat}) \tag{7.3.32}$$

$$\Delta L_{clm} = L_{itl} \log\left(\frac{\frac{V_{ds} - V_{dsat}}{L_{itl}} + \boldsymbol{E_M}}{E_{sat}}\right) \quad V_{ds} > V_{dsat} \tag{7.3.33a}$$

$$\Delta L_{clm} = 0 \qquad V_{ds} \le V_{dsat} \tag{7.3.33b}$$

$$L_{itl} = \sqrt{3X_J T_{ox}} \tag{7.3.34}$$

$$E_{sat} = 2\frac{\boldsymbol{vSAT}}{\mu_{eff}} \tag{7.3.35}$$

2). Moderate inversion and subthreshold regions (V_{gs}-V_{th} ≤ 0.1V)

In operation regions other than strong inversion (V_{gs}-$V_{th} \le 0.1$V), the following expression is used to calculate the noise density,

$$S_{id}(f) = \frac{S_{limit} S_{wi}}{S_{limit} + S_{wi}} \tag{7.3.36}$$

where S_{limit} is the flicker noise power density given by Eq. (7.3.29) at V_{gs}=V_{th}+0.1V, and S_{wi} is given by,

$$S_{wi} = \frac{\boldsymbol{NOIA} v_t I_{ds}^2}{W_{eff}' L_{eff} f^{\boldsymbol{EF}} 4\text{x}10^{36}} \tag{7.3.37}$$

7.4 Thermal Noise Models in BSIM3v3

There exist two choices for the thermal noise model. When ***noiMod*** is 1, the modified SPICE2 thermal noise model is used, and when ***noiMod*** is 2, the BSIM3 thermal noise model is used.

7.4.1 Modified SPICE2 thermal noise model (*noiMod=1*)

A modified version of the SPICE2 thermal noise model is included in BSIM3v3 as an option. The original SPICE2 model is

$$S_{id}(f) = \frac{8K_BT}{3} g_m \tag{7.4.1}$$

where g_m is the gate transconductance of the device.

Eq. (7.4.1) becomes inadequate in the linear region, especially when V_{ds}=0, where the transconductance is zero so that the calculated noise density is zero. However, the noise power density is not zero in reality. To resolve this problem, the SPICE2 noise model is modified into the following form in the BSIM3v3 implementation:

$$S_{id}(f) = \frac{8K_BT}{3}(g_m + g_{ds} + g_{mb}) \tag{7.4.2}$$

where g_{ds} and g_{mb} are the output conductance and the bulk transconductance.

7.4.2 BSIM3 thermal noise model (*noiMod=2*)

We will describe the model derivation briefly, and then give the thermal model equations implemented in BSIM3v3.

1. Model derivation

The derivation of the thermal noise model used in BSIM3v3 follows the steps described in [7.17]. As is well known, the power spectral density of the noise voltage, generated across a resistor of value *R*, is $4K_BTR$ [7.25]. If a small element in the MOSFET channel has a resistance, ΔR, the noise voltage power of this element is

$$\overline{(\Delta v_t)^2} = 4K_BT\Delta R\Delta f \tag{7.4.3}$$

Assuming the length of the small element of the channel is Δx, ΔR is

$$\Delta R = \frac{\Delta x}{W_{eff}\mu Q_{inv}} \tag{7.4.4}$$

where W_{eff} is the effective channel width, μ is the electron mobility, and Q_{inv} is the channel charge per unit area.

Substituting Eq. (7.4.4) into Eq. (7.4.3) gives

$$\overline{(\Delta v_t)^2} = \frac{4K_B T\Delta f\Delta x}{\mu W_{eff} Q_{inv}} \tag{7.4.5}$$

The current change caused by the voltage change Δv_t is given by

$$\Delta i_t = \frac{W_{eff}}{L_{eff}}\mu Q_{inv}\Delta v_t \tag{7.4.6}$$

The mean square value of Δi_t is

$$\overline{(\Delta i_t)^2} = [\frac{W_{eff}}{L_{eff}}\mu Q_{inv}]^2\,\overline{(\Delta v_t)^2} \tag{7.4.7}$$

Substituting Eq. (7.4.7) into Eq. (7.4.6), we have

$$\overline{(\Delta i_t)^2} = 4K_B T\frac{W_{eff}}{L_{eff}^2}\mu Q_{inv}\Delta x\Delta f \tag{7.4.8}$$

The total noise current power in a bandwidth Δf can be obtained by integrating the above expression along the channel,

$$\begin{aligned}\overline{\Delta i_t^2} &= 4K_B T\Delta f\frac{\mu}{L^2}\int_0^{L_{eff}} Q_{inv} W_{eff} dx \\ &= 4K_B T\frac{\mu}{L_{eff}^2} Q_{INV}\Delta f\end{aligned} \tag{7.4.9}$$

Where Q_{INV} is the total inversion layer charge in the channel.

The power spectral density of thermal noise in a MOSFET can then be expressed as,

$$S_{id}(f) = \frac{\overline{\Delta i_t^2}}{\Delta f} = 4K_BT\frac{\mu}{L_{eff}^2}Q_{INV} \tag{7.4.10}$$

2. Model implementation

The formula of the thermal noise model implemented in BSIM3v3 is based on Eq. (7.4.10) and is:

$$S_{id}(f) = \frac{4K_BT\mu_{eff}}{L_{eff}^2}Q_{INV} \tag{7.4.11}$$

where μ_{eff} is the effective mobility, and L_{eff} is the effective channel length. Q_{INV} is the total inversion charge in the channel, and is calculated from the charge expressions in the capacitance model.

As discussed in Chapter 5, BSIM3 supports several different capacitance (charge) model options. Thus, the detailed expression of the thermal noise power spectral density is slightly different according to the selection of the capacitance model option.

When ***capMod***=0, the total charge is given by:

1). Linear region ($V_{gs}>V_{th}$, $V_{ds}<V_{dsat'cv}$)

$$Q_{INV} = -W_{active}L_{active}C_{ox}\left(V_{gs} - V_{th} - \frac{A_{bulk}'}{2}V_{ds} + \frac{A_{bulk}'^2 V_{ds}^2}{12\left(V_{gs} - V_{th} - \frac{A_{bulk}'}{2}V_{ds}\right)}\right) \tag{7.4.12}$$

2). Saturation region ($V_{gs}>V_{th}$, $V_{ds}\geq V_{dsat'cv}$)

$$Q_{INV} = -\frac{2W_{active}L_{active}C_{ox}}{3}\left(V_{gs} - V_{th}\right) \tag{7.4.13}$$

When ***capMod***=1, the total charge is given as,

1). Linear region ($V_{gs}>V_{th}$, $V_{ds}<V_{dsat'cv}$)

$$Q_{INV} = -W_{active}L_{active}C_{ox}\left(V_{gsteff,cv} - \frac{A_{bulk}'}{2}V_{ds} + \frac{A_{bulk}'^2 V_{ds}^2}{12\left(V_{gsteff,cv} - \frac{A_{bulk}'}{2}V_{ds}\right)}\right) \tag{7.4.14}$$

2). Saturation region ($V_{gs}>V_{th}$, $V_{ds}\geq V_{dsat,cv}$)

$$Q_{INV} = -\frac{2W_{active}L_{active}C_{ox}}{3}V_{gsteff,cv} \tag{7.4.15}$$

When ***capMod***=2, the total channel charge is given by

$$Q_{INV} = -W_{active}L_{active}C_{ox}\left(\left(V_{gsteff,cv} - \frac{A_{bulk}'}{2}V_{cveff}\right) + \frac{A_{bulk}'^2 V_{cveff}^2}{12\left(V_{gsteff,cv} - \frac{A_{bulk}'}{2}V_{cveff}\right)}\right) \tag{7.4.16}$$

When ***capMod***=3, the total channel charge is calculated with

$$Q_{INV} = -W_{active}L_{active}C_{oxeff}\,[(V_{gsteff,cv} - \Phi\delta - \frac{A_{bulk}'}{2}V_{cveff}) + \frac{A_{bulk}'^2 V_{cveff}^2}{12(V_{gsteff,cv} - \Phi\delta - \frac{A_{bulk}'}{2}V_{cveff})}] \tag{7.4.17}$$

7.5 Helpful Hints

1. Options of the flicker and thermal noise models

As discussed in section 7.4, either the SPICE2 models or the BSIM3 models may be selected for the flicker noise and the thermal noise by setting ***noiMod*** to 1 or 2. However, users may want to use a combination of, say, the SPICE2 flicker noise model (because it is simpler than the BSIM3 1/f noise model) and the BSIM3 thermal noise model (because it is more physical and accurate than the SPICE2 model). Therefore, two more options are available in BSIM3v3 for the users to select different combinations of the noise models.

Table 7.5.1 gives the different combinations determined by the model parameter ***noiMod***.

Table 7.5.1 *noiMod* parameter for different noise models

noiMod	Flicker noise model	Thermal noise model
1	SPICE2	SPICE2
2	BSIM3v3	BSIM3v3
3	BSIM3v3	SPICE2
4	SPICE2	BSIM3v3

2. Charge models used in the thermal noise model

The users should be aware that the thermal noise calculation is dependent on user's selection of the capacitance model options. If ***capMod*** is selected by the user for the capacitance model, that charge model is used in the thermal noise calculation. The ***capMod***=3 inversion charge model will be used in BSIM3v3.2 to calculate the thermal noise if the ***capMod*** is not given and ***noiMod***=2 or 4 is selected in the model card.

3. Noise contribution of the drain/source resistances

BSIM3v3 includes the thermal noise and flicker noise in the intrinsic device and the noise from the external drain/source resistances. The latter is triggered by the model parameter R_{SH}. The default value for R_{SH} is zero, so the contribution from the drain/source external resistances is ignored if R_{SH} is not specified in the model card. The MOSFET small signal equivalent circuit with noise sources is given in Fig. 7.5.1.

In Fig. 7.5.1, i_d represents the noise contribution from the intrinsic MOSFET while i_{rd} and i_{rs} represent the noise contributions from the external source and drain resistances. The spectral noise power density of R_d and R_s can be calculated by the following equations:

$$S_{i\_s}(f) = \frac{\overline{i_{rs}^2}}{\Delta f} = \frac{4K_BT}{R_s} \tag{7.5.1}$$

$$S_{i\_d}(f) = \frac{\overline{i_{rd}^2}}{\Delta f} = \frac{4K_BT}{R_d} \tag{7.5.2}$$

The equations for R_s and R_d are given in section 8.2 when discussing the models of parasitic components.

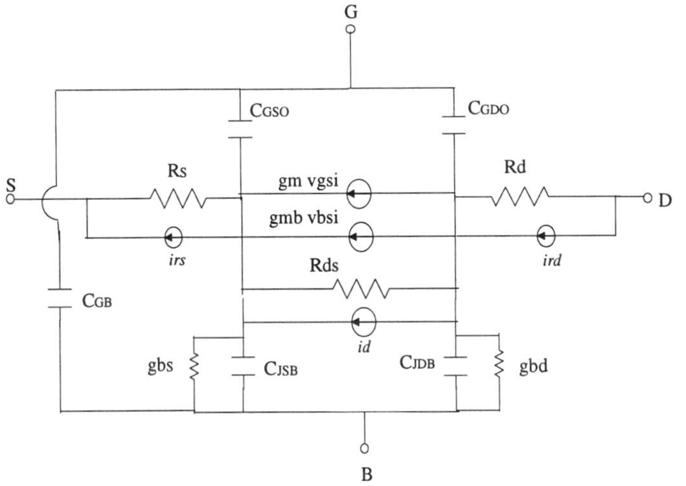

Fig. 7.5.1 MOSFET small-signal equivalent circuit with the three noise sources mentioned. Noise sources from extrinsic resistances at other terminals such as the gate and bulk are not included here.

It should be noted that careful extraction is needed to obtain the proper R_{SH} parameters. In addition to the measured DC *I-V* data the effect of R_{SH} on the noise characteristics needs to be checked, especially for analog and HF applications where the noise characteristics are important.

Noise contributions from the gate and substrate resistances are not included in the BSIM3 noise model.

4 Noise model parameters

Model parameters in the BSIM3v3 noise models are given in Table 7.5.2. The units for some of parameters in the table, such as K_F, N_{OIA}, N_{OIB}, and N_{OIC}, may look strange but they are correct. It is caused by the introduction of parameters E_F and A_F. By carrying such units for these parameters, the model can ensure correct units for the spectral noise power density.

Table 7.5.2 Noise model parameters

Symbols in Equation	Symbols in source code	Description	Default	Unit
noiMod	noimod	Parameter for noise models	1	none
A_F	af	Flicker noise exponent	1	none
E_F	ef	Flicker Frequency exponent	1	none
E_M	em	Saturation electrical field parameter	$4.1\text{x}10^7$	V/m
K_F	kf	Flicker noise coefficient	0	$s^{1\text{-}Ef}A^{2\text{-}Af}F$
N_{OIA}	noia	Noise parameter A	(nmos) 10^{20} (pmos) $9.9\text{x}10^{18}$	$s^{1\text{-}Ef}m^{-2}(ev)^{-1}$
N_{OIB}	noib	Noise parameter B	(nmos) $5\text{x}10^4$ (pmos) $2.4\text{x}10^3$	$s^{1\text{-}Ef}(ev)^{-1}$
N_{OIC}	noic	Noise parameter C	(nmos) $-1.4\text{x}10^{-12}$ (pmos) $1.4\text{x}10^{-12}$	$s^{1\text{-}Ef}m^{2}(ev)^{-1}$

The flicker noise exponent A_F typically falls in the range of 0.5 to 2. E_F can be from 0.8 to 1.2 depending on the technology. Reasonable values of K_F are in the range of 10^{-19} to 10^{-29}. MOSFETs fabricated with an experimental technology may have larger K_F values than MOSFETs fabricated with a mature technology because the latter has a better quality oxide and Si-SiO_2 interface.

5. Comparison with other flicker noise models

In some circuit simulators, such as HSPICE [7.26] and SPECTRE [7.27], users may find other flicker noise models. For example, both HSPICE and SPECTRE include the following flicker noise model:

$$S_{id}(f) = \frac{K_F I_{ds}{}^{A_F}}{C_{ox} L_{eff} W_{eff} f^{E_F}} \tag{7.5.3}$$

Also, HSPICE [a] introduces another flicker noise model with the spectral noise power density proportional to $g_m{}^2$:

a. Please note that HSPICE manual uses A_F instead of E_F in Eq. (7.5.4).

$$S_{id}(f) = \frac{K_F g_m^2}{C_{ox} L_{eff} W_{eff} f^{E_F}} \tag{7.5.4}$$

where g_m is the gate transconductance.

Here we compare the several models in their geometry, bias, and T_{OX} dependences.

1). The geometry dependence of the noise models

By comparing Eq. (7.5.3) with Eq. (7.3.1), it is obvious that the geometry dependence between these two models are different if the same A_F value is used. Eq. (7.3.1) is an empirically-based formula, so a prediction of correct geometry dependence of flicker noise is not expected when A_F =2. For Eq. (7.3.1) to obtain the correct geometry dependency, A_F in Eq. (7.3.1) may need to be set to around 1. This setting, however, may not model the V_g dependence accurately.

The comparison between Eq. (7.5.3) and the BSIM3 flicker noise model, Eq. (7.3.29), is a little bit more difficult because Eq. (7.3.29) is a complex expression. However, if we analyze Eq. (7.3.29) carefully, we can find that the geometry dependence of Eq. (7.5.3) and Eq. (7.3.29) is almost the same. If we set the noise exponent, A_F, to 2 in Eq. (7.5.3) according to the number fluctuation theory [7.5], and also note that $\frac{q^3 v_t I_{ds} \mu_{eff}}{L_{eff}^2}$ in Eq. (7.3.29) is proportional to $\frac{I_{ds}^2}{L_{eff} W_{eff}}$, we can derive similar geometry dependence in both Eq. (7.5.3) and Eq. (7.3.29). As for Eq. (7.5.4), it presents similar geometry dependence to that of Eq. (7.5.3) and Eq. (7.3.29).

2). The bias dependence of the noise models

Both Eq. (7.3.1) and Eq. (7.5.3) show that the spectral noise power density is proportional to $I_{ds}^{A_F}$. Eq. (7.5.4) suggests a g_m^2 dependence for the spectral noise power density. Eq. (7.3.29) shows yet another different bias dependence.

Eq. (7.3.1) shows a stronger V_{gs} dependence than Eq. (7.5.4), and the BSIM3 flicker noise model Eq. (7.3.29) presents a V_{gs} bias dependence that is weaker than Eq. (7.3.1) but stronger than Eq. (7.5.4).

3). The oxide thickness dependence of the noise models

The dependence of the flicker noise spectral power density on oxide thickness has been studied. It is concluded that the input referred noise power [a] at low drain bias has two T_{OX} dependencies [7.9]. At low V_{gs} bias condition, a T_{OX}^2 dependence may be dominant, and at high V_{gs} bias condition, the input referred noise power density presents a linear dependence on T_{OX}. It is obvious that Eq. (7.3.1) gives a T_{OX}^2 (assuming A_F=1) dependence of the input referred noise power density (a $T_{OX}^{1-A_F}$ dependence for drain current noise spectral power density). Eq. (7.5.4) only shows a T_{OX} dependence of the input referred noise power density (a T_{OX}^{-1} dependence for drain current noise spectral power density). However, the BSIM3 noise model has accounted for both the T_{OX} and T_{OX}^2 dependence of the input referred noise power density. Note that the second term in brackets of Eq. (7.3.29) contains a $1/T_{OX}$ term.

6. Understanding the A_F and K_F parameters in the different noise models

A_F and K_F are given as fitting parameters to improve the model accuracy. However, depending on the models they may have significantly different values and units.

A value around 1 for A_F is expected in Eq. (7.3.1) to follow reasonable geometry dependence. However, for the model given in Eq. (7.5.3) it may be difficult to extract the value for A_F. The bias dependence of Eq. (7.5.3) may be incorrect (too strong) if a value around 2 is used for A_F to ensure the correct geometry dependence, The geometry dependence may be incorrect if a value around 1 is used for A_F to ensure the reasonable bias dependence.

Depending on the value of A_F in the different models, the units of K_F are different. For example, a unit $s^{1-E_F}V^2F$ is given for K_F in Eq. (7.5.4). However, the unit of K_F in Eq. (7.3.1) is $s^{1-E_F}A^{2-A_F}F$.

7 Simulating the noise characteristics with circuit simulators

For the user's convenience, we give a test circuit in Fig. 7.5.2 for simulating the noise characteristics of a single MOSFET. A netlist in SPECTRE format is provided in Fig. 7.5.3. Fig. 7.5.4 shows the simulated result with ***noiMod***=1.

---

a. Input referred noise power density $S_{Vgn}=S_{id}/g_m^2$, where g_m is the transconductance of the device.

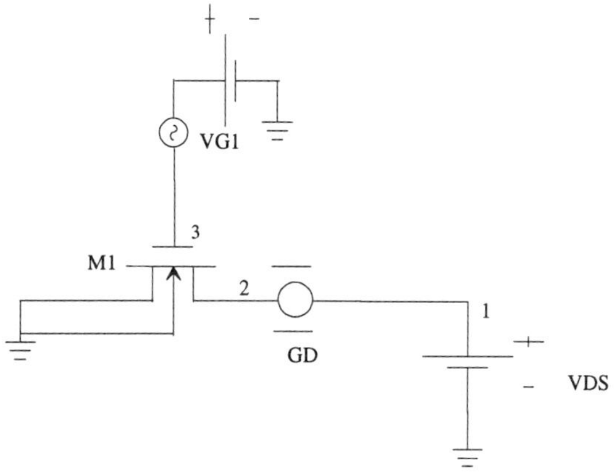

Fig. 7.5.2 A test circuit to simulate the noise characteristics of a single MOSFET.

```
*** noise simulation ***
simulator lang= spice
spectre options rawfmt=nutascii reltol=1e-3
VDS 1 0 vsource dc=1
VG1 3 0 vsource dc= 0. 86 mag= 1
GD 1 2 ccvs probe=vdd rm=1
do_noise_analysis 2 0 noise iprobe=vdd start= 1k stop=100Meg dec= 20
do_ac_analysis ac start=1 stop=100Meg dec=20
M1 2 3 0 0 nch w=1um l=1um
.model nch BSIM3v3
+ type =n
+Tnom=25.0
**** other model parameters need to be added to complete the netlist
.end
```

Fig. 7.5.3 Netlist to simulate the noise characteristics of a single MOSFET.

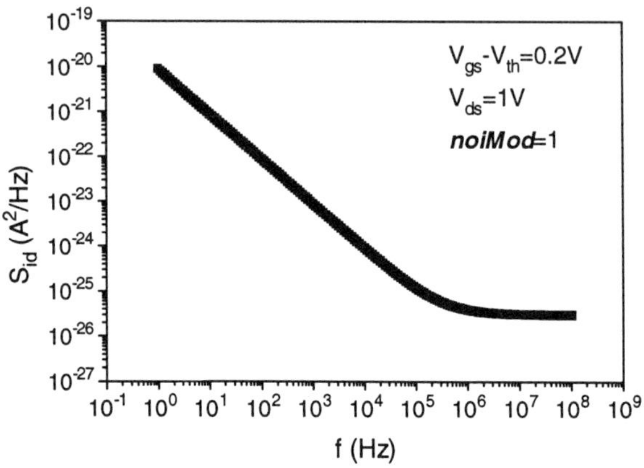

Fig. 7.5.4 The simulated noise characteristics of a MOSFET vs. frequency (1~10^8Hz).

References

[7.1] R. P. Jindal, and A. Van der Ziel, "Carrier fluctuation noise in a MOSFET channel due to traps in the oxide," *Solid-state Electronics*, vol.21 pp. 901-903, 1978.

[7.2] F. M. Klaassen, "Characterization of low 1/f noise in MOS transistors," *IEEE Trans. Electron Devices*, vol. ED-18, pp. 887-891, 1971.

[7.3] B. J. Gross and C. G. Sodini, "1/f noise in MOSEFTs with ultrathin gate dielectric," *IEDM Tech. Dig.* pp.881-884, 1992.

[7.4] T. G. M. Kleinpenning, "On 1/f trapping noise in MOST's," *IEEE Trans. Electron Devices*, vol. ED-37, pp. 2084-2089, 1990.

[7.5] L. K. J. Vandamme, X. Li, and D. Rigaud, "1/f noise in MOS devices, mobility or number fluctuations?," *IEEE Trans. Electron Devices*, vol. 41, pp. 1936-1945, 1994.

[7.6] F. N. Hooge, "1/f noise sources," *IEEE Trans. Electron Devices*, vol. 41, pp. 1926-1935, 1994.

[7.7] O. Jantsch, "Flicker (1/f) noise generated by a ransom walk of electrons in interface," *IEEE Trans. Electron Devices*, vol. 34, pp. 1100-1113, 1987.

[7.8] H. Mikeoshiba, "1/f noise in n-channel silicon-gate MOS transistors," *IEEE Trans., Electron Devices*, vol. ED-29, p.965, 1982.

[7.9] K. K. Hung, P. K. Ko, C. Hu, and Y. C. Cheng, "A unified model for the flicker noise in metal-oxide-semiconductor field-effect transistors," *IEEE Trans. Electron Devices*, vol. 37, pp. 654-665, 1990.

[7.10] Z. Celik-Butler and T. Y. Hsiang, "A thermal activation model for $1/f^{\gamma}$ noise on gate bias in N-MOSFETs", *Solid-state Electronics*, vol. 30, pp.419-423, 1987.

[7.11] C. Surya, and T. Y. Hsiang, "Theory and experiments on the $1/f^{\gamma}$ noise in P-channel metal-oxide-semiconductor field-effect transistors at low drain bias," *Physics Rev.*, vol B33, pp. 4898-4905, 1986.

[7.12] L. K. J, Vandamme, "Model for 1/f noise in MOS transistor based in linear region," *Solid-state Electronics*, vol. 23, p317, 1980.

[7.13] R. P. Jindal and A. Van der Ziel, "Phonon fluctuation model for flicker noise in elemental semiconductor," *J. Appl. Phys.* Vol. 52, p.2884, 1978.

[7.14] H. S. Park, A van der Ziel, and S. T. Liu, "Comparison of two 1/f noise models in MOSFETs," *Solid-state Electronics*, vol. 25, p.213, 1982.

[7.15] A. Van der Ziel, *Noise: Source, Characterization, and Measurements*, Prentice-Hall, Englewood Cliffs, N. J. 1970.

[7.16] A. G. Jordan, and N. A. Jordan, "Theory of the noise in metal oxide semiconductor devices," *IEEE Trans. Electron Devices*, vol. 12, pp. 148-156, 1965.

[7.17] Y. P. Tsividis, *Operation and modeling of the MOS transistor*, McGraw-Hill Book Company, New York, 1987.

[7.18] B. Wang, R. Hellums, and C. G. Sodini, "MOSFET thermal noise modeling for analog integrated circuits," *IEEE Journal of Solid-State Circuits*, vol.29, pp. 833-835, 1994.

[7.19] D. P. Triantis, A. N. Birbas, D. Kondis, "Thermal noise modeling for short channel MOSFETs," *IEEE Trans. Electron Devices*, vol. 43, pp. 1950-1955, 1996.

[7.20] S. Liu and L. W. Nagel, "Small signal MOSFET models for analog circuit design," *IEEE Journal of Solid-State Circuits,* vol. SC-17, pp.983-998, 1982.

[7.21] Y. Cheng et al., *BSIM3 version 3 User's Manual*, University of California, Berkeley, Memorandum No. UCB/ERL M97/2, 1997.

[7.22] K. K. Hung, P. K. Ko, C. Hu, and Y. C. Cheng, "A physical-based MOSFET noise model for circuit simulation," *IEEE Trans. Electron Devices*, vol. 37, pp. 1323-1333, 1990.

[7.23] G. Reimbold, "Modified 1/f trapping noise theory and experiments in MOS transistors biased from weak to strong inversion-influence of interface state," *IEEE Trans. Electron Devices*, vol. ED-31, p. 1190, 1984.

[7.24] S. C. Sun, and J. D. Plummer, "Electron mobility in inversion and accumulation layers on thermally oxidized silicon surfaces," *IEEE Trans. Electron Devices*, vol. ED-27, p. 1497, 1980.

[7.25] A. Ambrozy, *Electronic Noise*, McGraw-Hill, New York, 1982.

[7.26] *Star-Hspice user's manual*, Avanti Corporation, 1997.

[7.27] *Cadence Spectre User's manual*, Cadence Design Systems, 1996.

CHAPTER 8 Source/Drain Parasitics Model

We have discussed the modeling of the intrinsic part of a MOSFET in the previous chapters. In this chapter, we will discuss the modeling of the parasitic components that are also a part of the MOS device. These parasitics have become more important as the size of MOSFETs shrinks. This chapter will present the models of the parasitic components in BSIM3v3.

8.1 Parasitic Components in a MOSFET

As shown in Fig. 8.1.1, the four terminal MOSFET contains many parasitic components, such as the gate resistance R_g, gate/source overlap capacitance C_{gso}, gate/drain overlap capacitance C_{gdo}, gate/bulk overlap capacitance C_{gbo}, source series resistance R_s, drain series resistance R_d, source/bulk junction diode D_{sb}, drain/bulk junction diode D_{db}, and substrate resistances R_{sb}, R_{db} and R_{dsb}. These parasitic components can influence the device performance significantly. Therefore, accurate modeling of these parasitic components is essential for accurate circuit simulation.

The gate/source and gate/drain overlap capacitance models have been discussed in Chapter 5. In this chapter we will discuss the modeling of the source/drain series resistance, and D/B and S/B *p-n* junctions. The modeling of the gate and substrate resistances will be discussed later when covering RF modeling in Chapter 14.

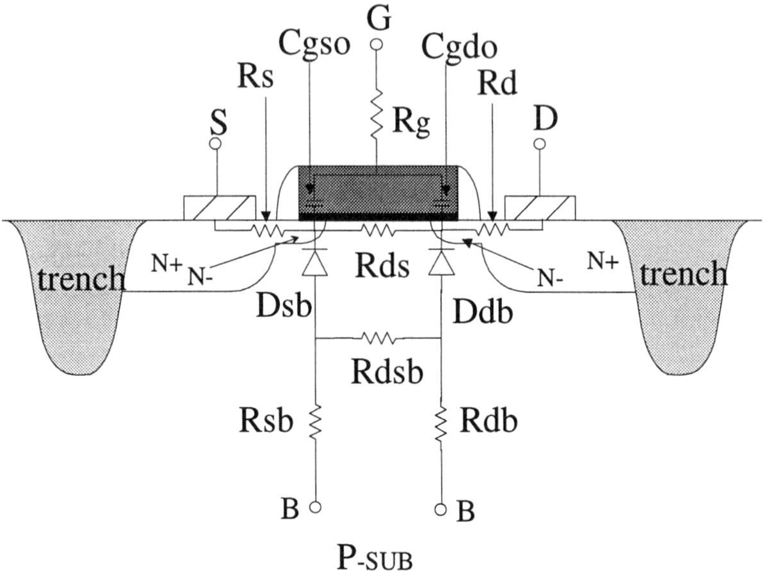

Fig. 8.1.1 MOSFET schematic cross section highlights the parasitic components.

8.2 Models of Parasitic Components in BSIM3v3

8.2.1 Source and drain series resistances

The parasitics at the source and drain regions are shown in Fig. 8.2.1. The modeling of the overlap capacitances of gate/drain and gate/source has been discussed in chapter 5. In this section, we examine the modeling of the source/drain series resistances.

The parasitic resistances of the source and drain regions are modeled in two different ways in BSIM3v3 [8.1, 8.2]. One is a traditional approach that has been used since SPICE2 [8.3], and the other was newly developed for BSIM3 [8.1, 8.2].

The approach used in SPICE2 introduces two internal nodes (S_i and D_i) at the source and drain, as shown in Fig. 8.2.2. The source and drain resistances are modeled with a parameter $\boldsymbol{R_{SH}}$ by assuming that the MOSFET is a symmetric structure:

$$R_s = \boldsymbol{R_{SH}} \times \boldsymbol{N_{RS}} \tag{8.2.1}$$

$$R_d = \boldsymbol{R_{SH}} \times \boldsymbol{N_{RD}} \tag{8.2.2}$$

where $\boldsymbol{R_{SH}}$ is a model parameter for the sheet resistance (in units of Ω/square), which has been introduced in section 7.5. $\boldsymbol{N_{RS}}$ and $\boldsymbol{N_{RD}}$ are the number of squares in the source and drain diffusion regions, respectively.

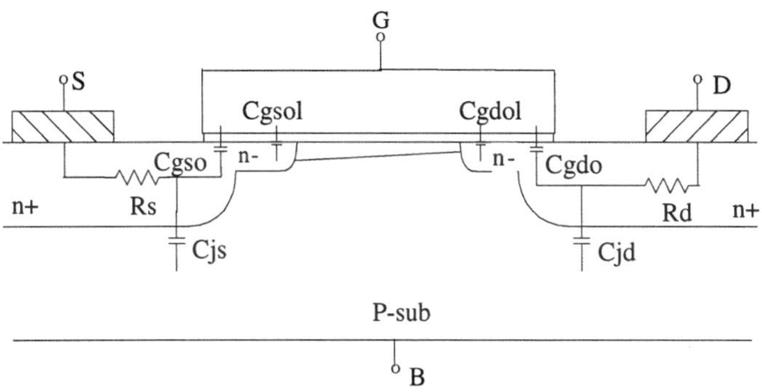

Fig. 8.2.1 Schematic cross section of a MOSFET including the parasitic components at the source and drain.

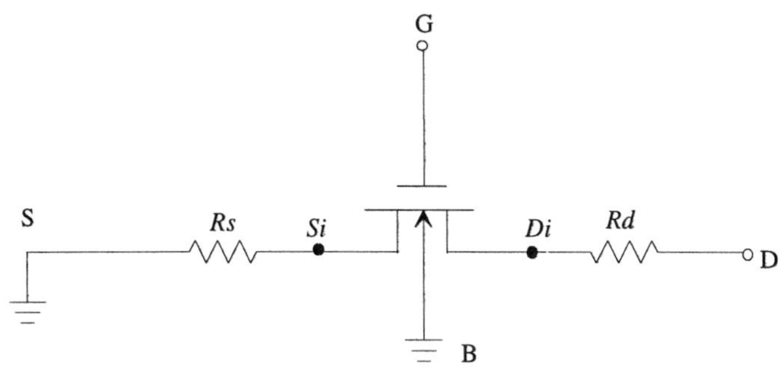

Fig. 8.2.2 Two nodes (S_i and D_i) are introduced to account for R_s and R_d.

The above approach is simple, but does not account for the geometry (channel width) and bias dependences of the series resistance. Also, because two additional nodes are introduced, the simulation time is increased.

To improve the simulation efficiency and model accuracy, another approach to model the influence of R_s and R_d has been developed in BSIM3 by including R_{ds}, the sum of R_s and R_d, in the *I-V* equation [8.1, 8.2]. The equations of the source/drain resistance R_{ds} and its influence on the DC current have been described in Chapter 4 without details of the derivation. We now derive Eq. (4.8.8).

For mathematical simplicity, BSIM3v3 assumes $R_s = R_d$. The total source and drain resistance R_{ds} is the sum of R_s and R_d, and is expressed in terms of the effective channel width W_{eff}:

$$R_{ds} = R_{dso} + \frac{R_{dsw}}{W_{eff}} \tag{8.2.3}$$

where R_{dso} is a width independent component.

The effective channel width W_{eff} can be written as,

$$W_{eff} = W_{drawn} - 2\Delta W = W_{drawn} - 2(\Delta W' + \Delta W_b) \tag{8.2.4}$$

where W_{drawn} is the channel width designated by the circuit designer. ΔW_b is the width change caused by the biases and $\Delta W'$ is the width change resulting from process-related issues (lithography, etch, and diffusion, etc.) on each side as illustrated in Fig. 8.2.3. For a wide channel width device, say $W > 10\mu m$, ΔW can be considered a constant independent of geometry (*W* and *L*) and bias. However, ΔW has been found to be a function of *W* and *L* when *W* and/or *L* is very small because photolithography and etch process are feature-size dependent. To model this, a geometry dependent $\Delta W'$ is included in BSIM3v3, as discussed in Chapter 4.

To model the bias dependence of the channel width, we need further analysis. Physically, when gate or body biases is applied, ΔW or the effective channel width is modulated. At a higher gate bias, ΔW is larger (or W_{eff} is smaller). At a higher body bias, ΔW is smaller (or W_{eff} is larger). A simple relationship is assumed [8.4]:

$$\Delta W_b = A(V_{gs} - V_{th}) = A[V_{gs} - V_{th}' + \gamma(\sqrt{\phi_s - V_{bs}} - \sqrt{\phi_s})] \tag{8.2.5}$$

where *A* is a constant, V_{th} is the threshold voltage of the device, V_{th}' is the V_{th} without the term of $\gamma(\sqrt{\phi_s - V_{bs}} - \sqrt{\phi_s})$, and γ is the body-effect coefficient.

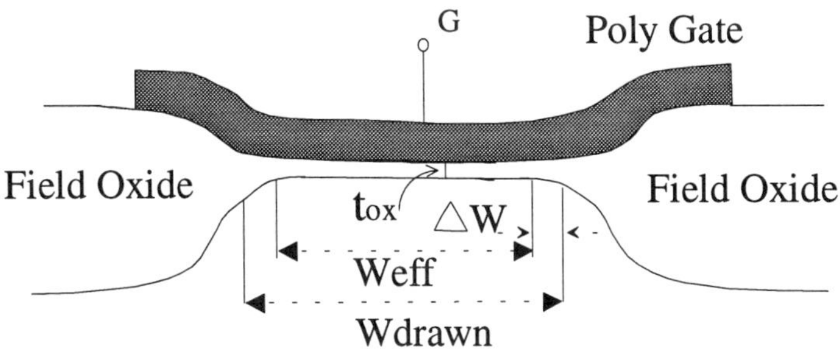

Fig. 8.2.3 Cross section of a MOSFET in the width direction.

Substituting Eq. (8.2.4) and Eq. (8.2.5) into Eq. (8.2.3), we have the following:

$$R_{ds} = R_{ds0} + \frac{R_{dsw}}{W_{drawn} - 2\Delta W' - A[V_{gs} - V_{th}' - \gamma(\sqrt{\phi_s - V_{bs}} - \sqrt{\phi_s})]} \quad (8.2.6)$$

The denominator of Eq. (8.2.6) may become zero under certain bias conditions, which is undesirable in circuit simulation. To avoid this, the first order Taylor expansion of Eq. (8.2.6) is used so that R_{ds} is given by:

$$R_{ds} = R_{ds0} + \frac{R_{dsw}[1 + A(V_{gs} - V_{th}') - B(\sqrt{\phi_s - V_{bs}} - \sqrt{\phi_s})]}{W_{eff}'} \quad (8.2.7)$$

where W_{eff}' is the effective channel width without the bias dependence given in Eq. (4.8.7). A and B are fitting parameters.

In BSIM3v3, Eq. (8.2.7) has been further modified to improve the accuracy and smoothness and to ease parameter extraction [8.5]. The constant R_{ds0} is merged into the width dependent term. The V_{gs}-V_{th}' term is replaced approximately with V_{gsteff} given in Eq. (4.2.15) to enhance the continuity of the model in the transition from subthreshold to strong inversion. The parameters $\boldsymbol{P_{RWG}}$ and $\boldsymbol{P_{RWB}}$ are used to represent A and B in Eq. (8.2.7). V_{bs} in Eq. (8.2.7) is replaced by V_{bseff} given in Eq. (3.4.26). Finally, the power exponent $\boldsymbol{W_R}$ is introduced to improve the model accuracy. Thus, the parasitic resistance R_{ds} in BSIM3v3 becomes [8.5]

$$R_{ds} = \frac{\boldsymbol{R_{DSW}}[1 + \boldsymbol{P_{RWG}} V_{gsteff} + \boldsymbol{P_{RWB}}(\sqrt{\phi_s - V_{bseff}} - \sqrt{\phi_s})]}{(W_{eff}')^{\boldsymbol{W_R}}} \quad (8.2.8)$$

We will further discuss the difference between the two modeling approaches of series resistance as well as their influence on the simulation results in section 8.3.

8.2.2 DC model of the source/drain diodes

In BSIM3v3.2, the source/drain diode *I-V* model supports a resistance-free model and a series resistance (current-limiting) model [8.6]. If the model parameter $\boldsymbol{I_{JTH}}$ is specified to be zero, the resistance-free diode model will be used in the simulation; otherwise the series resistance model will be used.

In both of the models mentioned above, the saturation current I_{sbs} of the S/B junction is calculated with

$$I_{sbs} = J_s \boldsymbol{A_S} + J_{ssw} \boldsymbol{P_S} \tag{8.2.9}$$

where J_s is the saturation current density of the source/bulk area junction and $\boldsymbol{A_S}$ is the area of the source junction. J_{ssw} is the saturation current density of the source/bulk sidewall junction and $\boldsymbol{P_S}$ is the perimeter length of the source junction.

J_s and J_{ssw} are functions of temperature and are described by:

$$J_s = \boldsymbol{J_{S0}} \exp\left[\frac{\dfrac{E_{g0}}{V_{tm0}} - \dfrac{E_g}{V_{tm}} + \boldsymbol{X_{TI}} \ln\left(\dfrac{T}{\boldsymbol{T_{NOM}}}\right)}{\boldsymbol{N_J}}\right] \tag{8.2.10}$$

$$J_{ssw} = \boldsymbol{J_{S0SW}} \exp\left[\frac{\dfrac{E_{g0}}{V_{tm0}} - \dfrac{E_g}{V_{tm}} + \boldsymbol{X_{TI}} \ln\left(\dfrac{T}{\boldsymbol{T_{NOM}}}\right)}{\boldsymbol{N_J}}\right] \tag{8.2.11}$$

where $\boldsymbol{X_{TI}}$ and $\boldsymbol{N_J}$ are the temperature exponent coefficient and emission coefficient of the junction diode, respectively. The two extracted model parameters $\boldsymbol{J_{S0}}$ and $\boldsymbol{J_{S0SW}}$ are the saturation currents of the S/B area and sidewall junctions at nominal temperature $\boldsymbol{T_{NOM}}$. If $\boldsymbol{J_{S0}}$ is not given it assumes the default value of 10^{-4}A/m^2. If $\boldsymbol{J_{S0SW}}$ is not given it is taken to be 0. E_{g0} and E_g in Eq. (8.2.10) and Eq. (8.2.11) are the energy band gaps at the nominal temperature $\boldsymbol{T_{NOM}}$ and the operating temperature T (in Kelvins) [8.6]:

$$E_{g0} = 1.16 - \frac{7.02 \times 10^{-4} \boldsymbol{T_{NOM}}^2}{\boldsymbol{T_{NOM}} + 1108.0} \tag{8.2.12}$$

$$E_g = 1.16 - \frac{7.02 \times 10^{-4} T^2}{T + 1108.0} \tag{8.2.13}$$

If the saturation current I_{sbs} given by Eq. (8.2.9) is not positive, the source/bulk diode current I_{bs} is calculated by:

$$I_{bs} = \boldsymbol{G_{MIN}} V_{bs} \tag{8.2.14}$$

Where V_{bs} is the bias at the S/B junction and $\boldsymbol{G_{MIN}}$ is a parallel junction conductance, which is introduced to improve the convergence of circuit simulation [8.8].

If the saturation current I_{sbs} of the S/B junction given by Eq. (8.2.9) is larger than zero, the following equations will be used to calculate the S/B junction current I_{bs}, depending on the value of $\boldsymbol{I_{JTH}}$ specified in the model card.

When $\boldsymbol{I_{JTH}}$ is equal to zero, the following resistance-free model is used:

$$I_{bs} = I_{sbs}[\exp(\frac{V_{bs}}{\boldsymbol{N_J} v_t}) - 1] + \boldsymbol{G_{MIN}} V_{bs} \tag{8.2.15}$$

where v_t is the thermal voltage.

If $\boldsymbol{I_{JTH}}$ is not zero, the following equation with a current limiting feature is used to calculate the diode current of the S/B junction by introducing a critical junction voltage V_{jsm}:

$$V_{jsm} = \boldsymbol{N_J} v_t \ln(\frac{\boldsymbol{I_{JTH}}}{I_{sbs}} + 1) \tag{8.2.16}$$

If $V_{bs} < V_{jsm}$

$$I_{bs} = I_{sbs}[\exp(\frac{V_{bs}}{\boldsymbol{N_J} v_t}) - 1] + \boldsymbol{G_{MIN}} V_{bs} \tag{8.2.17}$$

If $V_{bs} \geq V_{jsm}$

$$I_{bs} = \boldsymbol{I_{JTH}} + \frac{I_{JTH} + I_{sbs}}{N_{J}v_t}(V_{bs} - V_{jsm}) + \boldsymbol{G_{MIN}}V_{bs} \tag{8.2.18}$$

The *I-V* characteristics given by Eq. (8.2.17) and Eq. (8.2.18) smoothly changes from exponential to linear at $I_{bs}=\boldsymbol{I_{JTH}}$.

The current of the drain/bulk diode is modeled exactly the same way as that of the source/bulk diode with s replaced by d in the subscripts.

Fig. 8.2.4 gives an example of the calculated current characteristics of drain/bulk junction (in linear scale) using both the resistance free model and the series-resistance model. Fig. 8.2.5 gives the calculated current characteristics of the drain/bulk junction (in logarithmic scale) for both models.

8.2.3 Capacitance model of the source/bulk and drain/bulk diodes

As shown in Fig. 8.2.6, source/drain junction capacitance can be divided into three components: the bottom junction capacitance C_{jb}, the sidewall periphery junction capacitance C_{jpsw} (of the field oxide edge), and the gate-edge periphery junction capacitance C_{jpg}.

According to Fig. 8.2.6, the total source/bulk junction capacitance is:

$$C_{apbs} = C_{jbst} + C_{jbsswgt} + C_{jbsswt} \tag{8.2.19}$$

where C_{jbst} is the area capacitance of the source/bulk junction, $C_{jbsswgt}$ is the periphery capacitance of the source/bulk junction at the gate edge and C_{jbsswt} is the periphery capacitance of the source/bulk junction at the field oxide edge.

The area capacitance C_{jbst} can be calculated with

$$C_{jbst} = \boldsymbol{A_s}C_{jbs} \tag{8.2.20}$$

where C_{jbs} is the area capacitance per unit area, the equation of which will be given later in this section. $\boldsymbol{A_S}$ is the area of the source/bulk junction.

If the length of the periphery of the source/bulk junction $\boldsymbol{Ps}$ is larger than the effective channel width W_{eff}', $C_{jbsswgt}$ and C_{jbsswt} are [8.2]

$$C_{jbsswgt} = W_{eff}' C_{jbsswg} \tag{8.2.21}$$

$$C_{jbsswt} = (\boldsymbol{Ps} - W_{eff}')C_{jbssw} \tag{8.2.22}$$

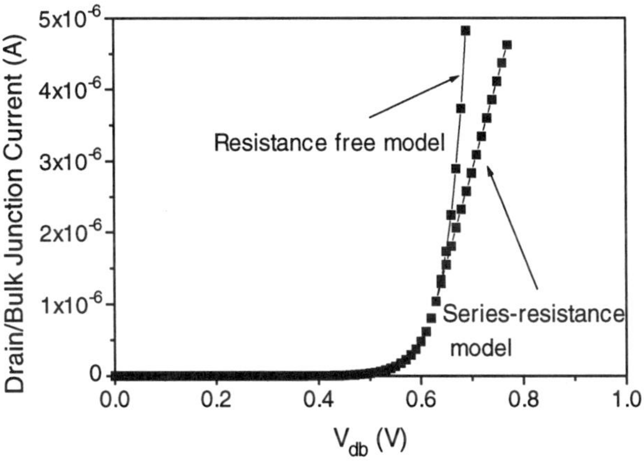

Fig. 8.2.4 Two *I-V* models of the drain/bulk junction in linear scale.

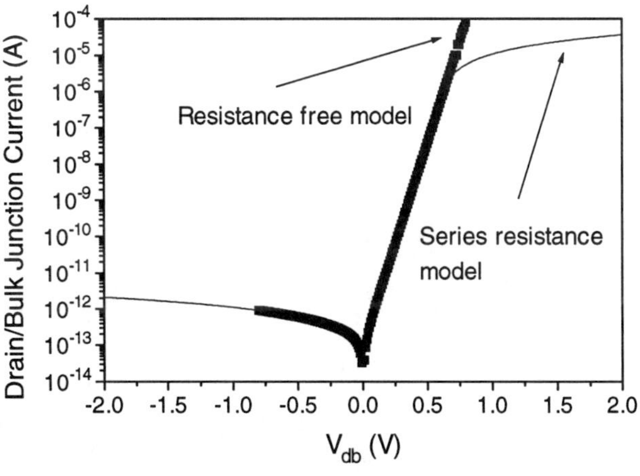

Fig. 8.2.5 Two *I-V* models of drain/bulk junction in logarithmic scale.

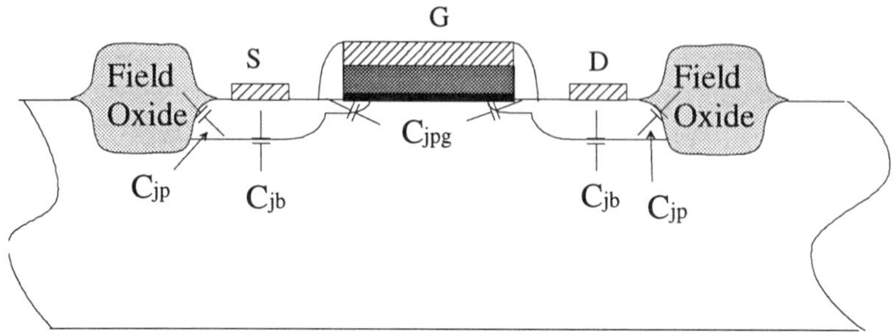

Fig. 8.2.6 Capacitance components of the source/drain junctions.

W_{eff}' is the effective channel width without bias dependence [8.5], $\boldsymbol{Ps}$ is the length of the periphery of the source/bulk junction, C_{jbsswg} is the gate edge periphery junction capacitance per unit length, and C_{jbssw} is the field-oxide edge periphery junction capacitance per unit length. Expressions for these capacitances are given later in this section.

Thus, the total junction capacitance can be calculated with

$$C_{apbs} = \boldsymbol{As} C_{jbs} + W_{eff}' C_{jbsswg} + (\boldsymbol{Ps} - W_{eff}') C_{jbssw} \tag{8.2.23}$$

If $\boldsymbol{Ps} \le W_{eff}$', only the gate edge periphery capacitance is considered and it is given by

$$C_{jsbswgt} = \boldsymbol{Ps} C_{jbsswg} \tag{8.2.24}$$

In this case, the total capacitance is given by

$$C_{apbs} = \boldsymbol{As} C_{jbs} + \boldsymbol{Ps} C_{jbsswg} \tag{8.2.25}$$

Nine model parameters, $\boldsymbol{C_J}$, $\boldsymbol{P_B}$, $\boldsymbol{M_J}$, $\boldsymbol{C_{JSW}}$, $\boldsymbol{P_{BSW}}$, $\boldsymbol{M_{JSB}}$, $\boldsymbol{C_{JSWG}}$, $\boldsymbol{P_{BSWG}}$, and $\boldsymbol{M_{JSWG}}$ are introduced in the junction capacitance model (the temperature effects of the capacitance are considered later). $\boldsymbol{C_J}$ is the unit area bottom capacitance at the zero bias, $\boldsymbol{P_B}$ is the built-in potential of the bottom junction, $\boldsymbol{M_J}$ is capacitance grading coefficient of the bottom junction, $\boldsymbol{C_{JSW}}$ is the unit length periphery capacitance at the field oxide edge at zero bias, $\boldsymbol{P_{BSW}}$ is the built-in potential of the sidewall junction at the field oxide edge, $\boldsymbol{M_{JSW}}$ is the capacitance grading coefficient of the sidewall junction at the field oxide edge, $\boldsymbol{C_{JSWG}}$ is the unit-length periphery capacitance at the gate edge at zero bias, $\boldsymbol{P_{BSWG}}$ is the built-in potential of the sidewall junction at the gate edge, and $\boldsymbol{M_{JSWG}}$ is capacitance grading coefficient of the sidewall junction at the gate edge.

Physically, $\boldsymbol{C_J}$, $\boldsymbol{C_{JSW}}$, and $\boldsymbol{C_{JSWG}}$ cannot be less than zero. If $\boldsymbol{C_J}$, $\boldsymbol{C_{JSW}}$, or $\boldsymbol{C_{JSWG}}$ is not larger than zero, the corresponding total capacitance such as C_{jbst}, C_{jbsswt}, and $C_{jbsswgt}$ is set to zero in the model implementation.

When $\boldsymbol{C_J}$ is larger than zero, C_{jbs} is calculated in the following way:

$$C_{jbs} = \boldsymbol{C_J}\left(1 - \frac{V_{bs}}{\boldsymbol{P_B}}\right)^{-\boldsymbol{M_J}} \quad V_{bs}<0 \tag{8.2.26}$$

$$C_{jbs} = \boldsymbol{C_J}\left(1 + \boldsymbol{M_J}\frac{V_{bs}}{\boldsymbol{P_B}}\right) \quad V_{bs}\geq 0 \tag{8.2.27}$$

If $\boldsymbol{C_{JSW}}$ is larger than zero, C_{jbssw} is calculated with the following equations:

$$C_{jbssw} = \boldsymbol{C_{JSW}}\left(1 - \frac{V_{bs}}{\boldsymbol{P_{BSW}}}\right)^{-\boldsymbol{M_{JSW}}} \quad V_{bs}<0 \tag{8.2.28}$$

$$C_{jbssw} = \boldsymbol{C_{JSW}}\left(1 + \boldsymbol{M_{JSW}}\frac{V_{bs}}{\boldsymbol{P_{BSW}}}\right) \quad V_{bs}\geq 0 \tag{8.2.29}$$

If $\boldsymbol{C_{JSWG}}$ is larger than zero, C_{jbsswg} is calculated with:

$$C_{jbsswg} = \boldsymbol{C_{JSWG}}\left(1 - \frac{V_{bs}}{\boldsymbol{P_{BSWG}}}\right)^{-\boldsymbol{M_{JSWG}}} \quad V_{bs}<0 \tag{8.2.30}$$

$$C_{jbsswg} = \boldsymbol{C_{JSWG}}\left(1 + \boldsymbol{M_{JSWG}}\frac{V_{bs}}{\boldsymbol{P_{BSWG}}}\right) \quad V_{bs}\geq 0 \tag{8.2.31}$$

The drain-bulk capacitance is modeled with the same equations after substituting s with d in the subscripts.

Fig. 8.2.7 shows an example of the calculated area junction capacitance per unit area as the bias changes. The bias dependence of the periphery junction capacitances are similar to this.

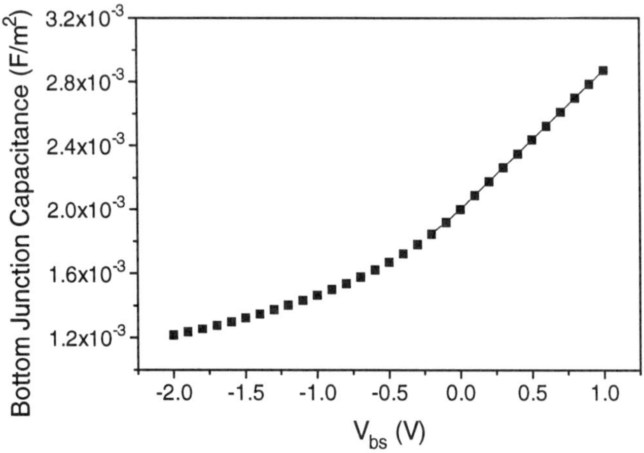

Fig. 8.2.7 Calculated source-bulk area junction capacitance versus body bias.

8.3 Helpful Hints

1. The difference between the R_s /R_d and the R_{ds} modeling approaches

We have discussed two different approaches to model the parasitic source and drain resistances. The R_s /R_d approach is straightforward but needs more simulation time because it introduces two additional circuit nodes. The R_{ds} model is the preferred BSIM3 model. It accounts for the geometry and bias dependences and can describe the parasitic source and drain resistance more accurately while requiring less simulation time. Depending on the applications, users can select one or the other model. For most digital and low frequency analog applications, the R_{ds} model given in Eq. (8.2.8) is recommended. However, for some applications that have very high resistances in the source/drain regions or that have different R_d and R_s, i.e. $R_d \neq R_s$, the R_{ds} model is not suitable. For example, in high voltage applications the voltage drops at the large source/drain series resistances needs to be accounted for with separate R_s and R_d components to evaluate their influence accurately. Another example is high frequency (RF) application. The R_{ds} model cannot accurately describe the noise characteristics and the input AC impedance of the device. In this case, external R_s and R_d terms should be introduced to model the high frequency characteristics correctly.

Furthermore, it should be pointed out that the two different source and drain resistance models can give different simulation results, even if we keep $R_s = R_d = 1/2R_{ds}$. The reason is that the *I-V* equation with the influence of R_{ds} in BSIM3v3 is a simplified analytical solution that may differ from the numerical solution of the R_s / R_d model. It is not a problem if the user extracts the model parameters from the measured data for either one of the resistance model approaches. However, if the user extracts the model parameters based on the R_{ds} model but then uses these model parameters in the simulations with external R_s/R_d (by assuming that $R_d = R_s = 1/2R_{ds}$ and the model parameter R_{DSW}=0), different simulation results will be obtained. Fig. 8.3.1 and Fig. 8.3.2 illustrate this difference.

Table 8.3.1 An example netlist to simulate a MOSFET with $R_d \neq R_s$.

```
subckt asymmetry_cir D G S B
parameters
+ wdrawn=10e-6  /* Designed device channel width */
+ ldrawn=0.5e-6  /* Designed device channel Length */
+ rshd=500      /* Sheet resistance in drain region */
+ rshs= 400     /* Sheet resistance in source region */
+ rcd = 20      /* Contact resistance in drain region */
+ rcs = 10      /* Contact resistance in source region */
+ hdiffd= 0.5u  /* Length of the drain highly diffused region */
+ hdiffs = 0.3u  /* Length of the source highly diffused region */
+ ls_perim_g= 10u /* Length of the source perimeter along the gate */
+ ld_perim_g = 10u /* Length of the drain perimeter along the gate */
+ rs = rcs + rshs*hdiffs/ls_perim_g /* total serial resistance at the source */
+ rd = rcd + rshd*hdiffd/ld_perim_g /* total serial resistance at the drain */
MI DI G SI B NFET W=WDRAWN  L=LDRAWN
RS S SI resistor r=  rs
RD D DI resistor r = rd
.MODEL NFET BSIM3V3
+ TYPE= N
+ VERSION = 3.1
+ LMIN = 3.5E-07
+ LMAX = 1
+ WMIN =3.5E-07
+ WMAX = 1
+ TNOM = 25
+ RDSW = 0
+ (other BSIM3v3 model parameters)
......
ends asymmetry_cir
```

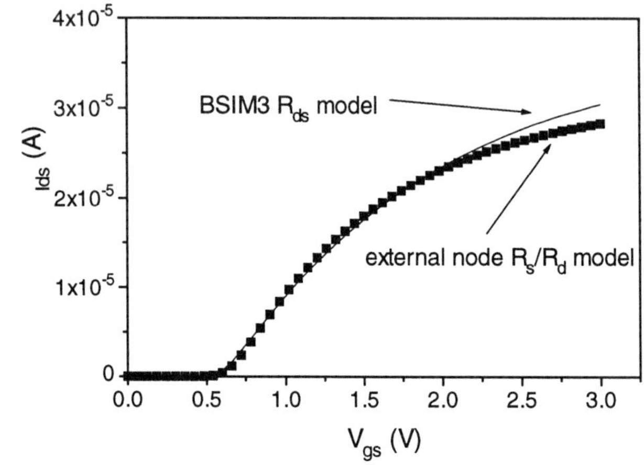

Fig. 8.3.1 Simulated I_{ds}-V_{gs} characteristics with the same set of model parameters except for R_{ds} and R_s/R_d. Solid line: R_{ds} model; Symbols: external node R_s/R_d model with $R_s = R_d = 1/2R_{ds}$ and the parameter $\boldsymbol{R_{DSW}}$ set to zero.

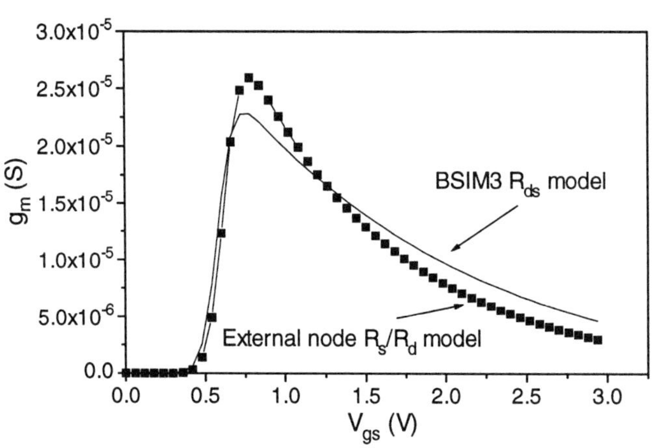

Fig. 8.3.2 Simulated g_m-V_{gs} characteristics with the same set of model parameters except for R_{ds} and R_s/R_d. Solid line: R_{ds} model; Symbols: external node R_s/R_d model with $R_s = R_d = 1/2\ R_{ds}$ and the parameter $\boldsymbol{R_{DSW}}$ set to zero.

In the two series resistance models discussed above, we have assumed that the devices are symmetric (R_s=R_d). For devices with asymmetric source and drain structures (R_d≠R_s), the simulation cannot be done directly with the BSIM3v3 model. However, we can use a sub-circuit approach to account for the influence of the asymmetric source and drain resistances. An example netlist in SPECTRE [8.8] format to simulate asymmetric device structures with the sub-circuit approach for BSIM3v3 model is given in Table 8.3.1. Similar sub-circuits can be created for other simulators such as HSPICE [8.9], ELDO [8.11], etc.

2. The source/bulk (S/B) and drain/bulk (D/B) junction diode models

All commercial circuit simulators have their own S/B and D/B diode models [8.9, 8.10, 8.11]. For example, HSPICE includes four options for the source and drain diode model, which can be selected with the model parameter ACM (ACM can be 0, 1, 2, or 3) [8.10]. Due to these simulator-related source and drain diode models, sometimes users may get different results from different simulators even if they use the same BSIM3v3 model for the intrinsic part of the MOSFET. (Simulator vendors offer BSIM3 model together with their own additional diode models.) Currently, the effort of standardizing the BSIM3v3 model is to standardize the BSIM3v3 model in both its intrinsic and parasitic parts [8.12]. However, some users who are familiar with a diode model from a specific simulator may insist on using the same diode model while adopting the rest of the BSIM3v3 model. To help users in that situation, we now give some suggestions on how to make the BSIM3v3 source/drain (S/D) diode model identical to some popular S/D diode models.

The diode models for the S/B and D/B junctions in BSIM3v3 have all the features of the HSPICE diode model with ACM=0 except that the saturation current in BSIM3v3 is not a given model parameter but is calculated from the model parameter J_{S0} [8.2, 8.10]. The BSIM3v3 diode model also contains the features of HSPICE diode models with ACM=2 and 3 [8.2, 8.10]. For example, it considers the periphery junction capacitances at the gate edge. However, BSIM3v3 does not implement the calculation for A_S, A_D, P_S, and P_D inside the model. Instead, the users can calculate the source/drain area and perimeter length according to their own definitions of A_S, A_D, P_S, and P_D (or following the same definitions of A_S, A_D, P_S, and P_D as those given in the HSPICE defined by the GEO parameter when ACM=3 [8.10]). By using the advanced feature provided in different simulators (such as .param function in HSPICE [8.10] and ELDO [8.11] and the parameters function in SPECTRE [8.8]), these instance parameters can be linked with the process and layout parameters. As an example here, we give a netlist in ELDO format in Table 8.3.2 to calculate A_S, A_D, P_S, and P_D defined by Eq. (8.3.1) to Eq. (8.3.4) with the geometry information shown in Fig. 8.3.3.

$$A_D = 2H_{dif}W_{eff} \tag{8.3.1}$$

$$P_D = 4H_{dif} + 2W_{eff} \tag{8.3.2}$$

$$A_S = 2H_{dif}W_{eff} \tag{8.3.3}$$

$$P_S = 4H_{dif} + 2W_{eff} \tag{8.3.4}$$

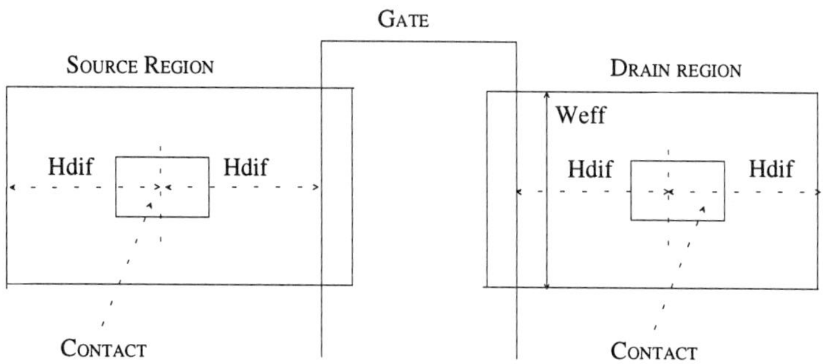

Fig. 8.3.3 A simple device layout plot showing geometry parameters used to define A_S, A_D, P_S, and P_D.

3. The G_{MIN} parameter

G_{MIN} is a small conductance added in parallel with every p-n junction to aid convergence. The default value for G_{MIN} is 10^{-12} mho, and users can change its value with the .options statement in the netlist [8.8]. However, the user cannot set G_{MIN} to zero.

4. The difference between the source/drain junction model and the stand-alone diode model

It should be noted that the diode model for the source/drain junctions in BSIM3v3 is not a complete one compared with the model for a stand-alone p-n junction diode model. The capacitance component contributed from the diffusion storage mechanism is not included in the model. Also, diode breakdown is not accounted for in the model [8.13]. These two mechanisms are not important in the normal operation of a MOSFET. However, sometimes they may be needed in some specific applications, such as to use the BSIM3v3 model in high voltage applications or to study the transient behavior when forward-biasing the source/drain junctions of the device. In those cases, one solution may be to add a full diode model to the intrinsic BSIM3v3 models with a sub-circuit approach by disabling the internal source/drain diode models in BSIM3v3.

Table 8.3.2 A netlist using the subcircuit to define A_S, A_D, P_S, and P_D parameters.

```
.subckt show_diode D G S B wdrawn=20u ldrawn=0.5u
.param hdif ={ 5e-6}
.param deltal ={ 5e-8}
.param deltaw ={ 3e-8}
.param Weff ={ wdrawn-deltaw}
.param Leff ={ ldrawn-deltal}
.param aseff = {2 * hdif * Weff }
.param pseff ={ 4 * hdif + 2 * Weff}
.param adeff ={ 2 * hdif * Weff}
.param pdeff ={ 4 * hdif + 2 * Weff}
MI D G S B dut_fet W={Weff} L={Leff}
+ AS={aseff} PS={pseff} AD={adeff} PD={pdeff}
.MODEL dut_fet NMOS
* BSIM3v3 model parameters
.ends show_diode
```

5. Source and drain diode model parameters

All of the instance and model parameters in the source and drain parasitic models are listed in Table 8.3.3.

Table 8.3.3 BSIM3v3.2 parasitic model parameters (*m* and *i* in the note column stand for instance and model parameters respectively).

Symbols in equations	Symbols in source code	Description	Default	Unit	Note
R_{SH}	rsh	Sheet resistance in source/ drain regions	0	Ω/square	m
A_S	as	Area of the source region	0	m^2	i
A_D	ad	Area of the drain region	0	m^2	i

P_S	ps	Perimeter of the source region	0	m	i
P_D	pd	Perimeter of the drain region	0	m	i
N_{RS}	nrs	Numbers of the squares in the source region	1	none	i
N_{RD}	nrd	Numbers of the squares in the drain region	1	none	i
J_{S0}	js	Saturation current density of bottom junction diode	10^{-4}	A/m^2	m
J_{S0SW}	jssw	Saturation current density of sidewall junction diode	0	A/m	m
N_J	nj	Emission coefficient of source/drain junctions	1	none	m
X_{TI}	xti	Temperature exponent coefficient of junction current	3.0	none	m
I_{JTH}	ijth	Diode limiting current	0.1	A	m
C_J	cj	Source/drain (S/D) bottom junction capacitance per unit area at zero bias	$5x10^{-4}$	F/m^2	m
M_J	mj	S/D bottom junction capacitance grading coefficient	0.5	none	m
P_B	pb	Bottom junction built-in potential	1.0	V	m
C_{JSW}	cjsw	S/D field oxide sidewall junction capacitance per unit length at zero bias	$5x10^{-10}$	F/m	m
M_{JSW}	mjsw	S/D field oxide sidewall junction capacitance grading coefficient	0.33	none	m
P_{BSW}	pbsw	Source/drain field oxide sidewall junction built-in potential	1.0	V	m
C_{JSWG}	cjswg	S/D gate edge sidewall junction capacitance per unit length at zero bias	Cjsw	F/m	m
M_{JSWG}	mjswg	S/D gate edge sidewall junction capacitance grading coefficient	Mjsw	none	m
P_{BSWG}	pbswg	Built-in potential of the source/drain gate edge sidewall junction	Pbsw	V	m

References

[8.1] J. H. Huang et al., *BSIM3 Manual (Version 2.0),* University of California, Berkeley, March 1994.

[8.2] Y. Cheng et al., *BSIM3 version 3.0 User's Manual,* University of California, Berkeley, 1995.

[8.3] L. W. Nagel, *SPICE2: A computer program to simulate semiconductor circuits*, ERL-M520, Electronics Research Laboratory, University of California, Berkeley, 1975.

[8.4] K. Chen et al., "Modeling of a MOSFET's parasitic resistance's narrow width and body bias effects for an IC simulator," *Solid-state Electronics*, vol. 39, pp.1405-1408, 1996.

[8.5] Y. Cheng et al., *BSIM3 version 3.1 User's Manual,* University of California, Berkeley, Memorandum No. UCB/ERL M97/2, 1997.

[8.6] W. Liu et al., *BSIM3 version 3.2 User's Manual,* University of California, Berkeley, 1998.

[8.7] S. M. Sze, *Physics of Semiconductor Devices*, New York: wiley, 1981.

[8.8] A. Vladimirescu, *The SPICE Book,* John Wiley & Sons, Inc., New York, 1994.

[8.9] *SPECTRE user's manual,* Cadence Design Systems, 1998.

[8.10] *Star-HSPICE user's manual,* Avanti Corporation, 1997.

[8.11] *ELDO user's manual,* Mentor Graphics, 1996.

[8.12] *Compact Model Workshop*, Burlington, Vermont, Aug, 1996.

[8.13] G. Massobrio and P. Antognetti, *Semiconductor Device Modeling with SPICE,* McGraw-Hill, Inc., New York, 1993.

CHAPTER 9

Temperature Dependence Model

In the previous chapters, we have discussed the DC and capacitance models at the nominal temperature. This chapter describes the analysis of the temperature dependence models. We briefly examine the parameters that vary with temperature. We then present the temperature dependence models of BSIM3v3.

9.1 Temperature Effects in a MOSFET

It is well known that a change in the operating temperature of a device affects its characteristics and hence the circuit performance. Accurate description of the temperature effects in devices is necessary to predict circuit behavior over a range of temperatures. A number of important model parameters such as mobility, threshold voltage, saturation velocity, parasitic series resistance, and source/drain junctions characteristics are temperature dependent. All of these temperature dependencies need to be modeled correctly.

1. Temperature dependence of mobility

Much research has been done to model the inversion charge mobility as a function of channel doping concentration, the gate and substrate voltages, and temperature [9.1, 9.2]. It is well known that phonon scattering, surface scattering, and coulombic scattering (including ionized impurity scattering and interface charge scattering) are the three major scattering mechanisms governing the characteristics of carrier mobility in the inversion layer. For temperatures above 250K, phonon scattering is the dominant mechanism.

2. Temperature dependence of the threshold voltage

Threshold voltage (V_{th}) is another important parameter that is sensitive to temperature. It increases as temperature decreases due to the Fermi-level and bandgap energy shifts. V_{th} depends linearly on the temperature over a wide range of temperatures for devices with long channel lengths [9.3]. Recent experiments show that V_{th} rolloff, i.e. the dependence of V_{th} on V_{ds} and L, is insensitive to temperature [9.4]. This can be explained by the fact that V_{th} rolloff results from the capacitive coupling of the drain and the channel - a temperature independent phenomenon.

3. Temperature dependence of the saturation velocity

It is known that the saturation velocity (υ_{sat}) is a weak function of temperature [9.5]. For simplicity, the temperature dependence of υ_{sat} is usually ignored in the compact modeling of MOSFETs.

4. Temperature dependence of the parasitic drain/source resistances

With the increasing current drive of MOSFETs and dropping supply voltages, the drain/source series resistance becomes a more important parameter. R_{ds} is composed of contact resistance, drain and source diffusion sheet resistance, and spreading resistance at the edge of the inversion layer due to current crowding. R_{ds} increases almost linearly with rising temperature.

5. Temperature dependence of the S/D diode characteristics

For the source/drain junctions in a MOSEFT, the temperature dependences of the saturation current and the junction capacitance at zero bias are important and need to be modeled.

It is known that the temperature dependence of the saturation current, I_s, of a p-n junction is determined by the temperature dependence of the intrinsic carrier density, n_i, or the energy band gap of the material, E_g [9.3, 9.6, 9.7]. The temperature dependence of the zero-bias junction capacitance C_{j0} is determined by the temperature dependences of the dielectric constant of silicon material, ε_{si}, and the junction built-in potential, V_{bi} [9.3, 9.7].

9.2 Temperature Dependence Models in BSIM3v3

1. Modeling the temperature dependence of mobility

Several empirical unified formulations have been suggested to describe the mobility as a function of process parameters and bias conditions [9.8, 9.9, 9.10, 9.11]. However, all of them contain a quantity, E_{eff}, that is not readily available for circuit simulation. It has been shown that E_{eff} may be expressed simply as $(V_{gs}+V_{th})/(6T_{ox})$ [9.12]. The effects of V_{bs} and doping concentration are reflected in the V_{th} term. In BSIM3v3, a second order polynomial with parameters U_a, U_b, and U_c, which are linear functions of temperature, is used to describe the temperature dependence of mobility [9.4, 9.13, 9.14]. For ***mobMod***=1, the mobility model, including temperature effects, becomes

$$\mu_{eff} = \frac{\mu_0 (T / T_{NOM})^{u_{te}}}{1 + (U_a(T) + U_c(T)V_{bseff})(\frac{V_{gsteff} + 2V_{th}}{T_{OX}}) + U_b(T)(\frac{V_{gsteff} + 2V_{th}}{T_{OX}})^2} \tag{9.2.1}$$

where

$$U_a(T) = U_A + U_{A1}(T / T_{NOM} - 1) \tag{9.2.2}$$

$$U_b(T) = U_B + U_{B1}(T / T_{NOM} - 1) \tag{9.2.3}$$

$$U_c(T) = U_C + U_{C1}(T / T_{NOM} - 1) \tag{9.2.4}$$

where parameters μ_0, U_A, U_{A1}, U_B, U_{B1}, U_C, U_{C1}, and U_{TE} can be extracted from the measured *I-V* data, and *T* is the temperature in Kelvin. T_{NOM} is the nominal temperature at which the model parameters μ_0, U_A, U_B, and U_C are extracted.

2. Modeling the temperature dependence of the threshold voltage

The following temperature model of V_{th} is used in BSIM3 [9.4, 9.13, 9.14]:

$$V_{th}(T) = V_{th}(T_{NOM}, L, V_{ds}) + (K_{T1} + \frac{K_{T1L}}{L} + K_{T2}V_{bs})(\frac{T}{T_{NOM}} - 1) \tag{9.2.5}$$

where $V_{th}(T_{NOM}, L, V_{ds})$ is the threshold voltage value at T_{NOM}. The expression for $V_{th}(T_{NOM}, L, V_{ds})$ has been given in Eq. (3.4.25) [9.14, 9.15]. The parameters K_{T1}, K_{T1L}, and K_{T2} are extracted from the experimental data. K_{T1L}/L is a minor term introduced to improve the fitting accuracy.

3. Modeling the temperature dependence of the saturation velocity

In BSIM3v3, the temperature dependence of υ_{sat} is modeled with the following [9.4, 9.13, 9.14]:

$$\upsilon_{sat}(T) = \boldsymbol{v_{SAT}} - \boldsymbol{A_T}\left(\frac{T}{\boldsymbol{T_{NOM}}} - 1\right) \tag{9.2.6}$$

where A_T is a parameter extracted from the measured data, and v_{SAT} is the saturation velocity at T_{NOM}.

4. Modeling the temperature dependence of the parasitic drain/source resistances

In BSIM3v3, R_{ds} and its temperature dependence is modeled as [9.4, 9.14]:

$$R_{ds}(T) = \frac{R_{dsw}(T)[1 + \boldsymbol{P_{RWG}} V_{gsteff} + \boldsymbol{P_{RWB}}(\sqrt{\phi_s - V_{bseff}} - \sqrt{\phi_s})]}{W_{eff}'^{\boldsymbol{W_R}}} \tag{9.2.7}$$

$$R_{dsw}(T) = \boldsymbol{R_{DSW}} + \boldsymbol{P_{RT}}\left(\frac{T}{\boldsymbol{T_{NOM}}} - 1\right) \tag{9.2.8}$$

where R_{DSW}, P_{RWG}, P_{RWB}, and W_R are extracted from the measured R_{ds} data at T_{NOM}. P_{RT} is extracted from the measured data at different temperatures. W_{eff}' is the effective channel width without consideration of bias dependence. V_{gsteff} is equal to V_{gs}-V_{th} in the strong inversion region [9.16, 9.17].

5. Modeling the temperature dependence of the S/B and D/B junctions

(1). Temperature dependence in the DC model

In BSIM3v3, the temperature dependence of the S/B and D/B junctions is described by the saturation current I_{sbs}. It is calculated as

$$I_{sbs} = J_s \boldsymbol{A_S} + J_{ssw} \boldsymbol{P_S} \tag{9.2.9}$$

$$I_{sbd} = J_s \boldsymbol{A_D} + J_{ssw} \boldsymbol{P_D} \tag{9.2.10}$$

where J_S is the saturation current density of the source or drain junction and $\boldsymbol{A_S}$ and $\boldsymbol{A_D}$ are the areas of the source and drain junctions. J_{SSW} is the saturation current density of the source or drain sidewall junction and $\boldsymbol{P_S}$ and $\boldsymbol{P_D}$ are the perimeters of the source and drain junctions. Both J_S and J_{SSW} are functions of temperature and are described by Eqs. (8.2.10) through (8.2.13).

Fig. 9.2.1 shows the current characteristics of the drain/bulk junction at several temperatures for the two different BSIM3v3.2 diode models discussed in Chapter 8.

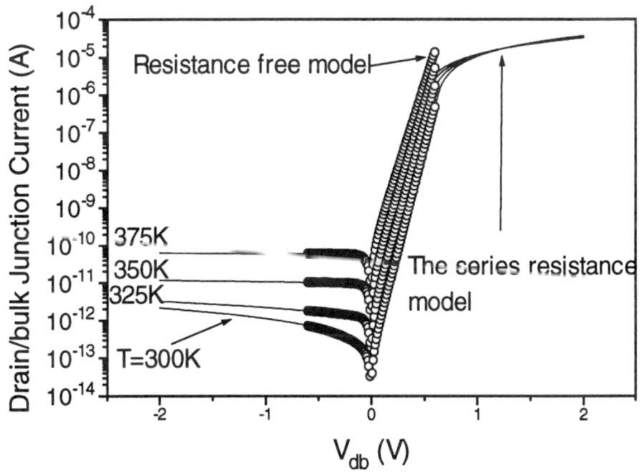

Fig. 9.2.1 *I-V* characteristics of the drain/bulk junction at different temperatures.

(2). Temperature dependence of the capacitance model

The temperature dependence of the source/drain junction capacitance is modeled by introducing the temperature-dependent zero-bias unit area/perimeter junction capacitances C_j(T), C_{jsw}(T), and C_{jswg}(T), and junction built-in potentials P_b(T), P_{bsw}(T), and P_{bswg}(T).

The temperature dependence of the zero-bias junction capacitance is modeled with the following equations [9.18]:

$$C_j(T) = \boldsymbol{C_J}[1 + \boldsymbol{T_{CJ}}(T - \boldsymbol{T_{NOM}})] \tag{9.2.11}$$

$$C_{jsw}(T) = \boldsymbol{C_{JSW}}[1 + \boldsymbol{T_{CJSW}}(T - \boldsymbol{T_{NOM}})] \tag{9.2.12}$$

$$C_{jswg}(T) = \boldsymbol{C_{JSWG}}[1 + \boldsymbol{T_{CJSWG}}(T - \boldsymbol{T_{NOM}})] \tag{9.2.13}$$

where C_j(T), C_{jsw}(T) and C_{jswg}(T) are zero-bias junction capacitance per unit area, the perimeter junction capacitance per unit length at the field-oxide edge, and the perimeter junction capacitance per unit length at the gate edge. $\boldsymbol{C_J}$, $\boldsymbol{C_{JSW}}$, and $\boldsymbol{C_{JSWG}}$ are the zero-bias capacitances at the nominal temperature $\boldsymbol{T_{NOM}}$. $\boldsymbol{T_{CJ}}$, $\boldsymbol{T_{CJSW}}$, and $\boldsymbol{T_{CJSWG}}$ are the model parameters for the temperature coefficients of C_j, C_{jsw}, and C_{jswg}.

The temperature dependence of the built-in potentials in the junction capacitances is modeled with the following equations [9.18]:

$$P_b(T) = \boldsymbol{P_B} - \boldsymbol{T_{PB}}(T - \boldsymbol{T_{NOM}}) \tag{9.2.14}$$

$$P_{bsw}(T) = \boldsymbol{P_{BSW}} - \boldsymbol{T_{PBSW}}(T - \boldsymbol{T_{NOM}}) \tag{9.2.15}$$

$$P_{bswg}(T) = \boldsymbol{P_{BSWG}} - \boldsymbol{T_{PBSWG}}(T - \boldsymbol{T_{NOM}}) \tag{9.2.16}$$

where P_b(T), P_{bsw}(T), and P_{bswg}(T) are the built-in potentials of the bottom junction, the periphery junction at the field-oxide edge, and the periphery junction at the gate edge at temperature T in Kelvin. $\boldsymbol{P_B}$, $\boldsymbol{P_{BSW}}$, and $\boldsymbol{P_{BSWG}}$ are the built-in potentials of the bottom junction, the periphery junction at the field-oxide edge, and the periphery junction at the gate edge at the nominal temperature $\boldsymbol{T_{NOM}}$. $\boldsymbol{T_{PB}}$, $\boldsymbol{T_{PBSW}}$, and $\boldsymbol{T_{PBSWG}}$ are the temperature coefficients of the built-in potentials.

C_{jbs}, the area capacitance of the source/bulk junction with temperature effects, is calculated by:

$$C_{jbs} = C_j(T)\left(1 - \frac{V_{bs}}{P_b(T)}\right)^{-\boldsymbol{M_J}} \qquad (V_{bs}<0) \tag{9.2.17}$$

$$C_{jbs} = C_j(T)\left(1 + \boldsymbol{M_J}\frac{V_{bs}}{P_b(T)}\right) \qquad (V_{bs}\geq 0) \tag{9.2.18}$$

C_{jbssw}, the periphery capacitance of the source/bulk junction at the field oxide edge with temperature effects, is calculated by:

$$C_{jbssw} = C_{jsw}(T)\left(1 - \frac{V_{bs}}{P_{bsw}(T)}\right)^{-\boldsymbol{M_{JSW}}} \qquad (V_{bs}<0) \tag{9.2.19}$$

$$C_{jbssw} = C_{jsw}(T)(1 + \boldsymbol{MJSW}\frac{V_{bs}}{P_{bsw}(T)}) \qquad (V_{bs} \geq 0) \qquad (9.2.20)$$

C_{jbsswg}, the periphery capacitance of the source/bulk junction at gate oxide edge with temperature effects, is calculated by:

$$C_{jbsswg} = C_{jswg}(T)(1 - \frac{V_{bs}}{P_{bswg}(T)})^{-\boldsymbol{MJSWG}} \qquad (V_{bs} < 0) \qquad (9.2.21)$$

$$C_{jbsswg} = C_{jswg}(T)(1 + \boldsymbol{MJSWG}\frac{V_{bs}}{P_{bswg}(T)}) \qquad (V_{bs} \geq 0) \qquad (9.2.22)$$

The equations for the temperature dependence of the drain/bulk junction are the same as the above except for the obvious change of the subscripts from "s" to "d" in Eq. (9.2.17) through Eq. (9.2.22).

Fig. 9.2.2 shows the minor influence of temperature change on the junction capacitance.

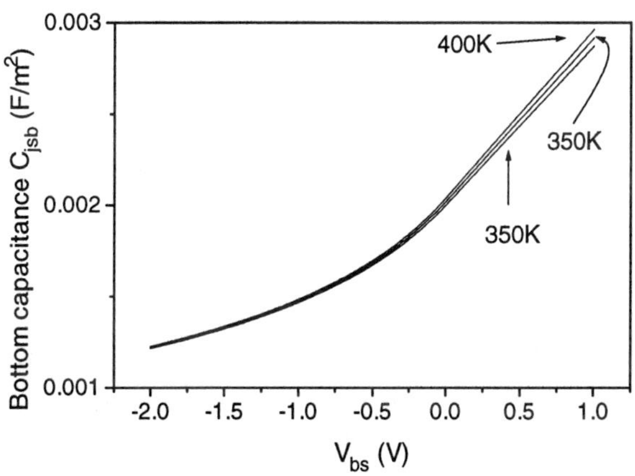

Fig. 9.2.2 The junction capacitance at several temperatures.

9.3 *Comparison of the Temperature-Effect Models with Measured Data*

The above models of the temperature dependence have been tested against measured data [9.4]. The MOSFETs used in the tests are from a 0.25μm CMOS technology with a T_{ox} of 4.5 nm. The BSIMPro model parameter extractor [9.19] is used to extract the model parameters.

The behavior of the threshold voltage (V_{th}) versus temperature for n- and p-channel devices with *W/L*=6μm/0.25μm are given in Figs. 9.3.1 and 9.3.2. The model can match the measured data well at different body bias conditions and temperatures. This shows that the temperature dependence of short channel effects can be well described by Eq. (9.2.5), and that the linear dependence of V_{th} on temperature still holds for devices with channel lengths down to, at least, quarter micron.

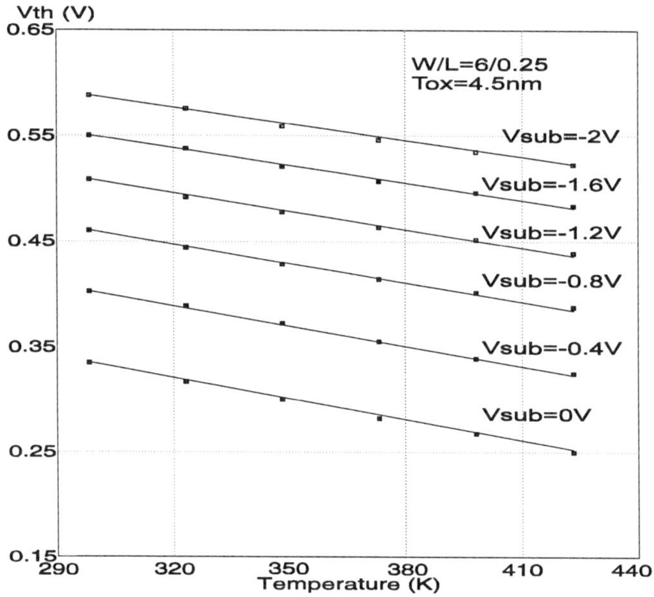

Fig. 9.3.1 V_{th}-T data of a *W/L*=6μm/0.25μm n-channel device at different V_{bs}. Data clearly indicates a linear dependence on *T*. After Cheng et al. [9.4].

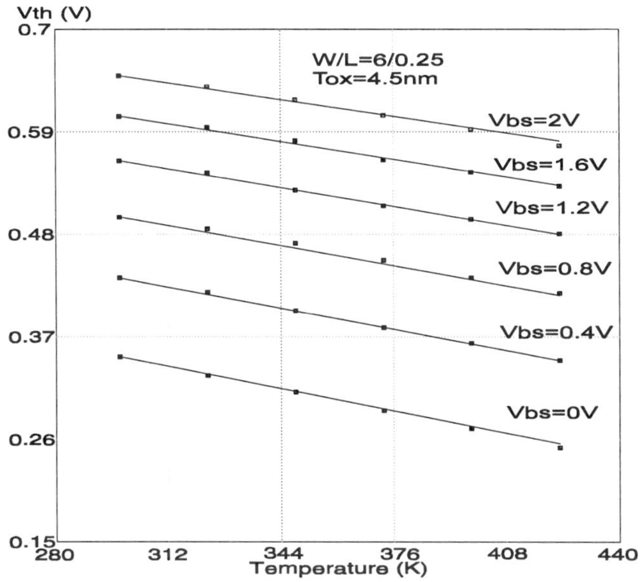

Fig. 9.3.2 p-channel devices also exhibit a linear temperature dependence of V_{th}. After Cheng et al. [9.4].

Figs. 9.3.3 and 9.3.4 show the curves of the threshold voltage versus channel length for different body bias conditions at 125°C. It can be seen that even at high temperatures the model can predict V_{th} at different V_{bs} for both n- and p-channel devices with different channel lengths.

The I_{ds}-V_{gs} characteristics are given in Figs. 9.3.5 and 9.3.6 for the n- and p-channel devices with W/L=6μm/0.25μm at V_{ds}=0.05V and V_{bs}=0V for different temperatures. It can be seen that the model can fit the measured data well, and the maximum error is less than 3.13% for nMOSFET and 2.89% for pMOSFET for a temperature range of 25°C to 150°C. It is well known that for short channel devices the I_{ds}-V_{gs} fit is determined by the accuracy of mobility and R_{ds} models. The good agreement between the model and the measured data of a 0.25μm device means that the temperature models of mobility and R_{ds} in BSIM3 are accurate.

In Fig. 9.3.7, the measured and modeled curves of the transconductance (g_m) versus V_{gs} at different temperatures are given. A good fit between the model and data can also be obtained for p-channel devices with 0.25 μm channel length [9.4].

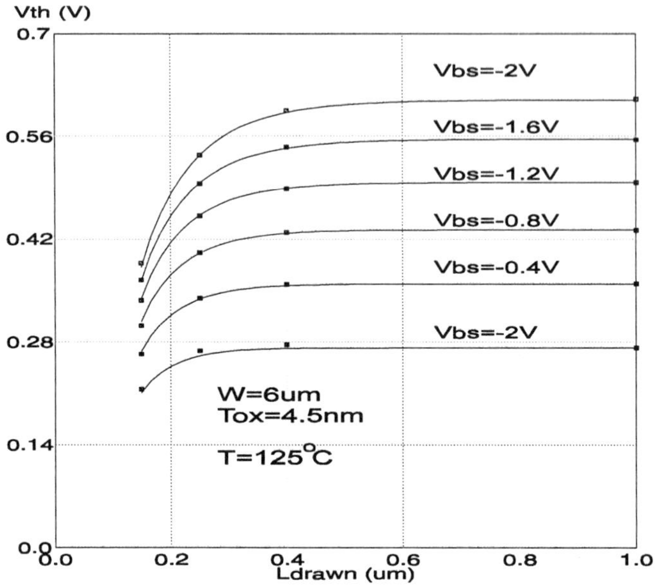

Fig. 9.3.3. V_{th} vs. channel length of n-MOSFETs for 125°C at different V_{bs} is well modeled by BSIM3. Solid lines: BSIM3v3; Symbols: measured data. After Cheng et al. [9.4].

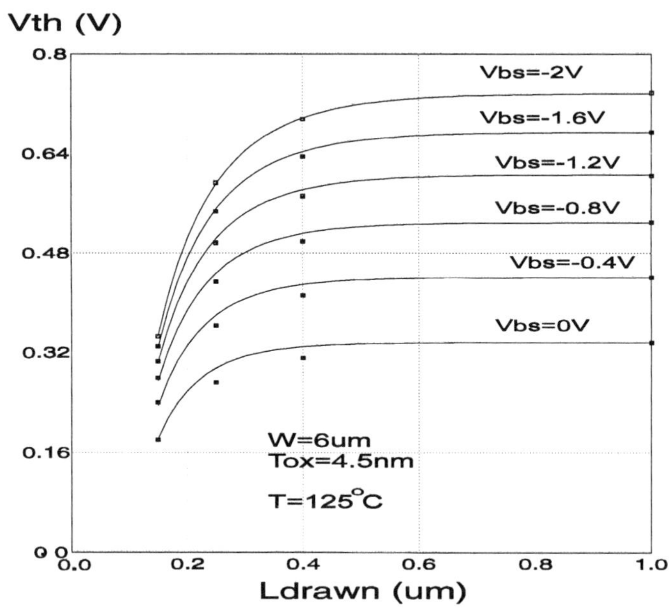

Fig. 9.3.4. Curves of V_{th} vs. L of p-MOSFETs for 125°C at different V_{bs}. After Cheng et al. [9.4].

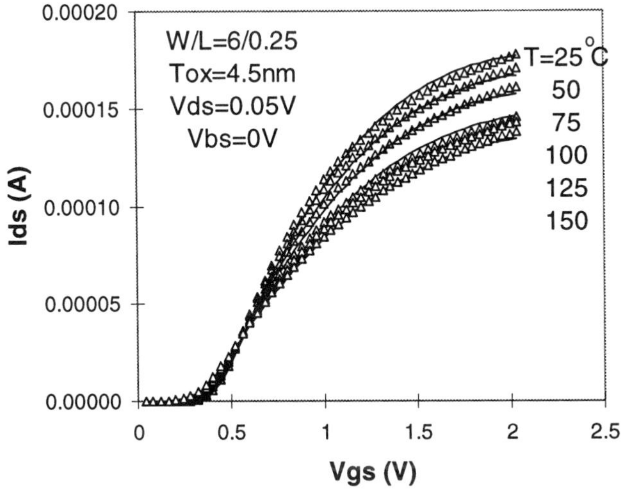

Fig. 9.3.5 I_{ds}-V_{gs} of an n-channel device of *W/L*=6μm/0.25μm at several temperatures. Solid lines: BSIM3v3; Symbols: data. After Cheng et al. [9.4].

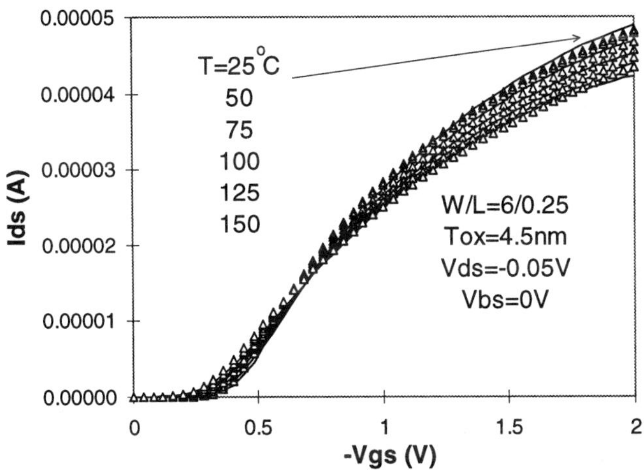

Fig. 9.3.6 I_{ds}-$V_{gs}$of a 0.25μm p-channel device at several temperatures. Solid lines: BSIM3v3; Symbols: data. After Cheng et al. [9.4].

In order to ensure that digital and analog simulations at different temperatures are accurate, the model accuracy in predicting the temperature dependence of I_{ds}-V_{ds}, and g_{ds}-V_{ds} characteristics needs to be verified. Figs. 9.3.8 and 9.3.9 show such a verification. The model agrees with the measured data with a maximum error of 2.15% for the n-MOSFET and 1.87% for the p-MOSFET without using global optimization during the parameter extraction. The temperature dependence of the saturation velocity given in Eq. (9.2.6) is very helpful for improving the accuracy of the model in the saturation region.

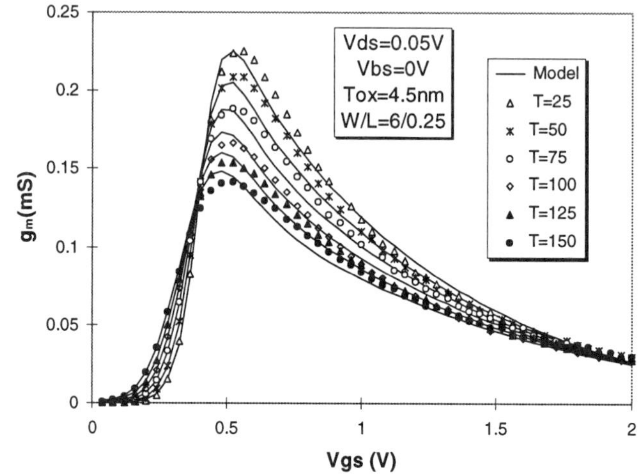

Fig. 9.3.7 g_m-V_{gs} curves of an n-channel device (W/L=6μm/0.25μm). After Cheng et al. [9.4].

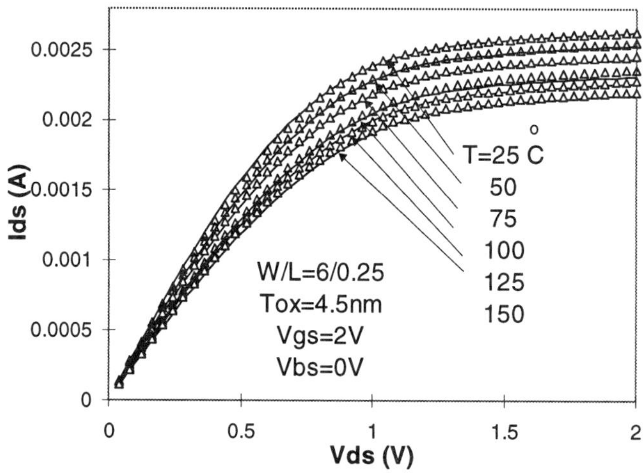

Fig. 9.3.8 I_{ds}-V_{ds} curves of an n-channel 0.25μm device at different temperatures. After Cheng et al. [9.4].

In Figs. 9.3.10 and 9.3.11, we present the modeled and measured g_{ds}-V_{ds} characteristics, which are important in analog circuit design. The figures show that the temperature dependence of g_{ds} can also be well described by the present model for different bias conditions and a wide temperature range.

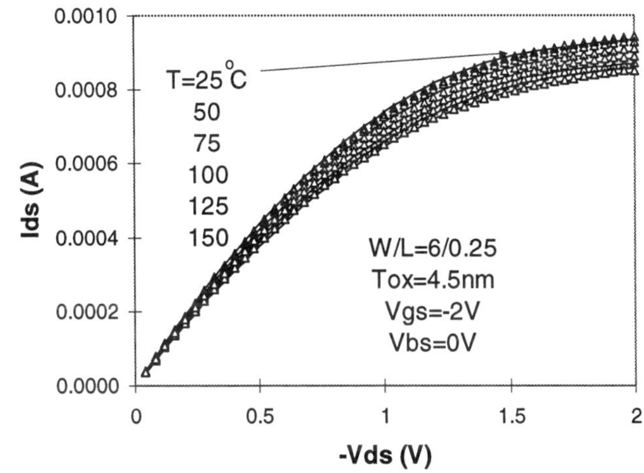

Fig. 9.3.9 I_{ds}-V_{ds} curves of a p-channel 0.25μm device at different temperatures. After Cheng et al. [9.4].

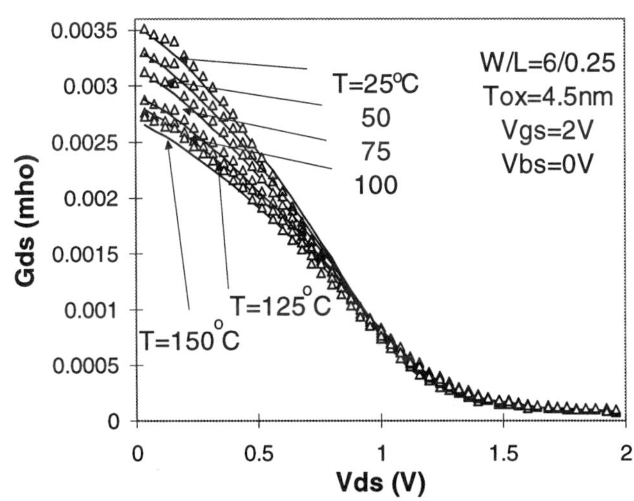

Fig. 9.3.10 g_{ds} -V_{ds} of an n-channel 0.25μm device at different temperatures. After Cheng et al. [9.4].

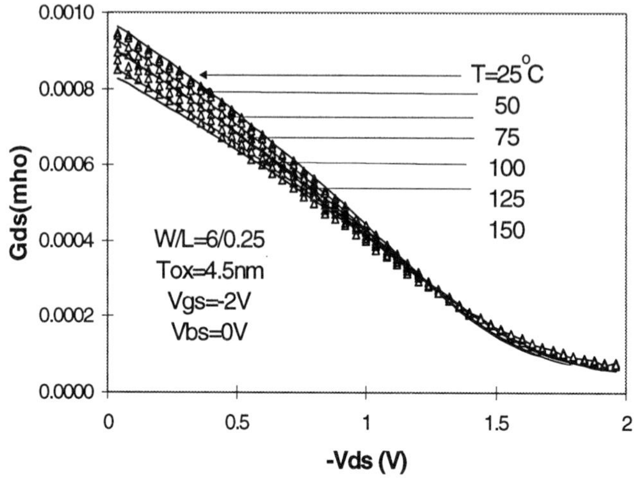

Fig. 9.3.11 g_{ds} -V_{ds} of a p-channel device at different temperatures. After Cheng et al. [9.4].

9.4 Helpful Hints

1. The empirical temperature dependence of V_{th}

In Eq. (9.2.5), the temperature dependence of V_{th} is accounted for in the second term. The first term is evaluated at the nominal temperature $\boldsymbol{T_{NOM}}$. Thus, the temperature dependences of the band-gap E_g, surface potential ϕ_s at the threshold, and the intrinsic carrier density n_i are all lumped together in the second term of Eq. (9.2.5) in a simple linear form. This approach is adopted because it can simplify the parameter extraction, and provide better fit to the measured data, which shows a strong linear relationship between V_{th} and temperature resulting from the overall contributions of these parameters to V_{th}. Parameters such as E_g, ϕ_s and n_i are calculated at $\boldsymbol{T_{NOM}}$ when Eq. (9.2.5) is used to evaluate the temperature dependence of V_{th}. To be consistent, and to simplify the model implementation, the ϕ_s appearing in the A_{bulk} and R_{ds} expressions is also calculated at $\boldsymbol{T_{NOM}}$. This does not influence the temperature dependence of the device significantly since the temperature dependence of such parameters as A_{bulk} is very weak, and the temperature effect of R_{ds} is accounted for in R_{DSW} given by Eq. (9.2.8).

This approach of modeling the temperature effects in a linear form significantly simplifies the parameter extraction. However, it may cause a problem when trying to use the model for statistical or predictive modeling, where the temperature dependence of all the parameters, including E_g, should be maintained. Without any change to the model implementation, a sub-circuit approach can be used to account for the temperature dependencies of these parameters by introducing the appropriate temperature-dependent equations within the sub-circuit.

2. The temperature dependence of E_g and n_i

As discussed above, the temperature effects of some physical parameters such as E_g, n_i, and ϕ_s are explicitly suppressed in favor of an empirical linear dependence of V_{th} on *T*. However, the temperature dependence of E_g and n_i are maintained in other parts of the model such as in evaluating the temperature dependence of the saturation current of source and drain junctions.

3. The operation temperature *T* and the nominal temperature T_{NOM}

T_{NOM} is the nominal, or reference, temperature at which the model parameters are extracted. It can be different in the model cards for different devices. *T* is defined as the operation temperature for the circuit being simulated. In most simulators, *T* must be kept the same for all the devices in the circuit being simulated.

4. The validation range of the temperature dependence model

The temperature range for the examples shown in this chapter is between 25°C and 150°C. However, at minimum the model is accurate from-50°C to 150°C. Additional temperature dependences need to be introduced for the models to be accurate at temperatures much lower than -50°C. For example, in this regime coulombic scattering may dominate over electron-phonon scattering and significantly change the temperature dependencies. Compact modeling for cryogenic temperature operation is a separate and challenging issue.

Another issue that the users need to be aware of is the temperature dependence model of the source/drain diode saturation current. The model given in Eq. (8.2.10) and Eq. (8.2.12) is based on the assumption that the diffusion component dominates the leakage current. However, it has been found that the diffusion leakage is actually not the dominant mechanism when the temperature is below 120°C (it is dominant in the temperature range above 120°C) [9.3]. Instead, the dominant mechanism of the junction leakage current in the temperature range of interest for normal circuit operation is generation-recombination [9.3, 9.20]. Thus, the saturation current is a linear function of the intrinsic carrier density. This is different from the quadratic dependence on n_i predicted by Eq. (8.2.10) and Eq. (8.2.12).

5. Parameters of the temperature effect model

The BSIM3v3 model parameters for the temperature effects are listed in Table 9.4.1.

Table 9.4.1 Temperature effect model parameters

Symbols in equation	Symbols in source code	Description	Default	Unit
T_{NOM}	tnom	Temperature at which parameters are extracted	27	^oC
P_{RT}	prt	Temperature coefficient for R_{dsw}	0.0	Ω-μm
μ_{TE}	ute	Mobility temperature exponent	-1.5	none
K_{T1}	kt1	Temperature coefficient for threshold voltage	-0.11	V
K_{T1L}	kt1l	Channel length sensitivity of temperature coefficient for threshold voltage	0.0	Vm
K_{T2}	kt2	Body-bias coefficient of the V_{th} temperature effect	0.022	none
U_{A1}	ua1	Temperature coefficient for Ua	4.31×10^{-9}	m/V
U_{B1}	ub1	Temperature coefficient for Ub	-7.61×10^{-18}	$(m/V)^2$
U_{C1}	uc1	Temperature coefficient for Uc	***mobMod***=1, 2: -5.6×10^{-11} ***mobMod***=3: -0.056	m/V^2 1/V
A_T	at	Temperature coefficient for saturation velocity	3.3×10^4	m/sec
N_J	nj	Emission coefficient of junction	1.0	none
X_{TI}	xti	Temperature exponent coefficient of junction current	3.0	none
T_{CJ}	tcj	Temperature coefficient of C_j	0.0	1/K

T_{CJSW}	tcjsw	Temperature coefficient of C_{jsw}	0.0	1/K
T_{CJSWG}	tcjswg	Temperature coefficient of C_{jswg}	0.0	1/K
T_{PB}	tpb	Temperature coefficient of P_b	0.0	V/K
T_{PBSW}	tpbsw	Temperature coefficient of P_{bsw}	0.0	V/K
T_{PBSWG}	tpbswg	Temperature coefficient of P_{bswg}	0.0	V/K

References

[9.1] C. L. Huang and G. Sh. Gildenblat, "Measurements and modeling of the n-Channel MOSFET inversion layer mobility and device characteristics in the temperature range 60-300K," *IEEE Trans. on Electron Devices*, vol. ED-37, pp. 1289-1300, 1990.

[9.2] M. S. Liang, J. Y. Choi, P. K. Ko and C. Hu, "Inversion-layer capacitance and mobility of very thin gate-oxide MOSFETs", *IEEE Trans. Electron Devices*, ED-33, p.409, 1986.

[9.3] S. M. Sze, *Physics of Semiconductor Devices*, New York: Wiley, 1981.

[9.4] Y. Cheng et al., "Modeling temperature effects of quarter micrometer MOSFETs in BSIM3v3 for circuit simulation," *Semiconductor Science and Technology*, Vol.12, pp. 1349-1354, 1997.

[9.5] N. Yasuda et al., "Analytical device model of SOI MOSFET's including self-heating," *Japan. J. Appl. Phys.*, Vol.30, pp.3677-3684, 1991.

[9.6] G. Massobrio and P. Antognetti, *Semiconductor Device Modeling with SPICE*, McGraw-Hill, Inc., New York, 1993.

[9.7] N. Arora, *MOSFET Models for VLSI Circuit Simulation*, Springer-Verlag, Wien New York, 1994.

[9.8] K. Lee et al, "Physical understanding of low field carrier mobility in silicon inversion layer," *IEEE Trans. Electron Devices*, ED-38, p. 1905, 1991.

[9.9] C. G. Sodini, P. K. Ko, and J. L. Moll, "The Effects of high fields on MOS device and circuit performance," *IEEE Trans. Electron Devices*, ED-31, p. 1386, 1984.

[9.10] E. A. Talkhan, I. R. Mansour, and A. I. Baroor, "Investigation of the effect of drift-filed-dependent mobility on MOSFET characteristics," *Part I and II, IEEE Trans. Electron Devices*, ED-19, p. 899, 1972.

[9.11] S. Takagi, A. Toriumi, M. Iwase, and H. Tango, "On the universality of inversion layer mobility in Si MOSFET's: part I - Effects of substrate Impurity Concentration", *IEEE Trans. Electron Devices*, Vol. ED-41, p. 2357, 1994.

[9.12] K. Chen et al., "MOSFET carrier mobility model based on gate oxide thickness, threshold and gate voltages", *Solid-State Electronics*, pp.1515-1518, Vol. 39, No. 10, October 1996.

[9.13] J. H. Huang et al., *BSIM3 Manual (Version 2.0)*, University of California, Berkeley, March 1994.

[9.14] Y. Cheng et al., *BSIM3 version 3.1 User's Manual*, University of California, Berkeley, Memorandum No. UCB/ERL M97/2, 1997.

[9.15] Y. Cheng et al., "Modeling of small size MOSFETs with reverse short channel and narrow width effects for circuit simulation", *Solid State Electronics*, vol. 41, (9), pp. 1227-1231, 1997.

[9.16] Y. Cheng et al., "A unified BSIM I-V mode for circuit simulation", *1995 International semiconductor device research symposium*, Charlottesville, pp. 312-313, Dec. 1995.

[9.17] Y. Cheng et al., "An investigation on the robustness, accuracy and simulation performance of a physics-based deep-submicronmeter BSIM model for analog/digital circuit simulation", *CICC'96*, pp. 321-324, May 1996.

[9.18] W. Liu et al., *BSIM3 version 3.2 User's Manual*, University of California, Berkeley.

[9.19] *BSIMpro Manual*, BTA Inc., Santa Clara, CA (http://www.btat.com).

[9.20] G. S. Gildenblat, *VLSI Electronics: Microstructure Science*, p. 11, vol. 18, 1989.

CHAPTER 10

Non-quasi Static (NQS) Model

In Chapter 5, the charge or CV models are derived based on the quasi-static (QS) approximation. The QS approximation breaks down when the signal changes occur on a time scale comparable to the device transit time. As we will discuss in this chapter, an non-quasi-static (NQS) model is desirable in some mixed signal IC and radio frequency (RF) applications. In this chapter, we first show some examples of where the NQS model is needed. Then we give the details of the derivation and implementation of the NQS model in BSIM3v3. Some test results of the BSIM3 NQS model are also given along with some helpful hints.

10.1 The Necessity of Modeling NQS Effects

As VLSI ICs become more performance-driven, it is sometimes necessary to predict the device performance for operation near the device transit time. However, as discussed in Chapter 5, most models available in SPICE use the QS approximation [10.1]. In a QS model, the channel charge is assumed to be a unique function of the instantaneous biases: i.e. the charge has to respond a change in voltages with infinite speed. Thus, the finite charging time of the carriers in the inversion layer is ignored. In reality the carriers in the channel do not respond to the signal immediately, and thus, the channel charge is not a unique function of the instantaneous terminal voltages (quasi-static) but a function of the history of the voltages (non-quasi-static). This problem may become pronounced in the RF applications, or when V_{gs} is close to V_{th}, or when long channel devices coexist with deep submicron devices as in many mixed signal circuits. In these circuits, the input signals may have rise or fall times comparable to, or even smaller than, the channel transit time. For long channel devices the channel transit time is roughly inversely proportional to (V_{gs}-V_{th}) and proportional to L^2. Because the carriers in these devices cannot follow the changes of the applied signal, the QS models may give inaccurate or anomalous simulation results that cannot be used to guide circuit design. For example, the most common QS model, which uses 40/60 drain/source charge partitioning [10.2], results in an unrealistic large drain current spike during a fast turn-on as shown in Fig. 10.1.1 [10.3].

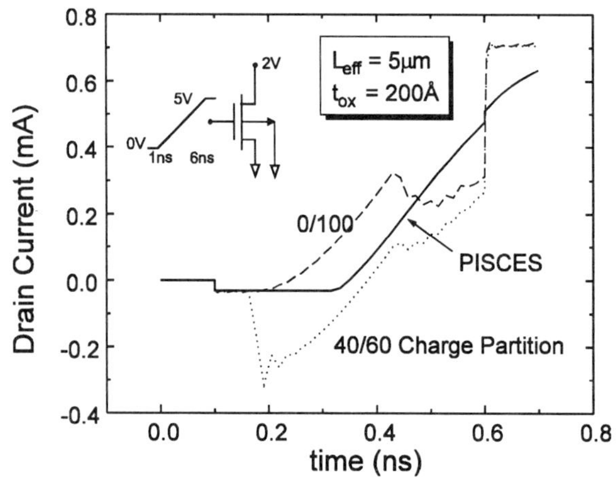

Fig. 10.1.1 NMOSFET drain current during a turn-on transient simulated by different QS models and PISCES 2-D simulation which does not use the QS assumption. After Chan et al. [10.3].

Besides affecting the accuracy of the simulation, the non-physical results can also cause oscillation and convergence problems in the numerical iterations. It is common among circuit designers to circumvent the convergence problem by using a 0/100 drain/source charge partitioning ratio [10.4], which attributes all transient charge to the source side. However, this non-physical solution merely shifts the current-spike problem to the source current as shown in Fig. 10.1.2, thus it only works when the source is grounded.

Moreover, none of these QS models can be used to accurately predict the high-frequency transadmittance of a MOSFET as pointed out in [10.5]. The result of modeling a 200μm long MOSFET in strong inversion saturation is completely different from two 100μm long MOSFET in series as shown in Fig. 10.1.3. It is common for circuit designers in high-frequency designs to break a long-channel MOSFET into N equal parts in series (N-lumped model) due to the lack of non-quasi-static models. The accuracy increases with N, at the expense of simulation time (of the order of 1.4N times longer [10.6]). However, this method becomes impractical when the device channel length is small because the short-channel effects in the sub-transistors may be activated.

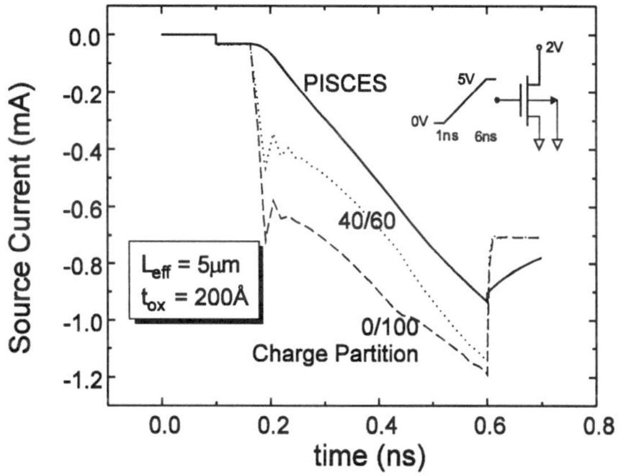

Fig. 10.1.2 NMOSFET source current during a turn-on transient simulated by the different QS models and PISCES 2-D simulation. After Chan et al. [10.3].

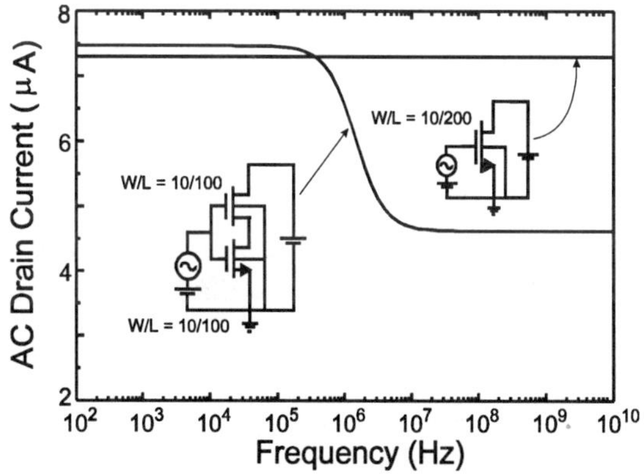

Fig. 10.1.3 AC drain current versus frequency for a 200μm long MOSFET. It is very different from that of two 100μm long devices in series, according to QS models. After Chan et al. [10.3].

It has been found that for RF applications the NQS model is necessary to fit the measured high frequency characteristics of devices with even short channel length where the operation frequency is above 1GHz [10.7, 10.8].

Therefore, a compact model that accounts for the NQS effect is highly desirable. Some non-quasi-static models based on solving the current continuity equation have been proposed [10.9, 10.10, 10.11]. They are complex and require long simulation times, making them unattractive for use in circuit simulation. In BSIM3, an NQS model based on the Elmore equivalent RC circuit is used [10.3]. It employs a physical relaxation time approach to account for the finite channel charging time. This NQS model is applicable for both large-signal transient and small signal AC analysis, as is discussed next.

10.2 The NQS Model in BSIM3v3

10.2.1 Physics basis and model derivation

As shown in Fig. 10.2.1(a), the channel of a MOSFET is analogous to a bias-dependent RC distributed transmission line [10.12]. In the QS approach, the gate capacitors are lumped to the intrinsic source and drain nodes (Fig. 10.2.1(b)). This ignores the fact that the charge build-up in the center portion of the channel does not follow a change in V_g as readily as at the source or drain edge of the channel. Breaking the transistor into N devices in series (Fig. 10.2.1(c)) gives a good approximation for the RC network but has the disadvantages mentioned in the previous section. A physical and efficient approach to model the NQS effect would be to formulate an estimate for the delay time through the channel RC network, and incorporate this time constant into the model equations.

One of the most widely used methods to approximate the RC delay was proposed by Elmore [10.13]. It is the mean, or the first moment, of the impulse response. Utilizing Elmore's approach, the RC distributed channel can be approximated by a simple RC equivalent which retains the lowest frequency pole of the original RC network. The new equivalent circuit is shown in Fig. 10.2.1(d). The Elmore resistance (R_{Elmore}) in strong inversion, calculated from the channel resistance, is given by [10.3]

$$R_{Elmore} \approx \frac{L_{eff}}{E_{LM}\mu W_{eff} Q_{ch}} \approx \frac{L_{eff}}{E_{LM}\mu W_{eff} C_{ox}(V_{gs} - V_{th})} \tag{10.2.1}$$

where Q_{ch} is the amount of channel inversion charge per area, and E_{LM} is the Elmore constant used to match the lowest frequency pole. The value of E_{LM} is found to be around 3 by matching the output of the equivalent circuits in Fig. 10.2.1 (a) and (d), and it is invariant with respect to W and L [10.3]. The time and frequency domain responses of the Elmore approximation, shown in Fig. 10.2.1 (d), and the original device with distributed channel, shown in Fig. 10.2.1 (a), are compared by SPICE simulation (Fig. 10.2.2 and Fig. 10.2.3). In the first case, a fast pulse is applied to the gate with both source and drain grounded and the gate current is measured. In the second case, a small signal voltage is applied to the gate and the resulting small signal voltages at different parts of the channel are measured. In both cases, a reasonable match between the equivalent circuit and the distributed RC network is observed.

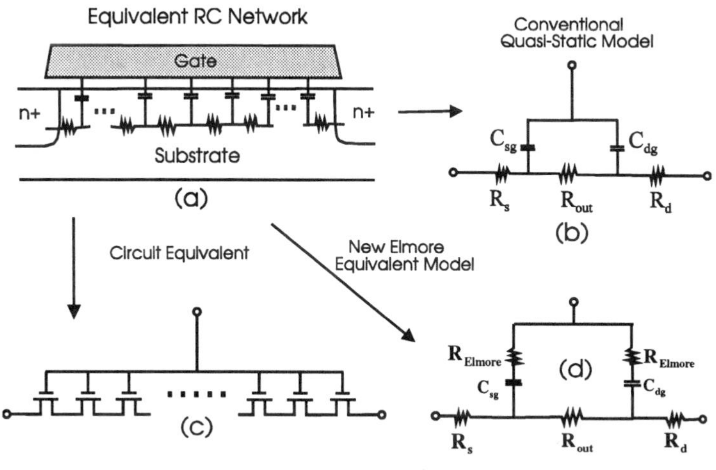

Fig. 10.2.1 Possible equivalent transient and ac small signal models for a MOS transistor. After Chan et al. [10.3].

However, direct implementation of the model shown in Fig. 10.2.1(d) requires the creation of two additional nodes, which increases the time to solve the Jacobian Matrix in SPICE by more than 70%. Also, the change in device topology may require modifications of the existing model equations. Therefore, a simpler way to incorporate the NQS effect is presented next.

The gate, drain, and source node currents can be described by the equation:

$$I_{G,D,S}(t) = I_{G,D,S}(t)\Big|_{DC} + G,D,S_{xpart}\frac{dQ_{ch}(t)}{dt} \tag{10.2.2}$$

where $I_{G,D,S}(t)$ are the gate, drain, and source currents, and $I_{G,D,S}(t)|_{DC}$ are the DC gate, drain, and source currents. $Q_{ch}(t)$ is the actual channel charge at a given time t, and G,D,S_{xpart} are the channel charge partitioning ratios [10.14, 10.15] for the gate, drain and source with

$$D_{xpart} + S_{xpart} = -G_{xpart} = 1 \tag{10.2.3}$$

In the 40/60 partitioning scheme, D_{xpart} varies from 0.5 at V_d=0V to 0.4 in the saturation region, and S_{xpart} varies from 0.5 to 0.6 respectively [10.1, 10.16]. However, because the 0.4/0.6 scheme covers a large voltage range and the error introduced by using a constant D_{xpart}=0.4 and S_{xpart}=0.6 is less than 5%, these values can be adopted to simplify the model.

In the QS approach, it is assumed that

$$\begin{aligned}\frac{dQ_{ch}(t)}{dt} &= \frac{dQ_{cheq}(t)}{dt} \\ &= \frac{dQ_{cheq}}{dV}\frac{dV}{dt}\end{aligned} \tag{10.2.4}$$

where $Q_{cheq}(t)$ is the equilibrium, or QS, channel charge under the instantaneous bias at a time t. The assumption of equilibrium at all times gives rise to the error in calculating the NQS currents. To account for the NQS current, a new state variable Q_{def} is introduced to keep track of the amount of deficit (or surplus) channel charge relative to the QS charge at a given time.

$$Q_{def}(t) = Q_{cheq}(t) - Q_{ch}(t) \tag{10.2.5}$$

and

$$\frac{dQ_{def}(t)}{dt} = \frac{dQ_{cheq}(t)}{dt} - \frac{dQ_{ch}(t)}{dt} \tag{10.2.6}$$

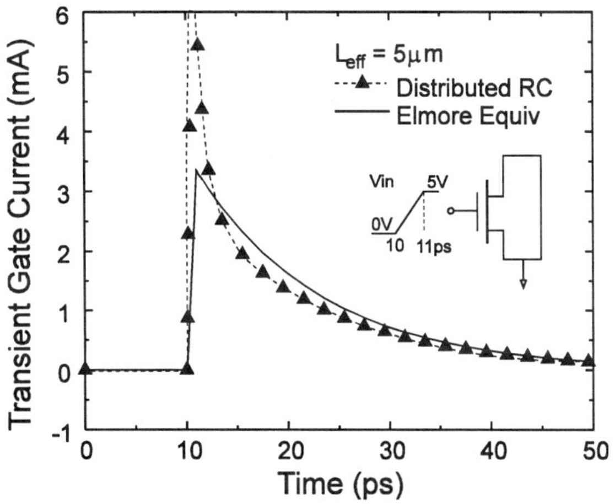

Fig. 10.2.2 Verification of the Elmore equivalent circuit in the time domain. It is a reasonable first order approximation to the RC network. After Chan et al. [10.3].

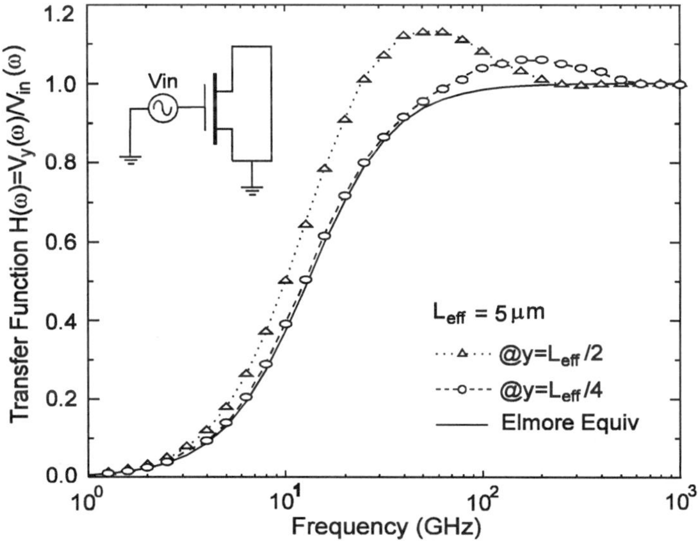

Fig.10.2.3 Verification of the Elmore equivalent circuit in the frequency domain, showing that good agreement is attained between the Elmore equivalent circuit and the distributed RC network. After Chan et al. [10.3].

Q_{def} is allowed to decay exponentially to zero after a step change in bias with a bias-dependent NQS relaxation time τ. Thus, the charging current can be approximated by

$$\frac{dQ_{ch}(t)}{dt} \approx \frac{Q_{def}(t)}{\tau} \tag{10.2.7}$$

$Q_{def}(t)$ can be calculated from Eq. (10.2.6) and a sub-circuit, shown in Fig. 10.2.4, has been introduced to obtain the solution. The subcircuit is a direct translation from Eq. (10.2.6). The node voltage gives the value of $Q_{def}(t)$. The total charging current is given by the current going through the resistor of value τ. With this approach, only one additional node is needed and the topology of the original transistor model is not affected.

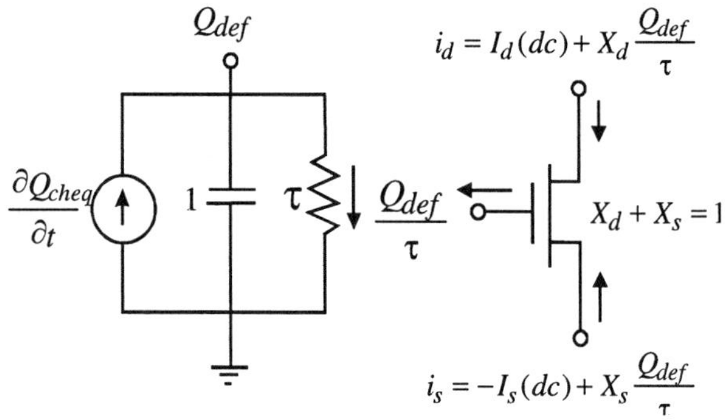

Fig. 10.2.4 BSIM3 implementation of the NQS model. A subcircuit is constructed to evaluate Q_{def}. The additional NQS currents calculated by the subcircuit are superimposed on the MOSFET DC currents. After Chan et al. [10.3].

The value of the channel relaxation time constant τ is composed of the terms related to the diffusion and drift currents (calculated from the RC Elmore equivalent circuit discussed above). The components of τ are given by

$$\tau_{diffusion} = \frac{q(L_{eff}/4)^2}{\mu kT} \tag{10.2.8}$$

$$\tau_{drift} = 0.5 R_{Elmore} C_{ox} W_{eff} L_{eff} \tag{10.2.9}$$

$$\frac{1}{\tau} = \frac{1}{\tau_{diffusion}} + \frac{1}{\tau_{drift}} \tag{10.2.10}$$

Fig. 10.2.5 compares the τ used in the new model with results obtained from a 2-D simulation under different biases. The model agrees with simulation results very well.

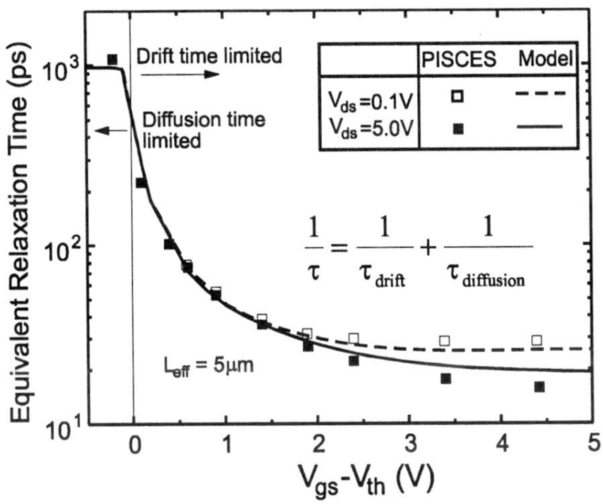

Fig. 10.2.5 Comparison between the relaxation time constant used in the NQS model and the value obtained from a 2-D simulation. After Chan et al. [10.3].

10.2.2 The BSIM3 NQS model

The NQS model discussed above has been implemented in BSIM3 [10.17, 10.18]. To improve the simulation performance and accuracy, this model is re-implemented in BSIM3v3.2 [10.19] with a new charge partitioning scheme that is physically consistent with that used in quasi-static capacitance models.

A model selector parameter, ***nqsMod***, is available for users to turn on the NQS model. In BSIM3v3.0 and BSIM3v3.1, ***nqsMod*** can be either an element (instance) or a model parameter. However, it is an instance parameter only in the official release of BSIM3v3.2 in Berkeley SPICE3, even though some simulator vendors still treat it as a model parameter in the implementations of their simulators. The NQS model is turned on when ***nqsMod*** =1.

In BSIM3v3.2, the capacitor C shown in Fig. 10.2.4 is multiplied by a scaling factor C_{fact} (with a typical value of $1\text{x}10^{-9}$) to improve the numerical accuracy of the computation. Fig. 10.2.6 gives the RC subcircuit of the NQS model implemented in BSIM3v3.2. Q_{def} now becomes

$$Q_{def}(t) = V_{def}(1 \cdot C_{fact}) \tag{10.2.11}$$

Eq. (10.2.10) gives the formula for τ. R_{elm} in Eq. (10.2.9) in strong inversion is calculated from the channel resistance.

$$R_{elm} = \frac{L_{eff}^{\,2}}{\boldsymbol{E_{LM}}\,\mu Q_{ch}} \approx \frac{L_{eff}^{\,2}}{\boldsymbol{E_{LM}}\,\mu Q_{cheq}} \tag{10.2.12a}$$

$$\mu = \mu_0 (T / \boldsymbol{T_{NOM}})^{\boldsymbol{U_{TE}}} \tag{10.2.12b}$$

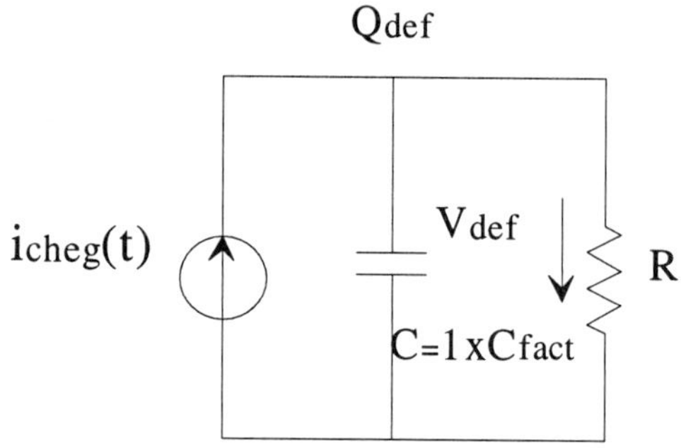

Fig. 10.2.6 The sub-circuit of NQS model implemented in BSIM3v3.2. After Liu et al. [10.19].

Note that the effective quasi-static (or equilibrium) channel charge Q_{cheq}(t), i.e the QS inversion channel charge Q_{inv} in Chapter 5, is used to approximate the actual channel charge Q_{ch}(t). The drift component of τ is formulated as

$$\tau_{drift} \approx R_{elm} C_{ox} W_{eff} L_{eff} \approx \frac{C_{ox} W_{eff} L_{eff}^{\,3}}{\boldsymbol{E_{LM}}\,\mu\, Q_{cheq}} \tag{10.2.13}$$

$\tau_{diffusion}$ is given by

$$\tau_{diffusion} = \frac{L_{eff}^{\,2}}{16 \mu v_t} \tag{10.2.14}$$

The terminal currents of D, G, and S are:

$$I_{D,G,S}(t) = I_{D,G,S}(DC) + \frac{\partial Q_{D,G,S}(t)}{\partial t} \tag{10.2.15}$$

Based on the relaxation time approach, the terminal charges and the corresponding charging currents can be formulated by

$$Q_{def}(t) = Q_{cheq}(t) - Q_{ch}(t) \tag{10.2.16}$$

$$\frac{\partial Q_{def}(t)}{dt} = \frac{\partial Q_{cheq}(t)}{dt} - \frac{\partial Q_{def}(t)}{\tau} \tag{10.2.17a}$$

$$\frac{\partial Q_{D,G,S}(t)}{dt} = D,G,S_{xpart}\frac{\partial Q_{def}(t)}{\tau} \tag{10.2.17b}$$

It is important for D_{xpart} and S_{xpart} to be consistent with the quasi-static charge partitioning factor $\boldsymbol{X_{PART}}$ and to be equal (D_{xpart} = S_{xpart}) at V_{ds}=0 when the transistor operation mode changes between the forward and reverse modes. Based on this consideration, D_{xpart} is formulated as

$$D_{xpart} = \frac{Q_D}{Q_D + Q_S} = \frac{Q_D}{Q_{cheq}} \tag{10.2.18}$$

D_{xpart} is now dependent on the bias conditions. The derivative of D_{xpart} can be obtained easily based on the quasi-static results:

$$\frac{dD_{xpart}}{dV_i} = \frac{S_{xpart}C_{di} - D_{xpart}C_{si}}{Q_{cheq}} \tag{10.2.19}$$

where i represents the four terminals (g,s,d,b), and C_{di} and C_{si} are the intrinsic capacitance calculated from the quasi-static analysis (see Chapter 5). The corresponding value of S_{xpart} can be derived from the fact that D_{xpart}+S_{xpart} =1.

To this point, the charge partitioning in strong inversion has been discussed. In the accumulation and depletion regions, the formula for D_{xpart} can be simplified. If $\boldsymbol{X_{PART}}$<0.5, D_{xpart}=0.4; if $\boldsymbol{X_{PART}}$ =0.5, D_{xpart}=0.5; if $\boldsymbol{X_{PART}}$>0.5, D_{xpart}=0 [10.19].

To derive the nodal conductance G_{tau}, note that $\tau = RC$. Then G_{tau} can be given by

$$G_{tau} = \frac{C_{fact}}{\tau} \tag{10.2.20}$$

τ is given by Eq. (10.2.10). Based on Eq. (10.2.17), the self-conductance due to NQS at the transistor node D can be derived as $\frac{dD_{xpart}}{dV_d} = G_{tau}V_{def} - D_{xpart}V_{def}\frac{dG_{tau}}{dV_d}$. The transconductance due to NQS at the node D is given by $D_{xpart}G_{tau}$. Other conductances needed to implement the NQS model can be also obtained in a similar manner [10.19].

10.3 Test Results of the NQS Model

The NQS model has been compared with 2-D simulations to verify the validity of the model [10.3]. Fig. 10.3.1 shows the simulated turn-on and turn-off transients simulated in the linear region (small V_d). Good agreement between the NQS model and 2-D simulation are observed in both cases. The simulation results in the saturation region (high V_d) are shown in Fig. 10.3.2. Again, very good agreement is achieved.

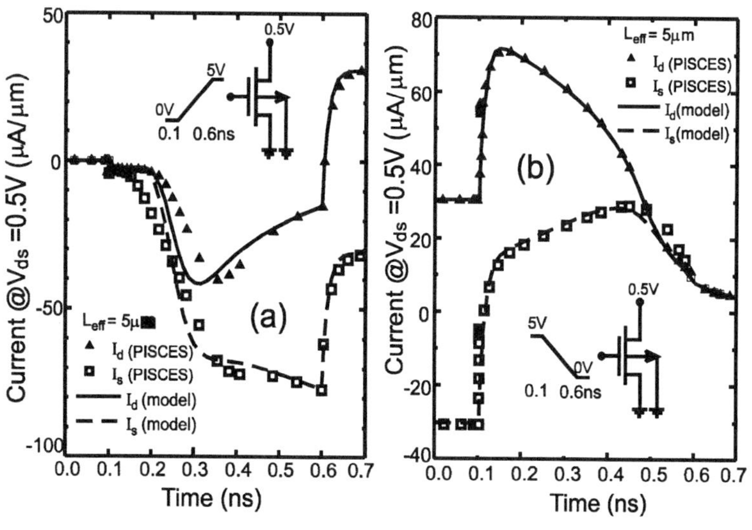

Fig. 10.3.1 Simulated turn-on/turn-off characteristics in the linear region. Excellent match between the model and 2-D simulation is observed. After Chan et al. [10.3].

Fig. 10.3.3 shows the high-frequency transadmittance test suggested in [10.5], where the real part of the transadmittance is plotted against the frequency of operation. The discrepancy between a single transistor and its equivalent N-lumped QS model is eliminated by the NQS model, which can predict the fall

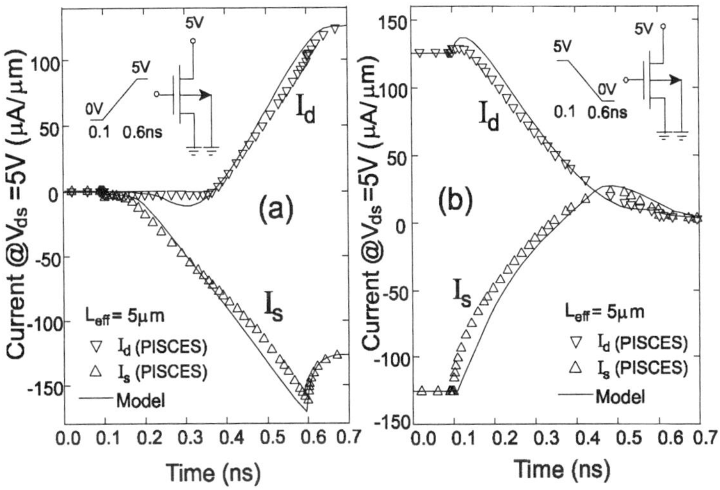

Fig. 10.3.2 Simulated turn-on/turn-off characteristics in the saturation region (high V_d). Very good agreement between the model and 2-D simulation is observed. After Chan et al. [10.3].

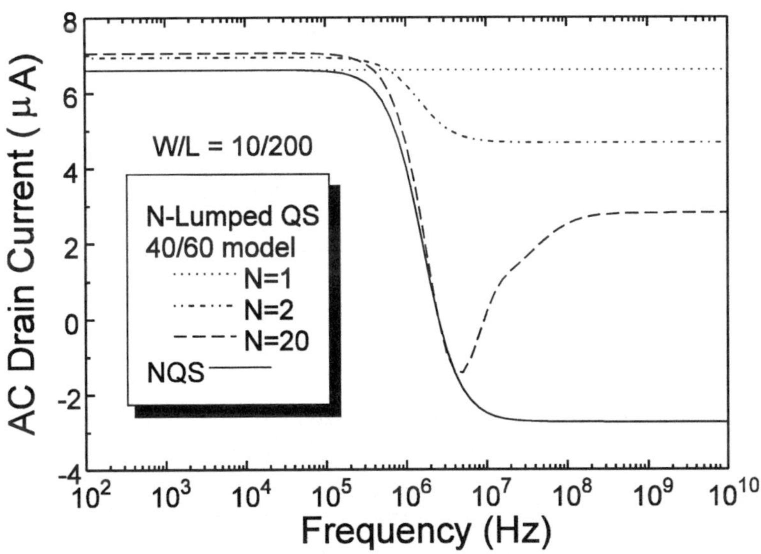

Fig. 10.3.3 Results of high-frequency transadmittance test. The NQS model is capable of predicting the transadmittance fall-off at high frequency. The N-lumped model asymptotically approaches the NQS model as N increases. After Chan et al. [10.3].

off of the transadmittance at high frequencies. It is also observed that the N-lumped model asymptotically approaches the NQS model as N increases.

Similar results can be observed in the magnitude and phase plots of a simple resistive-load inverter shown in Fig. 10.3.4.

As a practical example to illustrate the importance of the NQS effect in circuit design, a low voltage, high speed current output Digital/Analog converter (DAC) cell [10.20] is shown in Fig. 10.3.5. In this circuit, *M1* and *M1b* operate as current sources when they are turned on, and the output current appears as I_{out}, or I_{dump}. To obtain a high output resistance, *M1* must be a long channel device. Current switching was limited by the speed of voltage switching at node *1*. Fig. 10.3.6 shows the simulation results using the standard QS models and the NQS model. The NQS model indicates slower rise and settling times, limited by the NQS effect in the long channel *M1*. It illustrates an intrinsic limitation to the speed of this DAC circuit which is not apparent from simulation with the QS model. Table 10.3.1 compares the time required to simulate the DAC cell using different models. The overall simulation time penalty introduced by the NQS model is less than 30%.

Table 10.3.1:Time required for simulating the DAC cell with different model options.

	DC	0/100	40/60	NQS
Total time (sec)	1.777	1.863	1.906	2.243
# of iteration	732	729	781	848
# of time pt.	173	186	187	252
accepted time pt.	121	124	127	186

The high frequency (HF) small signal behaviors of the NQS model has also been studied through a comparison with 2-D device simulations [10.21]. The 2-D device simulator was used to determine the *I-V* characteristics, capacitances, and the Y parameters for a MOSEFT based on the doping profiles from a 0.5um CMOS process. In the study, comparisons with other models are also considered, such as (1) ***nqsMod***=0 (BSIM3v3 model without the NQS effect); (2) addition of an external gate resistance R_g with ***nqsMod***=0, and the value of R_g chosen to fit the phase of Y_{11}; and (3) a two-section lumped model with ***nqsMod***=0.

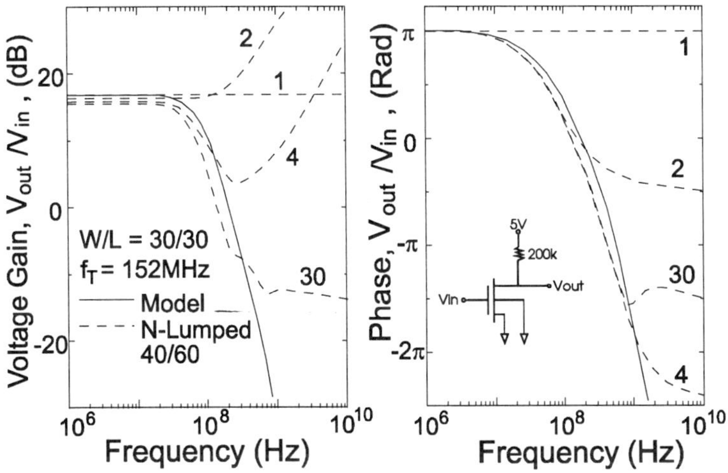

Fig. 10.3.4 Frequency response of an NMOS inverter with a resistive load simulated using the NQS model and the N-lumped model. The N-lumped model asymptotically approaches the NQS model as N increases. After Chan et al. [10.3].

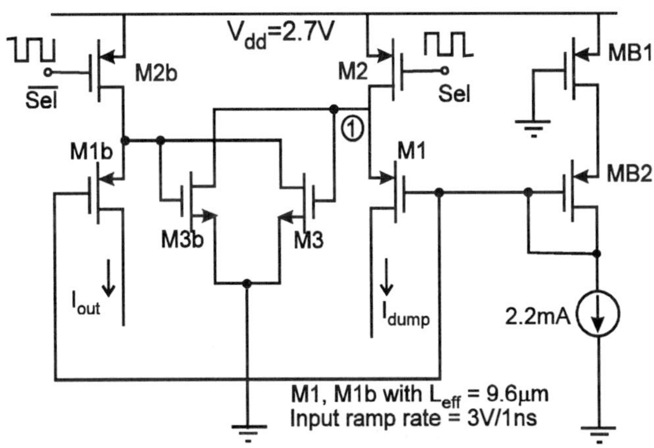

Fig. 10.3.5 A low voltage, high speed current output Digital/Analog Converter (DAC) cell is simulated with both the QS model and the NQS model. After Chan et al. [10.3].

Fig. 10.3.7 (a) and (b) show the magnitude and phase of Y_{11} as a function of frequency for W/L= 500μm/1.2μm at V_{gs}=0.8V and V_{ds}=3.0V. The results show that BSIM3v3 is reasonably well suited for small-signal analysis up to the frequency of 10GHz [10.21].

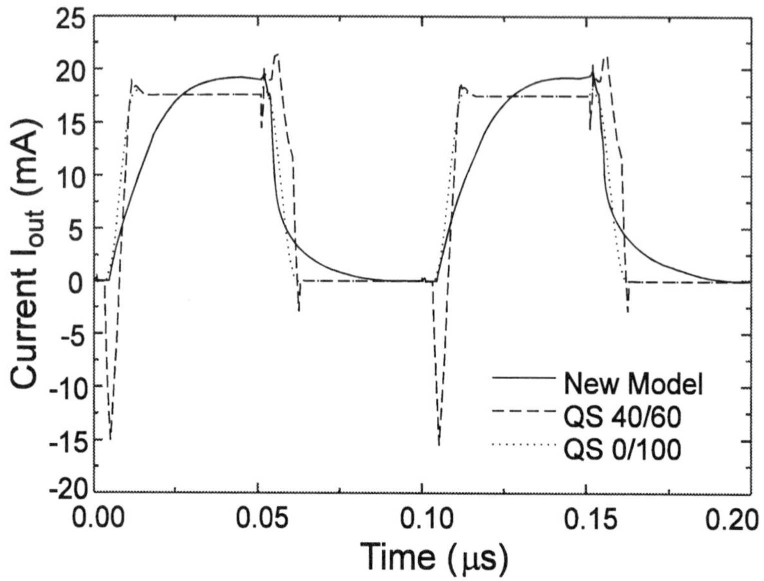

Fig. 10.3.6 Current output (I_{out}) of the DAC cell simulated by different models. The NQS model predicts a longer settling time compared with the QS models. After Chan et al. [10.3].

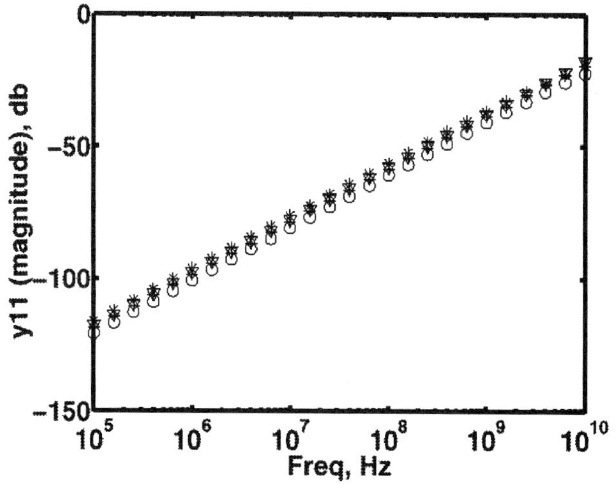

Fig. 10.3.7 (a) Magnitude of Y_{11} as a function of frequency for W/L=500μm/1.2μm at V_{gs}=0.8V and V_{ds}=3.0V with $\boldsymbol{E_{LM}}$=26 and R_g=2.5Ω. After Tin et al. [10.21].

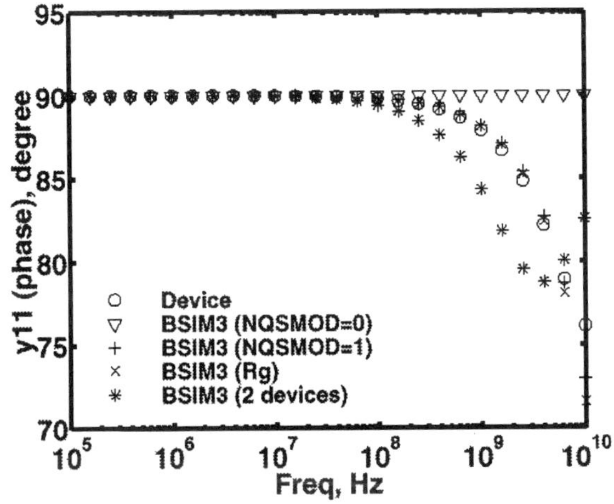

Fig. 10.3.7 (b) The phase of Y_{11} vs. frequency for W/L=500μm/1.2μm at V_{gs}=0.8V and V_{ds}=3.0V with E_{LM}=26 and R_g=2.5Ω. After Tin et al. [10.21].

10.4 Helpful Hints

1. NQS effects in short channel devices

In most practical cases, NQS effects are only important in circuits with long channel transistors driven by fast switching inputs. However, the NQS behavior has been recently observed even in short-channel devices [10.7, 10.8]. As the channel length of MOSFET's is reduced, the effect of velocity saturation cannot be neglected. Enhancement of the NQS model to include the effects of velocity saturation may be necessary.

When the MOSFET is operated in the velocity saturation regime the channel conductivity is reduced, increasing the value of τ as shown in Fig. 10.4.1. The error resulting from this effect when simulating a circuit using the NQS model is usually less than 20%. The error can be reduced by choosing an Elmore constant that provides a compromise between the linear and saturation regions. When a more accurate result is desired, an empirical model for the relaxation time given in Eq. (10.4.1) can be used [10.3]. The comparison between this model and 2-D simulation is shown in Fig. 10.4.1.

$$\tau'_{drift} = \begin{cases} \tau_{drift}\left(1+\frac{3}{8}\left(\frac{V_{ds}}{V_{dsat}}\right)^2\right) & ;\ \text{for } V_{ds} \le V_{dsat} \\ 1.375\ \tau_{drift} & ;\ \text{for } V_{ds} > V_{dsat} \end{cases} \qquad (10.4.1)$$

2. Special limits on the total drain current

The effects of the channel electric field in the current saturation region deserve special consideration. At high drain voltages, the electric field near the drain/ channel junction prevents a net carrier flow (dc current plus capacitive current) from the drain into the channel even during a fast turn-on. In this case, all the channel charge comes from the source, and the net drain current is never negative for NMOSFET's (positive for PMOSFET's) as seen in Fig. 10.3.2. This fact is modeled by forcing the drain current to be positive when the drain voltage is larger than the saturation drain voltage [10.3]. That is:

$$I_d \ge 0 \quad \text{for} \quad V_d \ge V_{dsat} \qquad (10.4.2)$$

During turn-off, another restriction is imposed due to the fact that the maximum drain current is limited by the number of carriers controlled by the drain and the maximum velocity with which they can move. Therefore the drain current must satisfy

$$I_d \le W_{eff} v_{sat} (Q_{cheq} - X_d Q_{def}) \qquad (10.4.3)$$

After incorporating the enhancements given by Eqs. (10.4.1)-(10.4.3) in the model, the simulation results in Fig. 10.4.2 shows a nearly perfect fit to 2-D simulation [10.3]. Comparison of Fig. 10.4.2 and Fig. 10.3.2 clearly demonstrates the benefit of adding the current limit.

3. Bulk charge in the NQS model

As shown in Fig. 10.2.4, the present BSIM3 NQS model adopts a simplifying approximation that considers only the charges at the source, drain, and gate, but assumes the bulk charging current to be zero. In other words, it ignores any bulk-charge dependence on biases [10.22]. This approximation is used to simplify the model implementation, and hence to improve the simulation efficiency. The body current can be included by partitioning Q_{def} between the gate and the body [10.3, 10.22]. For most applications the NQS effect from the bulk charge can be ignored, and it has little impact on small signal simulation. However, an NQS model that includes bulk charge would be desirable if it is physics-based and does not excessively increase the simulation time.

4. The approximation used in the present NQS model

In addition to ignoring the bulk charge, two other approximations are used in the NQS model. One uses C_{ox} instead of the sum of C_{dg} and C_{sg}, which are bias dependent, in Eq. (10.2.9) and Eq. (10.2.12) to reduce the complexity of calculating the derivatives of the capacitances. The other assumes $Q_{def}=Q_{ch}$ in the R_{elm} calculation to simplify the model implementation and reduce simulation time.

Examining the first approximation, we see that (C_{sg}+C_{dg}) varies between C_{ox} for small V_d and 0.75C_{ox} for V_{ds}>V_{dsat} [10.16]. So, the worst-case error shows up in the saturation region. To reduce this error, the Elmore constant ***ELM*** can be chosen to compromise between the linear and the saturation regions, as discussed in section 10.1. For example, the default value of ***ELM*** in the BSIM3v3 model is 5 instead of 3 although 3 is appropriate for the linear region.

The validity of the second approximation depends on the signal frequency applied to the devices. It works well as long as the rise time of the signal is slower than 5/f_T, where f_T is the cutoff frequency of the device. This should be sufficient for most current CMOS circuits. This approximation, however, may need to be re-examined before using this NQS model in RF applications operating close to f_T. Still, Fig. 10.3.7 shows good accuracy up to very high frequencies.

5. The necessity of introducing a higher order NQS model

The present BSIM3v3 NQS model can be used for fast transient and AC small signal analysis. It has been verified extensively with 2-D device simulation. Further verification is needed with measured data from devices and circuits, especially those for RF applications. The present BSIM3v3 NQS model is a first order approximation of the distributed RC network. Whether a higher-order NQS model is necessary depends on how well the present model works in real applications. Nevertheless, new NQS models with improved accuracy and simulation time would be welcome contributions to compact modeling because the prediction of device behavior near the cut-off frequency may become more important.

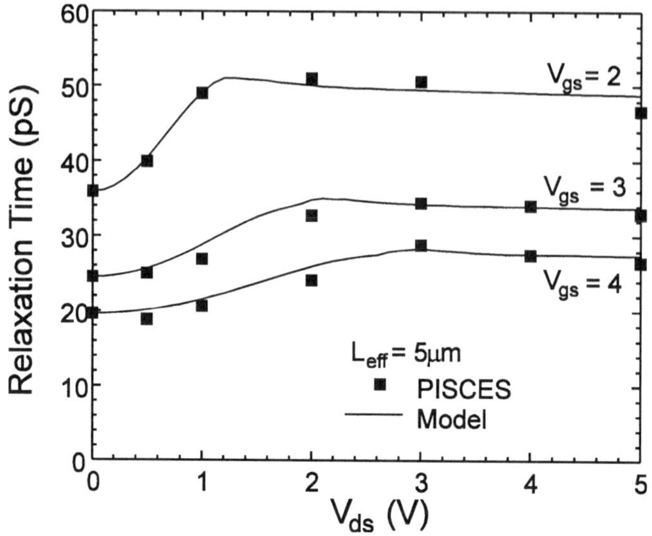

Fig. 10.4.1 Relaxation time constant as a function of drain bias with velocity saturation effects included. Eq. (10.4.1) is superimposed for comparison. After Chan et al. [10.3].

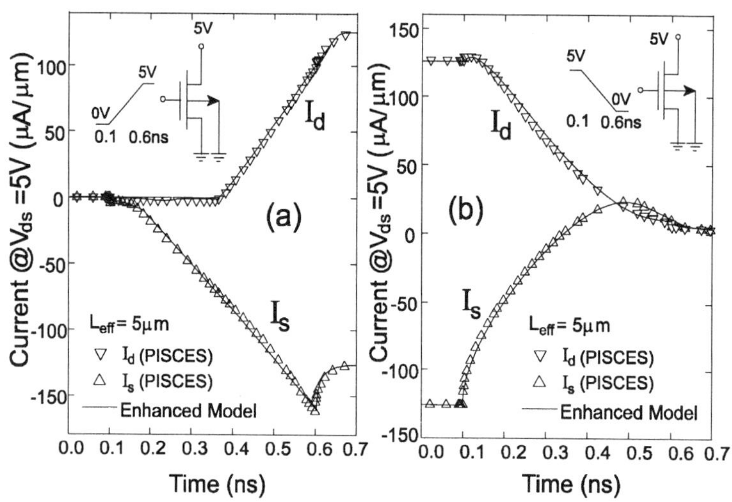

Fig. 10.4.2 Simulated current characteristics in the saturation region (high V_d) after current limits are included. Significant improvement to the model accuracy is observed. After Chan et al. [10.3].

6. The NQS model parameters

The model parameters of the NQS model are listed in Table 10.4.1.

Table 10.4.1 The Model parameters of the BSIM3 NQS model

Symbols in equation	Symbols in source code	Description	Default	Unit
nqsMod	nqsmod	NQS model selector	0	false
E_{LM}	elm	Elmore constant of the channel	5	none

References

[10.1] J. Paulous and D. A. Antoniadis, "Limitations of quasi-static capacitance models for the MOS transistors", *IEEE Electron Device Lett.*, vol. EDL-4, pp. 221-224, 1983.

[10.2] S. Y. Oh, D. E. Ward, R. W. Dutton, "Transient analysis of MOS transistors", *IEEE J. Solid-State Circuits*, vol. SC-15, no. 4, pp. 636-643, 1980.

[10.3] M. Chan et al. "A robust and physical BSIM3 non-quasi-static transient and AC small signal model for circuit simulation," *IEEE Trans. on Electron devices*, vol. ED-45, pp.834-841, 1998.

[10.4] P. Yang and P. K. Chatterjee, "SPICE modeling for small geometry MOSFET circuits", *IEEE Trans. Computer-Aided Des.*, vol. CAD-1, pp. 169-182, Oct. 1982.

[10.5] Y. P. Tsividis and G. Masetti, "Problems in the precision modeling of the MOS transistor for analog applications", *IEEE Trans. Computer-Aided Des.*, vol. CAD-3, pp. 72-79, Jan. 1984.

[10.6] T. L. Quarles, *SPICE 3 Implementation Guide*, Memorandum No. UCB/ ERL M89/42, April, 1989.

[10.7] R. Singh, A. Juge, R. Joly, and G. Morin, "An investigation into the non-quasi-static effects in MOS devices with an wafer S-parameter techniques, *Proc. IEEE Int. Conf. Microelectron Test Structures*, Barcelona, Mar. 1993.

[10.8] Y. Cheng et al., "RF modeling issues of deep-submicron MOSFETs for circuit design," *1998 International Conference of Solid-state and Integrated Circuit Technology*, pp.416-419, 1998.

[10.9] M. Bagheri, and Y. Tsividis, "A small signal dc-to-high frequency non-quasi-tatic model for the four-terminal MOSFET valid in all regions of operation", *IEEE Trans. Electron Devices*, vol. ED-32, no. 11, pp. 2383-2391, 1985.

[10.10] H. J. Park, P. K. Ko, and C. Hu, "A charge-conserving non-quasi-static MOSFET model for SPICE transient analysis", *IEDM 87 Technical Digest*, pp. 652-655, Dec. 1987.

[10.11] C. Turchetti, P. Mancini, and G. Masetti, "A CAD-oriented non-quasi-static approach for the transient analysis of MOS IC's", *IEEE Journal of Solid-State Circuits,* vol. SC-21, no. 5, pp. 827-836, 1986.

[10.12] Y. P. Tsividis, *Operation and Modeling of the MOS Transistor.* New York: McGraw-Hill, 1987.

[10.13] W. C. Elmore, "The transient response of damped linear networks with particular regard to wideband amplifiers", *J. Appl. Phys.*, vol. 19, no. 1, pp.55-63, 1948.

[10.14] J. G. Fossum, H. Jeong, and S. Veeraraghavan, "Significance of the channel-charge partition in the transient MOSFET model", *IEEE Trans. Electron Devices*, vol. ED-33, pp. 1621-1623, Oct. 1986.

[10.15] M. F. Sevat, "On the channel charge division in MOSFET modeling", *ICCAD Tech. Dig.*, Nov. 1987, pp. 208-210.

[10.16] B. J. Sheu and P. K. Ko, "Measurement and modeling of short-channel MOS transistor gate capacitances," *IEEE J. Solid-state Circuits,*. vol. SC-22, pp. 464-472, 1987.

[10.17] Y. Cheng et al., *BSIM3 version 3.0 User's Manual,* University of California, Berkeley, 1995.

[10.18] Y. Cheng et al., *BSIM3 version 3.1 User's Manual,* University of California, Berkeley, Memorandum No. UCB/ERL M97/2, 1997.

[10.19] W. Liu et al. *BSIM3 version 3.2 User's Manual, University of California, Berkeley, 1998.*

[10.20] T. Miki, Y. Nakamura, Y. Nishikawa, K. Okada, and Y. Horiba, "A 10bit 50MS/s CMOS D/A Converter with 2.7V power supply", *1992 Symp. on VLSI Circuits Dig. of Tech. Papers*, pp.92-93, 1992.

[10.21] S. F. Tin et al., "BSIM3 MOSFET model accuracy for RF circuit simulation," *Proceedings of RAWCON'98*, pp.351-354, 1998.

[10.22] W. Liu et al., "A CAD-compatible non-quasi-static MOSFET model," *IEDM Tech. Dig.*, pp. 151-154, 1996.

CHAPTER 11

BSIM3v3 Model Implementation

The importance of a MOSFET model with good accuracy, scalability, robustness, and simulation performance has been acknowledged by both circuit designers and device model developers [11.1,11.2]. It has been known that the discontinuity of model equations can result in non-convergence problems in circuit simulation [11.3]. Many model developers have been working on the improvement of the model equation continuity [11.4-11.12]. In addition, model implementation is a critical part of model development, and the robustness of model implementation is as important as the continuity of the model equations to ensure efficient circuit simulations. The enhancement of the continuity and smoothness of the BSIM3v3 model equations has been discussed in Chapter 4. In this chapter, we will discuss the robustness of the model achieved through careful model implementation.

11.1 General Structure of BSIM3v3 Model Implementation

Before analyzing the specific considerations of model implementation, we first give an introduction to the structure of implementation of BSIM3v3 in a typical circuit simulator such as SPICE3 [11.13]. The implementation of the BSIM3v3 model is realized with 21 different files that can be divided into the following five parts according to their functions: (1) data structures; (2) input routines; (3) output routines; (4) structure decomposition routines; and (5) processing routines.

1. Data structures

The BSIM3v3 model is described by a data structure that contains pointers to functions, which provide the specific operations of the BSIM3v3 model, and tables that describe the parameters of the BSIM3v3 model. This structure also contains other pointers to a variety of tables and size data that are needed by the model at the user-interface level and by high-level SPICE routines. BSIM3v3 requires two specific internal data structures, one for the global device model, the other for the instance parameters (for defining the individual devices in a circuit). Data placed in the model data structure is static. This data structure contains only the data that is universal to all the devices (of this type). Data placed in the instance data structure is also static and examined only by the code implementing the model. In addition to the basic model and instance data structures, there are several other data structures that must be defined to complete the description of the BSIM3v3 model to the higher levels of the simulator. For example, the BSIM3instSize field should be initialized to the size of the instance data structure; the BSIM3modSize field must be initialized to the size of the model data structure; static and initialized arrays should be defined. They describe the acceptable parameters and queries for the instances and models to specify parameters that are input parameters, output parameters, or input and output parameters. The definitions of these data structures and arrays can be found in files b3.c, bsim3itf.h, bsim3ext.h, and bsim3def.h.

2. Input routines

The input routines include two files called b3par.c and b3mpar.c. These routines are used by the front end to pass the input parameters to the device. The b3par.c file is for the function BSIM3Param that takes parameter values from the input parser and sets the appropriate field in the instance data structure of the device. The b3mpar.c file is for the function BSIM3mParam, which is very similar to the BSIM3Param function but provides values for model parameters instead of instance parameters.

3. Output routines

Output routines include two files called b3ask.c and b3mask.c. These routines are used by the simulator to obtain data from the BSIM3v3 model. In some sense, these routines are exactly the opposite of the two input routines described above.

4. Structure decomposition routines

Structure decomposition routines include three files called b3dest.c, b3del.c, and b3mdel.c. These routines are used to dismantle the data structures that have been built up. The b3dest.c file is for the function BSIM3destroy that follows the general pattern and loop through all instances and models to free all memory used by them. b3mdel.c is for the function BSIM3modDelete that is designed to delete BSIM3v3 from the circuit. This function is provided for future extensions and as such is never called by the present front end. The b3del.c file is for the function BSIM3delete that is used to delete the single specified instance from the circuit. This function is never called by the present front end.

5. Processing routines

Processing routines include ten files called b3set.c, b3temp.c, b3getic.c, b3check.c, b3ld.c, b3trunc.c, b3cvtest.c, b3acld.c, b3pzld.c, and b3noi.c.

B3set.c is for the function BSIM3setup that performs the first step of preparing BSIM3v3 for simulation. When this function is called, the devices are attached to the appropriate nodes and have most of their parameters set. At this point, space in the simulator state vector is reserved, incrementing it by the number of double precision values needed. All parameters defaulting to constant values are also set here.

b3temp.c is for the function BSIM3temp that completes the parameter preprocessing, such as binning parameter calculation and temperature setting to prepare BSIM3v3 for simulation at a certain temperature. All model and instance parameters should have their final default values assigned here. Every time a model or instance parameter or the circuit temperature is changed, this routine will be called.

b3check.c is for the function BSIM3checkModel, which examines whether the values of the parameters are invalid or unreasonable before they are loaded by the simulator for any calculation.

b3getic.c is for the function BSIM3getic that is used to convert node initial conditions to device initial conditions.

b3ld.c is for the function BSIM3load that is the most important function in the BSIM3v3 model implementation. This function is responsible for evaluating all instances at each iteration in the DC and transient analyses and for loading the Jacobian matrix and right hand side vector with the appropriate values.

b3trunc.c is for the function BSIM3trunc that is used to compute the truncation error for each device in the circuit. It reduces its timestep argument to the minimum of its previous value and the smallest timestep found for any of the instances it processes.

b3cvtest.c is for the function BSIM3convTest that performs the necessary convergence testing to determine whether each terminal current in each device has met the convergence requirements.

b3acld.c is for the function BSIM3acLoad that is a variation of bsim3Load. It is used when ac analysis is performed.

b3pzld.c is for the function BSIM3pzLoad, which is very similar to the BSIM3acLoad function, but evaluates the conductance at the complex frequency.

b3noi.c is for the function BSIM3noise that names and evaluates all of the noise sources.

Further information on the methodology and technique to implement a device model in SPICE3 can be found in [11.13]. The BSIM3v3 source code can be downloaded from the BSIM3 web site [11.14]. Next, we will discuss some practical and important issues that cannot be ignored in model implementation.

11.2 Robustness Consideration in the Implementation of BSIM3v3

The complete list of model parameters and equations can be found in Appendices A and B. Depending on the bias conditions, the device can work in different operating regimes. A good model should accurately describe the device behavior not only within each respective operation regime, but also ensure the accuracy and continuity of charge, current and their derivatives in all transition regions, e.g., between weak inversion and strong inversion regions, and between linear and saturation regions.

As we discussed in Chapter 4, BSIM3v3 employs unified channel charge and mobility expressions to achieve model continuity from weak inversion to strong inversion, and from the linear to saturation regions. This ensures continuity of the current, conductances, and capacitances in all operation regimes. [11.9, 11.10].

However, some discontinuity problems may still arise if the model implementation is not performed carefully though the model equation itself does not introduce discontinuities. This is due to the following reasons. First, the model includes many different physical effects, and some of them are present only in some bias regions (e. g. the polysilicon gate depletion model is valid only in the bias range at which the band bending in the polysilicon is less than 1.12V [11.11]). Specific care in the implementation is needed to ensure smooth transition of equations for these physical effects in different regions. Second, divide by zero, square root domain, or overflow/underflow problems may happen when certain "bad" parameter values are used or extreme bias conditions are encountered during the Newton Raphson iteration. Careful considerations are needed in the model implementation to avoid these problems. In general [11.2], all exponential and divisions must be limited, and hard limits of parameters must be avoided in the model implementation to make the model more robust in the simulation.

In the implementation of BSIM3v3.1, which was released in Dec. 1996 [11.12], all identified discontinuities due to implementation were eliminated. We will give some details next.

1. Limiting the exponential and divisions

Most model equations include some terms of the form $1/f=1/(1+Cx)$, where C is a constant and x is a variable. It is clear that $1/f$ will encounter divide-by-zero problem when x=-1/C. Generally, Cx is expected to be larger than -1 for a device model to maintain reasonable physical behavior. However, divide-by-zero may happen during the simulation if some "bad" values of model parameters or unusual, out-of-norm, operation bias conditions are encountered.

To avoid any potential problems, a function F in the form of $(1-\beta x)/(1-\alpha x)$ is introduced to replace f in the model implementation:

$$F = f = 1 + C\,x \quad x > x_0 \tag{11.2.1}$$

$$F = \frac{1-\beta x}{1-\alpha x} \qquad x \le x_o \qquad (11.2.2)$$

The expressions of α and β can be found by the conditions of F=f and dF/dx=df/dx at the boundary $x= x_o$ so that the continuity of F and its first derivative can be ensured by this implementation solution to avoid the potential divide-by-zero problem. As shown in Fig. 11.2.1, when $x>x_o$, F follows f exactly; when $x\le x_o$, which is far beyond the values of interest in the simulation, F approaches the function given in Eq. (11.2.2). It can be seen that the value of f is limited to β/α as $|x|$ increases, so that divide by zero and overflow problems can be avoided. The value x_o (x_o >-1/C) should be carefully selected according to some reasonable criteria. For example, this point cannot be too close to the -1/C point to avoid too sharp a transition of dF/dx at x_o. Also this point cannot be too far from -1/C; otherwise F would not be a good approximation of f.

The F function with the form discussed above has been used in the implementation of BSIM3v3.1 for the terms related to parameters such as D_{VT2}, D_{VT2W}, N_{FACTOR}, C_{DSC}, C_{DSCB}, and C_{DSCD} [11.12]. Another function similar to F is also introduced for the terms related to parameters A_{bulk0}, A_{bulk}, W_{eff}, P_{RWG}, P_{RWB}, K_{ETA}, U_A, U_B, U_C, P_{DIBLCB}, P_{VAG}, E_{TA0}, and E_{TAB} [11.12] based on the idea that all functions should not have overflow/underflow problem and their first derivatives must be continuous.

As an example, Fig. 11.3.2 shows the simulated I-V characteristics of BSIM3v3.0 (released in Oct. 1995 [11.11]) and BSIM3v3.1 for a parameter set with which BSIM3v3.0 had a discontinuity at a large V_{gs}. It can be seen that BSIM3v3.1 removes the discontinuity and gives a very smooth transition. It should be pointed out that a very unreasonable parameter set is used in the simulation in Fig. 11.2.2 to emphasize the problem, and x_o is set to -0.1/C (not the value used in the final implementation of BSIM3v3.1) so that the results of BSIM3v3.0 and BSIM3v3.1 deviate clearly at V_{gs}>2V. This deviation depends significantly on the selection of x_o. The point of the illustration is that, even though unreasonable bias voltages or model parameter values are used, there will be no discontinuity in BSIM3v3.1 and higher versions.

Similarly, bounds for all exponential terms have been set in the implementation to avoid any overflow/underflow problem.

2. Smoothing functions for parameters that have hard limits

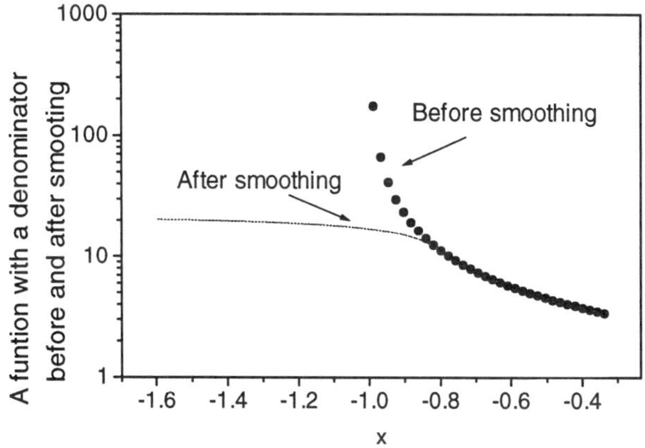

Fig. 11.2.1 A function with a denominator before and after introducing the value-limiting function. x_0=-0.8/*C*, x has been normalized by 1/*C* in this plot.

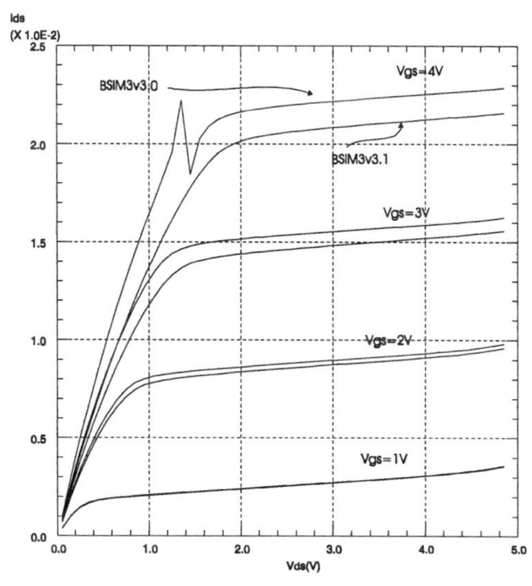

Fig. 11.2.2 I_d-V_d characteristics of BSIM3v3.0 and BSIM3v3.1. The discontinuity that is caused by the "bad" values of the mobility parameters in BSIM3v3.0 has been removed in BSIM3v3.1. After Cheng et al. [11.15].

There is another type of smoothing function used in BSIM3v3.1 to remove the discontinuity caused by the hard limits of some parameters. This type of smoothing function can be continuous to infinite order. As an example, the different implementations of the λ term in BSIM3v3.0 and BSIM3v3.1 are discussed here. The λ term has the following form:

$$\lambda = A_1 V_{gsteff} + A_2 \quad (11.2.3)$$

The implementation in BSIM3v3.0 set a hard limit to the value of λ at 1. This maximum value of λ is determined by the physical meaning of λ. But a piecewise hard limit would result in a discontinuity of the transconductance at the V_{gs} point at which λ=1. For some model parameter files this V_{gs} is less than V_{dd}, the power supply voltage, and the problem becomes obvious. In BSIM3v3.1, a smoothing function is introduced in the implementation for the λ term. λ is calculated as follows:

when A_1>0,

$$\lambda = 1 - \frac{T_1 + T_2}{2} \quad (11.2.4)$$

$$T_1 = 1 - A_2 - A_1 V_{gsteff} - \chi \quad (11.2.5)$$

$$T_2 = \sqrt{T_1^2 + 4\chi(1 - A_2)} \quad (11.2.6)$$

where χ is $1\text{x}10^{-4}$.

when $A_1 \le 0$

$$\lambda = \frac{T_1 + T_2}{2} \quad (11.2.7)$$

$$T_1 = A_2 + A_1 V_{gsteff} - \chi \quad (11.2.8)$$

$$T_2 = \sqrt{T_1^2 + 4\chi A_2} \quad (11.2.9)$$

With the above implementation, λ is given by Eq. (11.2.3) when λ<1, as shown in Fig. 11.2.3, and approaches 1 smoothly as V_{gs} increases. The previous implementation caused a discontinuity for the derivative of λ versus V_{gs}. In Fig. 11.2.4, we show the g_m-V_{gs} characteristics produced by BSIM3v3.1 and BSIM3v3.0. It can be seen that the a discontinuity is produced by BSIM3v3.0, and has been removed in BSIM3v3.1. A similar smoothing function has been used for the terms related to the polysilicon gate depletion effect.

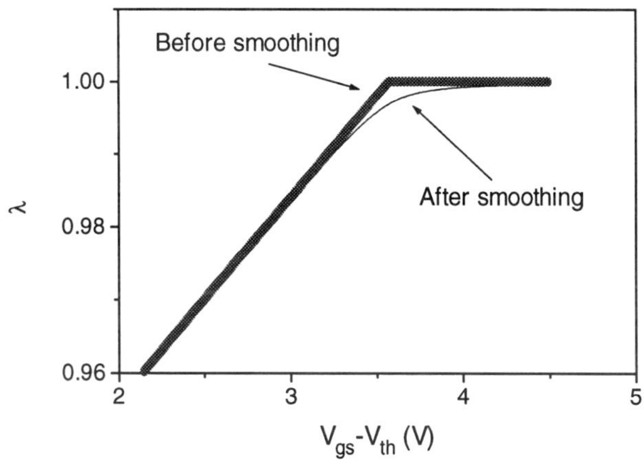

Fig. 11.2.3 (a) λ with and without smoothing.

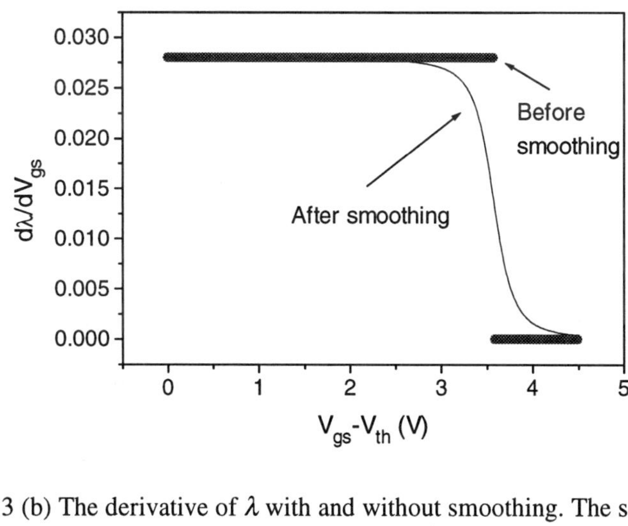

Fig. 11.2.3 (b) The derivative of λ with and without smoothing. The smoothing function can ensure the continuity of the derivatives of λ.

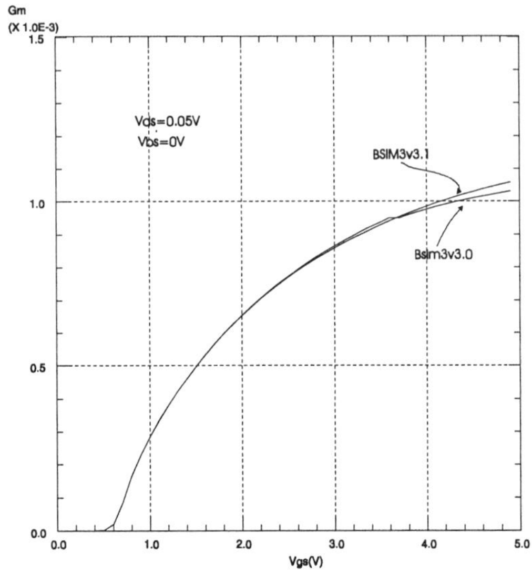

Fig.11.2.4 Simulated results of g_m-V_{gs} characteristics with BSIM3v3.0 and BSIM3v3.1. The discontinuity caused by the hard limit of λ in the implementation of BSIM3v3.0 has been removed in BSIM3v3.1. After Cheng et al. [11.15].

3. Parameter checking

It has been recognized that a good model should have the ability of outputting some warning or error messages when "bad" or unreasonable values of model parameters are used. This is convenient so that users can double-check their model parameters before the simulation starts. In BSIM3v3.1 such a function has been introduced so that the simulator, before doing the simulation, can check whether the values of the parameters are within the reasonable range specified by the model developer. Parameter checking is divided into three categories in BSIM3v3 implementation, "Fatal error abort", "Warning message and clamping", and "Warning message only". The following parameters are checked in BSIM3v3.1 and BSIM3v3.2: L_{eff}, L_{active}, W_{eff}, W_{active}, N_{LX}, N_{CH}, V_{BSC}, T_{OX}, D_{VT0}, D_{VT0W}, D_{VT1}, D_{VT1W}, W_0, N_{FACTOR}, C_{DSC}, C_{DSCD}, E_{T0}, B_1, U_0, D_{ELTA}, A_1, A_2, R_{DSW}, $V_{SATTEMP}$, P_{CLM}, P_{DIBLC1}, P_{DIBLC2}, C_{LC}, P_{SCBE2}, M_{OIN}, A_{CDE}, N_{OFF}, I_{JTH}, T_{OXM} [11.11]. For certain parameters such as oxide thickness T_{OX} and junction depth X_J, the simulator outputs "fatal error" and quits the simulation if the users input any parameters outside the specified range, as listed in Table 11.2.1. For other parameters such as A_1, A_2, etc., the simulator outputs a "warning" message if some unsuitable parameters are detected, and also the simulator sets clamping values for those parameters, as shown in Table 11.2.2.

Table 11.2.1. Conditions of fatal errors for some parameters

Conditions	Temperature dependence	Size Dependence
N_{LX}<-L_{eff}	No	Yes (all)
$T_{OX}\leq 0$	No	No
$T_{OXM}\leq 0$	No	No
$N_{CH}\leq 0$	No	Yes
$X_J\leq 0$	No	Yes
$D_{VT1}<0$	No	Yes
$D_{VT1W}<0$	No	Yes
$W_0=-W_{eff}$(DC)	No	Yes (all)
$D_{SUB}<0$	No	Yes
$B_1=-W_{eff}$(DC)	No	Yes (all)
$U_{0temp}\leq 0$	Yes	Yes
$D_{ELTA}<0$	No	Yes
$V_{SATtemp}\leq 0$	Yes	Yes
$P_{CLM}\leq 0$	No	Yes
$D_{ROUT}<0$	No	Yes
$W_{eff}\leq 0$	No	Yes
$L_{eff}\leq 0$	No	Yes
$N_{SUB}\leq 0$	No	Yes
$N_{GATE}<0$	No	Yes
$N_{GATE}>1\text{x}10^{25}\text{cm}^{-3}$	No	Yes
$C_{LC}<0$	No	Yes
$I_{JTH}<0$	No	No

Table 11.2.2. Conditions of warning messages with clamping of some parameters

Conditions	Clamping	Temperature Dependence	Size Dependence	Bypass
$A_2<0.01$	$A_2=0.01$	No	Yes	Yes
$A_2>1$	$A_2=1, A_1=0$	No	Yes	Yes
$R_{DSW}<0$	$R_{DSW}=0$, $R_{ds}=0$	Yes	Yes	Yes
$0<R_{ds}<0.001$	$R_{ds}=0$	Yes	Yes	Yes
$C_{GDO}<0$	$C_{GDO}=0$	No	No	Yes
$C_{GSO}<0$	$C_{GSO}=0$	No	No	Yes
$C_{GBO}<0$	$C_{GBO}=0$	No	No	Yes
$P_B<0.1$	$P_B=0.1$	No	No	No
$P_{BSW}<0.1$	$P_{BSW}=0.1$	No	No	No
$P_{BSWG}<0.1$	$P_{BSWG}=0.1$	No	No	No

Table 11.2.3 lists the parameters for which the model outputs warning messages only (no clamping) if they are in the value region shown in the table. The checking of some parameters in Table 11.2 and 11.2.3 can be turned off or on by setting the model parameter ***paramChk***=0 or 1.

Table 11.2.3 Conditions of warning message only for some parameters

Conditions	Temperature Dependence	Size Dependence	Bypass
N_{LX}<0	No	Yes	Yes
T_{OX}<1x10^{-9} m	No	No	Yes
N_{CH}≤1x10^{15}cm^{-3}	No	Yes	Yes
N_{CH} ≥1x10^{21}cm^{-3}	No	Yes	Yes
N_{SUB}≤1x10^{14}cm^{-3}	No	Yes	Yes
N_{SUB} ≥1x10^{21}cm^{-3}	No	Yes	Yes
|1x10^{-6}/ (W_{eff}(DC)+W_O)|>10	No	Yes (all)	Yes
D_{VT0}<0	No	Yes	Yes
|B_0/ (W_{eff}(DC)+B_1)|>10	No	Yes (all)	Yes
N_{FACTOR}<0	No	Yes	Yes
C_{DSC}<0	No	Yes	Yes
C_{DSCD}≤0	No	Yes	Yes
E_{TA0}<0	No	Yes	Yes
v_{SAT}<1x10^{3}m	Yes	Yes	Yes
L_{eff}<5x10^{-8}m	No	Yes	Yes
W_{eff}<1x10^{-7}m	No	Yes	Yes
P_{DIBLC1}<0	No	Yes	Yes
P_{DIBLc2}<0	No	Yes	Yes
N_{GATE}>0 but N_{GATE}≤1x10^{18}cm^{-3}	No	Yes	Yes
P_{SCBE2} ≤ 0	No	Yes	No
M_{OIN} <5.0 or M_{OIN} >25.0	No	Yes	No
A_{CDE} <0.4 or A_{CDE} >1.6	No	Yes	No
N_{OFF} <0.1 or N_{OFF}>4.0	No	Yes	No
V_{OFFCV}<-0.5 or V_{OFFCV}>0.5	No	Yes	No
P_D< W_{eff}'	No	No	No
P_S<W_{eff}'	No	No	No

11.3 Testing of Model Implementation

To check the general robustness and performance of BSIM3v3.1 in circuit simulation, two sets of benchmark tests have been performed. One set of tests was performed with both BSIM3v3.0 and BSIM3v3.1 to check the improvement of model implementation in BSIM3v3.1 and the closeness of BSIM3v3.1 to BSIM3v3.0, using the same parameter set. In BSIM3v3.0, no smoothing functions were used in the implementation, and no considerations were given for problems such as divide-by-zero. In BSIM3v3.1, all considerations discussed in section 11.2 have been incorporated. Another set of tests was performed with BSIM3v3 and previous BSIM models to compare the simulation efficiency and convergence performance. SPICE3f5 and SPECTRE were used in this comparison.

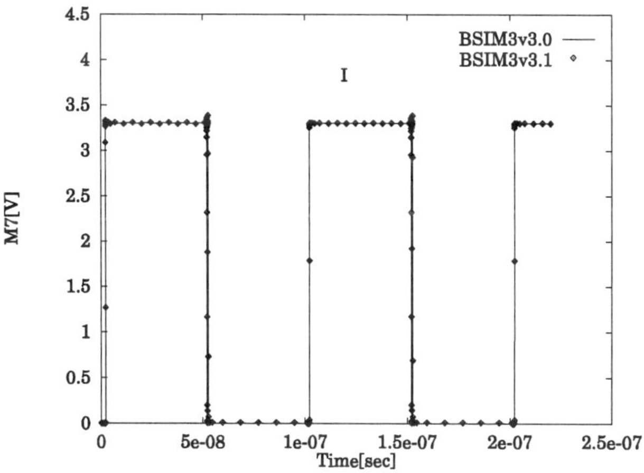

Fig. 11.3.1 Waveform of a 4 bit multiplier simulated with BSIM3v3.0 and BSIM3v3.1. The smoothing functions in BSIM3v3.1 did not alter the simulation results significantly.

The SPICE3f5 simulation results with BSIM3v3.1 and BSIM3v3.0 for two circuits are shown in Figs 11.3.1 and 11.3.2. The circuit simulated in Fig. 11.3.1 is a 4 bit multiplier. The circuit simulated in Fig 11.3.2 is a 204 stage ring oscillator. The parameter set used in the simulation is extracted from a commercial CMOS technology. It can be seen that curves from BSIM3v3.1 and BSIM3v3.0 coincide with each other, meaning that the smoothing functions did not alter the results significantly. The simulation performance of these two circuits with BSIM3v3.1 and BSIM3v3.0 is summarized in Table 11.3.1. Compared with the results from BSIM3v3.0, BSIM3v3.1 can reduce the iteration number and CPU time. This can be a significant advantage of BSIM3v3.1 (and higher versions) when simulating circuits that have converge problems.

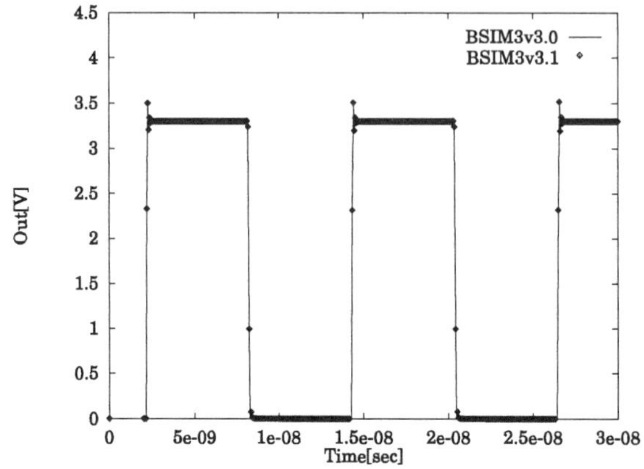

Fig. 11.3.2 Simulation results using BSIM3v3.0 and BSIM3v3.1 for a 204-stage ring oscillator.

The comparisons between BSIM3v3.1 and earlier BSIM models are also given in Table 11.3.1. The results simulated by BSIM1, BSIM2, BSIM3v3.0, and BSIM3v3.1 for 9 circuits are listed. According to these test results, the simulation time (and hence the time/iteration) of BSIM3v3.1 for some circuits may be longer than previous BSIM models because of the complex smoothing functions that may be used heavily during the iterations. However, the convergence performance has been improved since the iteration number is reduced. The iteration number is a very important measure for the quality of a model in the circuit simulation. This implies that BSIM3v3.1 may be superior in simulating more complex circuits.

Table 11.3.1 Simulation results of BSIM models with Spectre

Circuit	Number of MOSFET	Model	Number of Iteration	CPU Time (s)	Time/per iteration
Ring Oscillator	812	BSIM1	6237	1285	0.206
		BSIM2	6929	1738	0.205
		BSIM3v3.0	5237	1185	0.226
		BSIM3v3.1	5160	1108	0.214
ASIC-DRAM	1311	BSIM1	1087	218	0.201
		BSIM2	1469	326	0.257
		BSIM3v3.0	1038	281	0.271
		BSIM3v3.1	1032	244	0.236

Circuit	Number of MOSFET	Model	Number of Iteration	CPU Time (s)	Time/per iteration
4bit Multi-plier	782	BSIM1	1054	134	0.127
		BSIM2	1167	222	0.140
		BSIM3v3.0	835	159	0.190
		BSIM3v3.1	835	136	0.163
8bit Multi-plier	2492	BSIM1	1735	728	0.414
		BSIM2	1756	1212	0.466
		BSIM3v3.0	1234	627	0.508
		BSIM3v3.1	1226	582	0.475
16bit Multi-plier	8340	BSIM1	2942	4469	1.52
		BSIM2	3095	8081	1.60
		BSIM3v3.0	1851	3506	1.89
		BSIM3v3.1	1842	3050	1.65
24bit Multi-plier	17320	BSIM1	3825	14808	3.87
		BSIM2	3931	22478	3.92
		BSIM3v3.0	2626	10783	4.11
		BSIM3v3.1	2626	11175	4.26
32bit Multi-plier	29556	BSIM1	4773	38839	8.14
		BSIM2	5387	66230	8.66
		BSIM3v3.0	3319	28186	8.49
		BSIM3v3.1	3304	29488	8.92
SRAM1	31360	BSIM1	1794	5892	3.04
		BSIM2	2025	13096	3.10
		BSIM3v3.0	1757	7205	4.10
		BSIM3v3.1	1706	6884	4.04
SRAM2	51196	BSIM1	1816	8294	4.56
		BSIM2	2053	23988	5.26
		BSIM3v3.0	1774	11601	6.54
		BSIM3v3.1	1702	11159	6.56

11.4 Model Selectors of BSIM3v3

Several model selectors are introduced in BSIM3v3 for users to select options of the model. Besides ***Level*** (which is a model selector reserved for use by simulator vendors), 7 model selectors are introduced in BSIM3v3. They are ***mobMod, capMod, noiMod, nqsMod, Version, binUnit,*** and ***paramChk.***

mobMod is a model selector for the mobility model. When ***mobMod*** =1, the mobility model given in Eq. (4.3.5) is used. When ***mobMod*** =2, the mobility model given in Eq. (4.3.6) is used. When ***mobMod*** =3, the mobility model given in Eq. (4.3.7) is used.

Please note the different units of the parameter U_C in different mobility model options. The values of U_C can be significantly different in magnitude in ***mobMod*** =1 and 3, for example, as shown by the default values given in Appendix A. The default value of ***mobMod*** is 1.

capMod is a capacitance model selector. When ***capMod***=0, the BSIM1-like long channel capacitance model is used. When ***capMod***=1, the short channel capacitance model is used. When ***capMod***=2, the short channel capacitance model with $V_{gsteff,cv}$ and V_{cveff} (see Chapter 5) is selected. When ***capMod***=3, the short channel capacitance model with quantization effects is selected. The default value of ***capMod*** is 3 in BSIM3v3.2.

noiMod is a noise model selector. When ***noiMod*** =1, the SPICE2 flicker and thermal noise models are used. When ***noiMod***=2, the BSIM3 flicker noise and thermal noise models are used. When ***noiMod***=3, the BSIM3 flicker noise model and SPICE2 thermal noise model are used. When ***noiMod***=4, the SPICE2 flicker noise model and BSIM3v3 thermal noise model are used. The default value of ***noiMod*** is 1.

nqsMod is a selector for the NQS model. When ***nqsMod***=1, the NQS model is activated. The default value of ***nqsMod*** is 0.

Version is a model parameter introduced in BSIM3v3 for the convenience of simulator vendors to implement different versions of BSIM3v3, such as BSIM3v3.0, BSIM3v3.1, and BSIM3v3.2. When ***Version***=3.0, the BSIM3v3.0 version is selected. When ***Version***=3.1, the BSIM3v3.1 version is used. When ***Version***=3.2, the BSIM3v3.2 version is selected. The default value of ***Version*** is 3.2.

binUnit is a bin unit selector. In general, BSIM3v3 uses MKS units for most parameters. However, if the MKS unit is used for binning parameter calculation, the magnitude for L, W, and P dependent parameters varies greatly as will be shown later. Therefore, ***binUnit*** is introduced so that micrometer units can be used for W_{eff} and L_{eff} in the binning equations.

Appendix A gives a list of all BSIM3v3 model parameters which can and cannot be binned. All model parameters which can be binned are calculated with the following equation:

$$P_i = P_{I0} + P_{IL} / L_{eff} + P_{IW} / W_{eff}' + P_{IP} / (W_{eff}' L_{eff}) \tag{11.4.1}$$

where W_{eff}' is the effective channel width without bias dependence.

Let us take the parameters $\boldsymbol{K_1}$ and $\boldsymbol{K_2}$ as examples to understand the above equation. For the model parameter $\boldsymbol{K_1}$, P_{I0} will be $\boldsymbol{K_1}$, P_{IL} will be $\boldsymbol{L_{K1}}$, P_{IW} will be $\boldsymbol{W_{K1}}$, and P_{IP} will be $\boldsymbol{P_{K1}}$. The final value of $\boldsymbol{K_1}$ used in circuit simulation will be P_i calculated with Eq. (11.4.1). For the $\boldsymbol{K_2}$ case, P_{I0} will be $\boldsymbol{K_2}$, P_{IL} will be $\boldsymbol{L_{K2}}$, P_{IW} will be $\boldsymbol{W_{K2}}$, and P_{IP} will be $\boldsymbol{P_{K2}}$. The final value of $\boldsymbol{K_2}$ used in circuit simulation will be P_i calculated with Eq. (11.4.1).

Next we give an example to show how to use ***binUnit.*** If ***binUnit*** =1, the units of L_{eff} and W_{eff}' used in the binning equation above have the units of micrometers. Otherwise, they are in meters. For example, take a device with L_{eff}=0.5μm and W_{eff}' =10μm. If ***binUnit***= 1 and the values of the parameters $\boldsymbol{v_{SAT}}$, $\boldsymbol{Lv_{SAT}}$, $\boldsymbol{Wv_{SAT}}$, and $\boldsymbol{Pv_{SAT}}$ are 1x10^5, 1x10^4, 2x10^4, and 3x10^4 for saturation velocity $\boldsymbol{v_{SAT}}$, respectively. The final value of $\boldsymbol{v_{SAT}}$ used in the simulation is:

$$v_{SAT} = 1\text{x}10^5 + 1\text{x}10^4/0.5 + 2\text{x}10^4/10 + 3\text{x}10^4\ /(0.5\text{x}10) = 1.2810^5\ (\text{m/s})$$

However, to get the same final value of $\boldsymbol{v_{SAT}}$ by using the meter unit for W_{eff}' and L_{eff} (***binUnit*** = 0), the values of the parameters $\boldsymbol{v_{SAT}}$, $\boldsymbol{Lv_{SAT}}$, $\boldsymbol{Wv_{SAT}}$, and $\boldsymbol{Pv_{SAT}}$ should be 1x10^5, 1x10^{-2}, 2x10^{-2}, and 3x10^{-8}, respectively. Thus,

$$v_{SAT}=1\text{x}10^5+1\text{x}10^{-2}/(0.5\text{x}10^{-6})+2\text{x}10^{-2}/(10\text{x}10^{-6})+3\text{x}10^{-8}/(0.5\text{x}10^{-6}\text{x}10\text{x}10^{-6})=1.2810^5(\text{m/s})$$

The values for $\boldsymbol{Lv_{SAT}}$, $\boldsymbol{Wv_{SAT}}$, and $\boldsymbol{Pv_{SAT}}$ are significantly different, in the order of magnitude, from $\boldsymbol{v_{SAT}}$ when ***binUnit***=0, but they are of the same order when ***binUnit*** =1. The default value of ***binUnit*** is 1.

paramChk is a selector to determine whether the pre-checking of some parameters is to be performed. When ***paramChk*** is 0, the pre-checking of some parameters listed in Table 11.2.2 and 11.2.3 is bypassed. The default value of ***paramChk*** is 1.

11.5 Helpful Hints

1. V_{th} implementation

Depending on the model parameters provided in the model card by the user, the threshold voltage V_{th} may be calculated in different ways. If V_{THO}, K_1, and K_2 are given, the calculation of V_{th} follows Eq. (3.4.25). If V_{THO} or K_1 and K_2 are not given, the calculation of V_{th} becomes a little complex. If V_{THO} is not specified and K_1 is specified in the model parameters, V_{th} is calculated using the following:

$$V_{THO} = V_{FB} + \phi_s + K_{1OX}\sqrt{\phi_s} \tag{11.5.1}$$

where V_{FB}=-1.0 if not specified as an input model parameter, and

$$\phi_s = 2v_t \ln\left(\frac{N_{CH}}{n_i}\right) \tag{11.5.2}$$

Please note that v_t and n_i are calculated at T_{NOM}, as already discussed in Chapter 8.

If K_1 and K_2 are not given, but γ_1 and γ_2 are given, they are calculated using

$$K_1 = \gamma_2 - 2K_2\sqrt{\phi_s - V_{BM}} \tag{11.5.3}$$

$$K_2 = \frac{(\gamma_1 - \gamma_2)(\sqrt{\phi_s - V_{BX}} - \sqrt{\phi_s})}{2\sqrt{\phi_s}(\sqrt{\phi_s - V_{BM}} - \sqrt{\phi_s}) + V_{BM}} \tag{11.5.4}$$

where V_{BX} is the body bias at which the width of the depletion region equals the doping depth X_T in the channel. V_{BM} is the maximum applied body bias.

If γ_1 is not given, it is calculated using

$$\gamma_1 = \frac{\sqrt{2q\varepsilon_{si}N_{CH}}}{C_{ox}} \tag{11.5.5}$$

If γ_2 is not given, it is calculated using

$$\gamma_2 = \frac{\sqrt{2q\varepsilon_{si}N_{SUB}}}{C_{ox}} \tag{11.5.6}$$

if γ_1 is given, but N_{CH} is not given, N_{CH} is calculated from:

$$N_{CH} = \frac{\gamma_1^2 C_{ox}^2}{2q\varepsilon_{si}} \qquad (11.5.7)$$

If both γ_1 and N_{CH} are not given, N_{CH} defaults to $1.7\text{x}10^{23}\text{m}^{-3}$ and γ_1 is calculated from N_{CH}.

2. The default calculation of V_{BX} and V_{BI}

If V_{BX} is not given, it is calculated using

$$V_{BX} = \phi_s - \frac{qN_{CH}X_T^2}{2\varepsilon_{si}} \qquad (11.5.8)$$

where X_T is the effective doping depth in the channel with a default value of $1.55\text{x}10^{-7}$ m.

$$V_{BI} = v_{t0}\ln\left(\frac{N_{CH}N_{DS}}{n_i^2}\right) \qquad (11.5.9)$$

where $N_{DS}=1\text{x}10^{20}\text{cm}^{-3}$.

3. The difference between V_{FB} in *I-V* model and v_{fb} in *C-V* model

The expression for the flat-band voltage may be different in the *I-V* model and the *C-V* model in order to increase the model flexibility. Because the flat-band voltage in compact modeling, unlike the threshold voltage, is not directly extracted from the measured data, using different parameters for the flat-band voltage may help to improve the accuracy of both the *I-V* and *C-V* models.

In the *I-V* model, the parameter for the flat-band voltage is V_{FB}, which becomes a user input parameter in BSIM3v3.2. If it is not given, it is calculated using the following equation if V_{THO} is specified

$$V_{FB} = V_{THO} - \phi_s - K_{1OX}\sqrt{\phi_s} \qquad (11.5.10)$$

Otherwise, V_{FB} =-1.

In the *C-V* models of BSIM3v3, the parameter for the flat-band voltage is V_{FBCV} (***capMod***=0) or v_{fb} (***capMod***=1, 2, or 3). For ***capMod***=0, V_{FBCV} is a user-defined model parameter. When ***capMod*** is set to 1, 2, or 3, v_{fb} is calculated according to the following equations.

For ***capMod*** =2, if ***Version*** <3.2, v_{fb} is calculated from

$$v_{fb} = V_{th} - \phi_s - K_{1OX}\sqrt{\phi_s - V_{bseff}} \tag{11.5.11}$$

where V_{th} is calculated with Eq. (3.4.25) including the bias dependence.

For ***capMod***=1 or 3, or ***capMod*** =2 with ***Version***=3.2, v_{fb} is calculated by

$$v_{fb} = V_{th}' - \phi_s - K_{1OX}\sqrt{\phi_s - V_{bseff}} \tag{11.5.12}$$

where V_{th}' is calculated using the threshold voltage expression without the bias dependence:

$$\begin{aligned} V_{th}' = {} & V_{tho} + K_{1OX}\left(\sqrt{1+\frac{N_{LX}}{L_{eff}}}-1\right)\sqrt{\phi_s} + K_3\frac{T_{OX}}{W_{eff}' + W_0}\phi_s \\ & - D_{VT0}\left(\exp(-D_{VT1}\frac{L_{eff}}{2l_{t0}}) + 2\exp(-D_{VT1}\frac{L_{eff}}{l_{t0}})\right)(V_{BI} - \phi_s) \\ & - D_{VT0W}\left(\exp(-D_{VT1W}\frac{W_{eff}' L_{eff}}{2l_{tw0}}) + 2\exp(-D_{VT1W}\frac{W_{eff}' L_{eff}}{l_{tw0}})\right)(V_{BI} - \phi_s) \end{aligned} \tag{11.5.13}$$

where l_{t0} and l_{tw0} are given in Chapter 3.

4. The default calculation of overlap capacitance

If C_{GSO} is not given in the model card, but D_{LC} is given and D_{LC}>0, C_{GSO} is calculated with

$$C_{GSO} = D_{LC}C_{ox} - C_{GSL} \tag{11.5.14a}$$

If C_{GSO}, given in Eq. (11.5.14a), is negative,

$$C_{GSO} = 0 \tag{11.5.14b}$$

If C_{GSO} is not given in the model card and D_{LC} is not given or is negative,

$$C_{GSO} = 0.6\, X_J C_{ox} \tag{11.5.14c}$$

Similarly, if C_{GDO} is not given in the model card but D_{LC} is given and positive, C_{GDO} is calculated with

$$C_{GDO} = D_{LC} C_{ox} - C_{GDL} \tag{11.5.15a}$$

If C_{GDO}, given in Eq. (11.5.15a), is negative,

$$C_{GDO} = 0 \tag{11.5.15b}$$

If C_{GDO} is not given in the model card and D_{LC} is not given or is negative,

$$C_{GDO} = 0.6\, X_J C_{ox} \tag{11.5.15c}$$

If C_F is not given,

$$C_F = \frac{2\varepsilon_{ox}}{\pi} \ln\left(1 + \frac{4\times 10^{-7}}{T_{OX}}\right) \tag{11.5.16}$$

5. Understanding the T_{OXM} parameter

In BSIM3v3.2, the T_{OXM} parameter is introduced to account for the oxide thickness dependence of the threshold voltage while keeping the backward compatibility with previous versions of BSIM3v3. T_{OXM} is the gate oxide thickness at which the parameters are extracted as a nominal value of T_{OX}. This parameter is useful if users wish to use the BSIM3v3 model to predict the statistical behavior of circuits.

6. The units of some parameters in the simulation

Because of historical reasons, some users like to use the CGS system for the units of some parameters such as mobility and doping concentration. Unit conversion has been provided in the BSIM3v3 model implementation for some parameters such as the doping concentration in the channel and gate, and the mobility parameter μ_0. The model parameter N_{CH} and N_{GATE} can be entered either in m^{-3} or in cm^{-3}, and μ_0 can be given the values either in cm^2/Vs or m^2/Vs. Similarly for the corresponding binning parameters, they can also be entered in units composed of either m or cm. However, the parameter N_{SUB} must be entered in cm^{-3} unit.

7. The difference between W_{eff} and W_{eff}' in the I-V model

W_{eff} given in Eq. (4.8.3) is the complete equation of the effective channel width with bias dependence, while W_{eff}' given in Eq. (4.8.6) can be considered as the effective channel width without bias dependence. W_{eff}' is used in the calculation of the threshold voltage, R_{ds}, A_{bulk}, junction capacitance, etc. W_{eff} is used in the calculation of I-V (Eq. (4.6.2)), V_{dsat} (Eq. (4.6.10)), and V_{ASAT} (Eq. (4.6.4)).

8. The activation of the substrate current in the model

In the model implementation, no selector for the substrate current is used. The substrate current calculation is activated if the parameters $\boldsymbol{A_{LPHA0}}$ or $\boldsymbol{B_{ETA0}}$ have a value larger than zero given in the model card.

9 The activation of polysilicon gate depletion

Similarly, no model selector for polysilicon gate depletion is used in the BSIM3v3 implementation. The polysilicon gate depletion effect is activated when $\boldsymbol{N_{GATE}}$ with the value of larger than $1\text{x}10^{18}$ but less than $1\text{x}10^{25}\text{cm}^{-3}$ is given if V_{gs} is larger than $v_{fb} + \phi_s$. In that case, all V_{gs} in the model equations will be replaced by V_{gs_eff}:

$$V_{gs\_eff} = V_{FB} + \phi_s + \frac{q\varepsilon_{si} N_{GATE} T_{OX}^2}{\varepsilon_{ox}^2}\left(\sqrt{1 + \frac{2\varepsilon_{ox}^2 (V_{gs} - V_{FB} - \phi_s)}{q\varepsilon_{si} N_{GATE} T_{OX}^2}} - 1\right) \quad (11.5.17)$$

It should be pointed out that the V_{gs_eff} given in Eq. (11.5.17) is used in the calculation of the polysilicon gate depletion effect for both I-V and C-V models.

References

[11.1] Y. Tsividis and K. Suyama, "MOSFET modeling for analog circuit CAD: problems and prospects," *CICC Tech. Dig.*, pp. 14.1.1-14.1.6, 1993.

[11.2] *Compact Model Workshop*, Sunnyvale, CA, Aug., 1995.

[11.3] R. Kielkowski, *inside Spice*, McGraw-Hill, Inc. New York, 1994.

[11.4] N. D. Arora et al., "PCIM: a physically based continuous short-channel IGFET model for circuit simulation," *IEEE Trans. Electron Devices*, vol.ED-41, pp. 988-997, 1994.

[11.5] R. M. D. A. Velghe et al., "Compact MOS modeling for analog circuit simulation", *IEDM Tech. Dig.*, pp.485-488, Dec. 1993.

[11.6] C. C. Enz et al., "An analytical MOS transistor model valid in all regions of operation and dedicated to low voltage and low-current applications", *J. Analog Integrated Circuit and Signal Processing*, Vol. 8, pp.83-114, 1995.

[11.7] M. Shur, T. A. Fjeldly, T. Ytterdal, and K. Lee, "An unified MOSFET model," *Solid-State Electronics*, 35, pp.1795-1802, 1992.

[11.8] Y. Cheng et al., "A unified BSIM I-V mode for circuit simulation", *1995 International semiconductor devices research symposium*, Charlottesville, pp. 603-606, 1995.

[11.9] Y. Cheng et al., "An investigation on the robustness, accuracy and simulation performance of a physics-based deep-submicrometer BSIM model for analog/digital circuit simulation", CICC *Tech. Dig*, pp. 321-324, May 1996.

[11.10] Y. Cheng et al., "A physical and scalable BSIM3v3 I-V model for analog/ digital circuit simulation", *IEEE Trans. Electron Devices*, vol. 44, pp.277-287, Feb. 1997.

[11.11] Y. Cheng et al., *BSIM3 version 3.0 User's Manual*, University of California, Berkeley, 1995.

[11.12] Y. Cheng et al., *BSIM3 version 3.1 User's Manual*, University of California, Berkeley, Memorandum No. UCB/ERL M97/2, 1997.

[11.13] T. L. Quarles, *Adding device to SPICE3*, University of California, Berkeley, Memorandum No. UCB/ERL M89/45, 1989.

[11.14] http://www-device.eecs.berkeley.edu/~bsim3.

[11.15] Y. Cheng et al., *Compact Model Workshop*, Burlington, Vermont, Aug., 1996

CHAPTER 12

Model Testing

The general requirements for a MOSFET model include robustness, accuracy, and scalability [12.1-12.8]. In the past two decades, many MOSFET models have been developed by different companies and universities [12.9-12.17]. Recently, the need for a good standard model has been widely recognized. The Compact Model Council (CMC) [12.5] and the semiconductor industry in general are making efforts towards MOSFET model standardization [12.3-12.8]. In order to objectively study the model performance, CMC proposed a comprehensive set of tests to evaluate the models both qualitatively and quantitatively. The tests were compiled by experts in the modeling and circuit design fields [12.1, 12.2, 12.5]. These tests can identify certain problems and flaws of the models so that the users can be aware of the weaknesses of the models, and model developers can have guidance for future model development or improvement.

In this chapter, we discuss the benchmarking tests for qualifying a compact model for use in circuit design.

12.1 Requirements for a MOSFET Model in Circuit Simulation

Basically, a MOSFET model should meet the following criteria [12.1, 12.2, 12.5]:

(1) It should include most or all of the important physical effects in modern MOSFETs.

(2) The model should meet the requirements for accuracy and continuity of the I-V equations and charge conservation.

(3) It should give accurate values and ensure the continuity (with respect to any terminal voltage) of all small signal quantities such as transconductance g_m, g_{mb}, g_{ds} and all capacitances.

(4) It should ensure the continuity of g_m/I_d, an important quantity for analog circuit design, when V_{gs} is varied.

(5) It should give good results even when the device operates non-quasi-statically, or at least it should degrade gracefully for such operation, as frequency is increased.

(6) It should give accurate predictions for both thermal and $1/f$ noise in the triode and saturation regions.

(7) It should meet the above requirement over the weak, moderate and strong inversion regions, including $V_{bs}\neq 0$.

(8) It should meet all of the above requirements over the temperature range of interest.

(9) It should ensure the symmetry of model at $V_{ds}=0$ if the device itself is symmetric.

(10) It should pass the Gummel slope ratio test and the treetop curve test [12.5].

(11) It should do all of the above for any combination of channel width and length.

(12) One set of model parameters should be sufficient for all device channel lengths and widths.

(13) The model should provide warning or stop the simulation when the model is used outside its limits of validity.

(14) It should have as few parameters as possible, and those parameters should be linked as closely as possible to the device and process parameters.

(15) It should be conducive to an efficient parameter extraction method.

(16) It should be compact and computationally efficient.

12.2 Benchmark Tests

To examine the model behavior according to the above criteria, a series of benchmark tests have been suggested by the Compact Model Council [12.2,12.5]. They can be divided into two categories, qualitative and quantitative tests [12.2, 12.5]. The purpose of the qualitative tests is to check the general behavior of a model without comparison to the experimental data. The purpose of the quantitative tests is to check the accuracy and scalability of the model against measured data.

1. Qualitative tests:

The Compact Model Council suggested a set of qualitative tests for the *I-V* model [12.2, 12.5], but some of them are similar to each other, and some of them can be better classified as quantitative tests. Here, we list 15 qualitative tests with discussions of 8 of them.

(1) Triode-to-saturation characteristics (around V_{th}) of I_{ds} and g_{ds} [12.2,12.5]

This test checks the output characteristics of a model in the region around V_{th}, for devices with different *W/L* ratios such as wide/long and wide/short. The bias conditions are $V_{bs} = 0$ and $V_{bs}= -V_{dd}$, $V_{gs}= V_{th}-\Delta V$, V_{th} and $V_{th}+\Delta V$, where ΔV is a small voltage increment such as 0.15V. V_{ds} is swept from 0 to V_{dd} in 0.05 volt steps. Both I_{ds} and g_{ds} (output conductance) are plotted on both linear and logarithmic scales. A good model should show smooth transitions from the triode to saturation region. Negative g_{ds} should not be seen in any region, nor should there be any kinks, glitches, or discontinuities.

(2) Triode-to-saturation characteristics (in strong inversion) of I_{ds} and g_{ds}

This test checks the output characteristics of a model in the strong inversion region for devices with different *W/L* ratios such as wide/long and wide/short. The bias conditions are V_{bs}=0 and V_{bs}=-V_{dd}, with at least 3 values of V_{gs} equally spaced between some voltage (higher than V_{th}) and V_{dd}, and V_{ds} is swept from 0 to V_{dd} in 0.05 volt steps. Both I_{ds} and g_{ds} (output conductance) are plotted on linear and logarithmic scales. A good model should show smooth transitions from the triode to saturation region. Negative g_{ds} should not be seen in any region, nor should there be any kinks, glitches, or discontinuities.

(3) Strong inversion characteristics in a linear plot of I_d and g_m [12.2, 12.5]

This test checks strong inversion characteristics of a model for devices with different W/L ratios. The bias conditions are a fixed V_{ds} such as 0.1V and several V_{bs} such as 0, $-V_{dd}/2$, and $-V_{dd}$. V_{gs} is swept from 0 to V_{dd} in small steps such as 0.05V. Both I_d and g_m should be plotted on a linear coordinate. A good model will show smooth transition from below to above threshold with no kinks, glitches, or discontinuities.

(4) Subthreshold characteristics on a logarithmic scale, Log(I_d) and Log(g_m) [12.2, 12.5]

Similar to the above test, this test checks subthreshold characteristics of the model for devices with different W/L ratios. The test conditions are $V_{ds}=V_{dd}$, V_{sb}=0, $-V_{dd}/2$, and $-V_{dd}$, and V_{gs} swept from 0 to V_{dd} in small voltage steps. Both I_d and g_m should be plotted on a logarithmic coordinate. A good model will show smooth transition from below to above threshold with no kinks, glitches, or discontinuities.

(5) g_m/I_d characteristics [12.1, 12.2, 12.5]

This test plots the transconductance-current ratio, g_m/I_d (an important quantity for analog circuit design), versus V_{gs} or Log(I_d). A good model will show smooth transition from below to above threshold with no kinks, glitches, or discontinuities. The g_m/I_d peaks in the subthreshold region, but is not exactly constant in this region.

(6) Gummel symmetry test [12.2, 12.5]

This test is to check the symmetry of a model at V_{ds}=0. I_d must be an odd function of V_x, that is $I_d(V_x)=-I_d(-V_x)$ must hold if the device is symmetric. By plotting $g_o=dI_d/dV_x$, both on a large scale and on a fine grid about V_x=0, the symmetry can be examined. A model fails this test if there is discontinuity in the derivative of the curve of g_o at V_x=0. A model passes this test if g_o varies smoothly and continually from negative to positive V_x, with a slope of zero at V_x=0.

(7) Gummel slope ratio test [12.2, 12.5]

Asymptotic behavior for low V_{ds} operation can be well approximated by simple theoretical models. This leads to a simple and useful aid to evaluate whether a MOSFET model exhibits the correct behavior. The Gummel slope ratio test can be described in the following way. Consider two points, $I_1=I_d(V_{db}=V_1)$ and $I_2=I_d(V_{db}=V_2)$ for two small values V_1 and V_2 of V_{db} and $V_{sb}=0$.

Typically, these two points are the first two non-zero points on an output curve. The slope ratio S_r is defined as: $S_r=[(I_2+I_1)(V_2-V_1)]/[(I_2-I_1)(V_2+V_1)]$ and is the ratio of the slope of the line through the origin and the midpoint of the line connecting (V_1, I_1) and (V_2, I_2) to the slope of the line joining (V_1, I_1) and (V_2, I_2). For large V_{gb}, the transistor is biased in the triode, or linear, region, and I_d is close to linear in V_{db}, so S_r should approach unity. For small V_{gb} in subthreshold region, I_d is nearly proportional to $1\text{-}\exp(-V_{ds}/v_t)$, where v_t is the thermal voltage. S_r should thus reach an asymptote that is determined by the temperature and the values of V_1 and V_2. A model passes this test if it approaches the expected asymptotes. A model fails this test if it does not approach the asymptotes, and if it displays kinks and glitches in the S_r curve.

(8) Gummel treetop curve test [12.2, 12.5]

The subthreshold slope is an important parameter of a MOSFET, and lends itself for testing asymptotic behavior of MOSFET models. This test examines how a model agrees with the expected theoretical behavior. For a long channel MOSFET with a uniformly doped substrate, to a good approximation, g_m/I_d of a model should asymptotically reach a value that depends on V_{gb} in subthreshold operation. For large V_{gb} this value approaches $1/nv_t$, where v_t is the thermal voltage and $n>1$ is the subthreshold swing ideality factor. A model passes this test if it closely follows the treetop curve for subthreshold operation, and if it does not display any kinks or glitches.

(9) Additional qualitative tests:

Besides those discussed above, 7 additional qualitative tests are suggested.

(a) The V_{th}-L test. Examine whether the model can describe the short channel effects such as DIBL and V_{th} roll-off by checking the characteristics of threshold voltage versus device channel lengths.

(b) The V_{th}-W test. Examine whether the model can describe the narrow width effects by checking the characteristics of threshold voltage versus device channel widths.

(c) The I_{dsat}-L test. Examine whether the model can predict the saturation current, a very important parameter in circuit design and statistical modeling, by checking the characteristics of the saturation current versus the device channel lengths.

(d) Thermal noise test. Examine whether the model can predict the thermal noise characteristics correctly by simulating the thermal noise at a frequency low enough to avoid any influence from the capacitances [12.2].

(e) Flicker noise test. Test if the model can predict the flicker noise reasonably by checking the scalability of the noise model versus geometry at frequencies where the 1/f noise is dominant [12.2].

(f) High frequency AC test. Test if the model can describe the NQS effect by checking the AC frequency response of the drain current in a wide frequency range up to the GHz regime [12.2].

(g) Capacitance characteristics test. Examine if the capacitance model shows reasonable behavior by plotting the capacitance characteristics versus V_{gs} at a fixed V_{ds} and V_{bs}, and versus V_{ds} at a fixed V_{gs} and V_{bs}.

2. Quantitative tests:

In reality, all of the qualitative tests discussed above can be also used as quantitative tests as long as measurement data is available. Here, we list the basic quantitative tests for the I-V model:

(1) Triode-saturation characteristics of I_{ds} and g_{ds} versus V_{ds} (around V_{th})

(2) Triode-saturation characteristics of I_{ds} and g_{ds} versus V_{ds} in strong inversion

(3) Strong inversion characteristics of I_{ds} and g_m in both linear and saturation regions

(4) Subthreshold characteristics of Log(I_{ds}) and Log(g_m) in both linear and saturation regions

(5) g_m/I_{ds} characteristics at different V_{ds}, and different V_{bs}

(6) Characteristics of V_{th} versus channel length at different V_{bs}

(7) Characteristics of V_{th} versus channel width at different V_{bs}

(8) Saturation current I_{dsat} versus channel length at different V_{gs}

12.3 Benchmark Test Results

BSIM3v3 has been examined extensively with the above quantitative and qualitative tests by both model developers and users [12.18-12.22]. We show some test results here to further clarify the tests.

The MOSFETs used in the tests are from 4 different CMOS technologies with T_{ox} of 12.8nm, 11nm, 9nm, and 6.5 nm. The device geometry ranges from 0.25μm to 6μm for channel length, and from 0.6μm to 20μm for the width. The BSIMPro model parameter extractor [12.23] is used to extract the model parameters for the quantitative tests and to plot the results. Unless otherwise indicated in the figures, symbols represent measured data and solid lines represent the BSM3v3 model.

(1) Triode-to-saturation characteristics (around V_{th} point) of I_{ds} and g_{ds} [12.2, 12.5]

The device threshold voltage is 0.35V at V_{bs}=0V, and 0.54V at V_{bs}=-1V. Four different gate biases are used from 0.1-0.7 V (with 0.2V steps) to ensure that the device can work in the moderate inversion region at both V_{bs}=0 and -1V. Fig. 12.3.1 gives the g_{ds}-V_{ds} characteristics of the device with W_{drawn}/L_{drawn} of $6\mu m/0.25\mu m$ at V_{bs}=0V in linear scale. A good and smooth fit of g_{ds} can be seen.

To examine the model continuity, Figs. 12.3.2 and 12.3.3 show the I_{ds} -V_{ds} and g_{ds}-V_{ds} characteristics of the same device at V_{bs}=0V in logarithmic scale. The results show that the model can describe the current and conductance characteristics continuously and smoothly from the linear to saturation regions in the moderate inversion regime.

Furthermore, Fig. 12.3.4 gives the comparison of BSIM3v2 and BSIM3v3. It can be seen that BSIM3v3 has removed the negative conductance problems in BSIM3v2 [12.24, 12.25]. To test the model behavior with body bias, Figs 12.3.5 and 12.3.6 show the I_{ds} -V_{ds} and g_{ds}-V_{ds} characteristics of the same device at V_{bs}=-1V in logarithmic scale. Again, the model can match the measured data for both current and conductance, and is smooth from the linear to saturation regions.

(2) Triode-to-saturation characteristics (in strong inversion) of I_{ds} and g_{ds}

The modeled and measured I_{ds}-V_{ds} characteristics in the strong inversion region are shown in Fig. 12.3.7 for a device with W_{drawn}/L_{drawn}=10μm/0.4μm at different V_{gs} bias conditions. It can be seen in Fig. 12.3.7 that the model can fit the measured data well over the whole operation range, and the maximum error is 1.84%.

The characteristics of drain output conductance g_{ds} and resistance R_{out} are shown in Figs. 12.3.8 and 12.3.9 respectively for a device with W_{drawn} L_{drawn}=10μm/0.4μm. A very good and smooth fit of g_{ds} can be seen in Fig. 12.3.8. Furthermore, in Fig.12.3.9, the model can describe the R_{out} characteristics well in different gate bias conditions, which is a special feature of the BSIM3v3 model [12.24, 12.26].

(3) Strong inversion characteristics, I_d and g_m [12.2, 12.5]

Fig.12.3.10 shows the I_{ds}-V_{gs} characteristics of a device with W_{drawn} L_{drawn}=$20\mu m/0.4\mu m$ at V_{ds}=50mV and different body biases in linear scale. A good fit can be observed between the measured data and model results. The figure shows that the model can describe the current characteristics at different bias condition satisfactorily. Fig. 12.3.11 gives the g_m-V_{gs} characteristics of a device with W_{drawn}/L_{drawn}=$20\mu m/0.4\mu m$ at different V_{bs} conditions in linear scale. The model can match the measured data well.

(4) Subthreshold characteristics, Log(I_d) and Log(g_m) [12.2, 12.5]

Figs. 12.3.12 and 12.3.13 show the characteristics of I_{ds}-V_{gs} and g_m-V_{gs} of a device with W_{drawn}/L_{drawn}=20μm/0.4μm at V_{ds}=50mV and different body biases in logarithmic scale. It can be seen that the model can fit the measured data very well in the subthreshold regime and guarantee a continuous and smooth transition from the subthreshold to strong inversion regimes.

(5) g_m/I_d characteristics [12.1, 12.2, 12.5]

As discussed in [12.1], g_m/I_{ds} characteristic of the model is a very important aspect of a model used in analog circuit design. It can be seen from Figs. 12.3.14 and 12.3.15 that the model produces smooth g_m/I_{ds} characteristics and fit the data well at different V_{ds} and V_{bs} bias conditions. A further test of the g_m/I_{ds} characteristic of the model will be described in the Gummel slope ratio test.

(6) Gummel symmetry test [12.2, 12.5]

The results of the Gummel symmetry test are given in Figs. 12.3.16, 12.3.17, and 12.3.18 to show the model behavior clearly in different operation regimes. All of the simulations are performed using Berkeley SPICE3e2 with the netlist provided by the Compact Model Council [12.5]. Fig. 12.3.16 gives the test results in the subthreshold region, and shows that g_o of the model is symmetrical at V_{ds}=0. It is well known that the *I-V* characteristics of the model is also symmetric, but is not shown here. The test results of the model in strong inversion are given in Figs. 12.3.17 and 12.3.18, with different gate bias conditions. Fig. 12.3.17 shows the model performance in the strong inversion region using large gate voltage steps to show the model behavior over the entire bias range. Fig. 12.3.18 shows the model performance in the strong inversion region with small gate voltage steps to observe the model symmetry more clearly. Both results shows that the model can guarantee the symmetry of current and g_{ds} at V_{ds}=0.

(7) Gummel slope ratio test [12.2, 12.5]

Fig. 12.3.19 gives the results of the Gummel slope ratio test for devices with W/L=10μm/10μm and 10μm/0.5μm at V_1=0.01V and V_2=0.02V respectively. S_r, defined earlier, should reach 1.31 for small V_{gs} in the subthreshold region, and approaches 1 as V_{gs} increases in strong inversion. As shown in Fig. 12.3.19, the S_r characteristics of long channel devices can indeed tend to 1.31 approximately in the subthreshold region and becomes 1 when V_{gs} increases in strong inversion. Since this test is mainly to examine if the model can follow the fundamental device physics, that is, the 1-exp(-V_{ds}/v_t) behavior in subthreshold, it demonstrates that the *I-V* model in BSIM3v3 is physics-based in subthreshold region.

(8) Gummel treetop curve test [12.2, 12.5]

Fig. 12.3.20 shows the Gummel treetop curves generated by BSIM3v3 for a device of W/L=10μm/10μm. The diode current has been turned off and the value of G_{min} has been reduced in the simulation to eliminate their influence on the test results. It can be seen that the g_m/I_d curves simulated by the model can follow the so called treetop asymptotic behaviors. No kinks and glitches are observed.

(9) Additional quantitative tests:

Figs. 12.3.21 and 12.3.22 shows the measured and modeled threshold voltage characteristics for devices with different channel lengths and widths. It can be seen that the short channel and narrow width effects can be well described by the model at different body biases. Some verification results on the threshold voltage are given in section 3.5.

The saturation current (I_{dsat}) vs. device channel length is given in Fig.12.3.23, which shows that I_{dsat} can be described accurately by the model at different gate biases.

In Figs. 12.3.24, 12.3.25, and 12.3.26, I_{ds}-V_{ds}, I_{ds}-V_{gs} and g_{ds}-V_{ds} characteristics of devices of different W/L are shown. The maximum error in I_{ds} across different device geometries is less than 5%. Only one set of models parameters is used for all the W's and L's.

Detailed test results on the temperature effect are given in Chapter 9. The test results on the NQS model are given in Chapter 10.

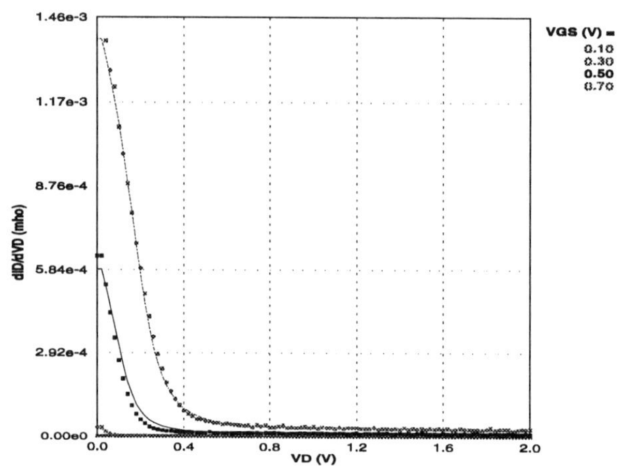

Fig. 12.3.1 Measured (symbols) and modeled (lines) g_{ds}-V_{ds} characteristics of a device with W_{drawn}/L_{drawn}=6μm/0.25μm at different gate voltages near V_{th} and V_{bs}=0V.

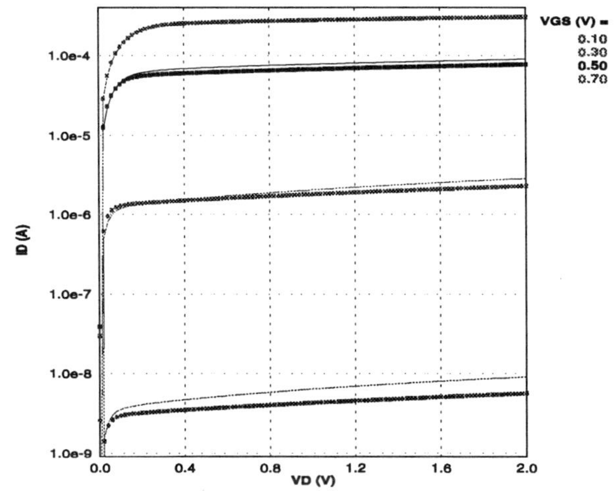

Fig. 12.3.2 Measured (symbols) and modeled (lines) I_{ds}-V_{ds} characteristics (in logarithmic scale) of a device with W_{drawn}/L_{drawn}=6μm/0.25μm at different gate voltages near V_{th} and V_{bs}=0V.

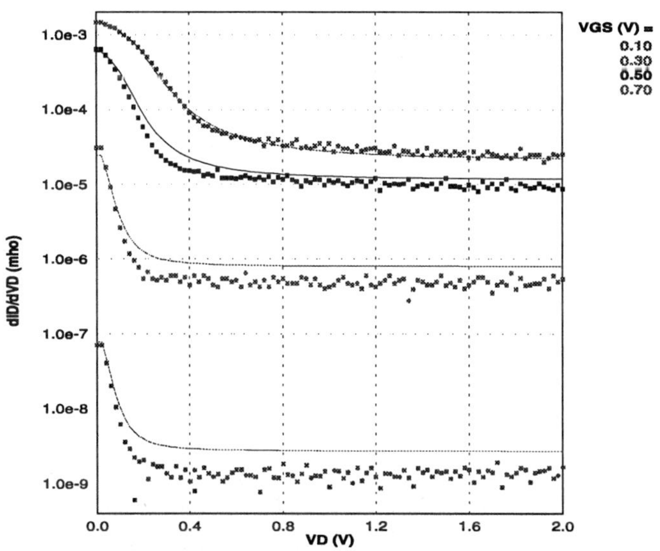

Fig. 12.3.3 Measured and modeled g_{ds}-V_{ds} characteristics (in logarithmic scale) of a device with W_{drawn}/L_{drawn}=6μm/0.25μm at different gate voltages near V_{th} and V_{bs}=0V.

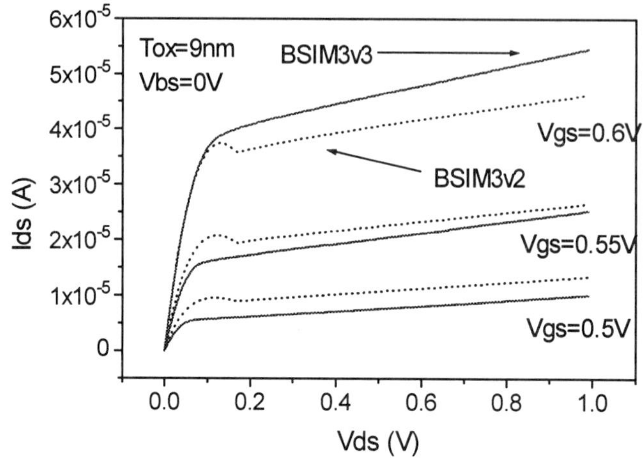

Fig. 12.3.4 BSIM3v3 removes the negative conductance that existed in BSIM3v2.

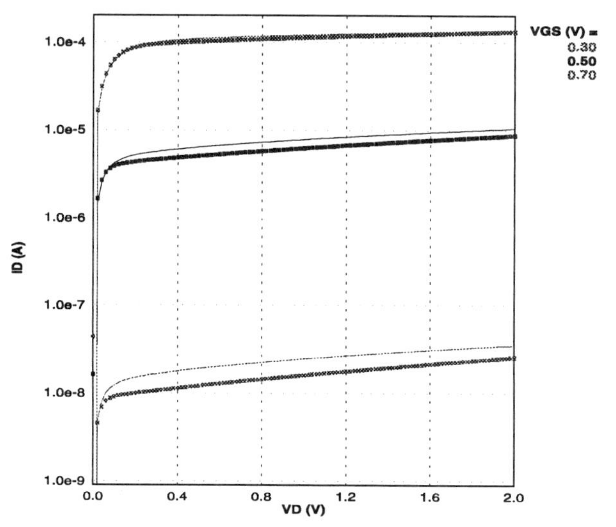

Fig. 12.3.5 Measured and modeled I_{ds}-V_{ds} characteristics (in logarithmic scale) of a device with W_{drawn}/L_{drawn}=6μm/0.25μm at different gate voltages near V_{th} and V_{bs}=-1V.

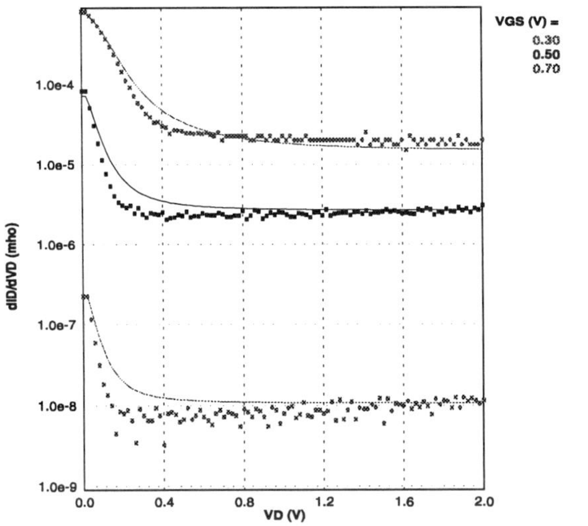

Fig. 12.3.6 Measured and modeled g_{ds}-V_{ds} characteristics (in logarithmic scale) of a device with W_{drawn}/L_{drawn}=6μm/0.25μm at different gate voltages near V_{th} and V_{bs}=-1V.

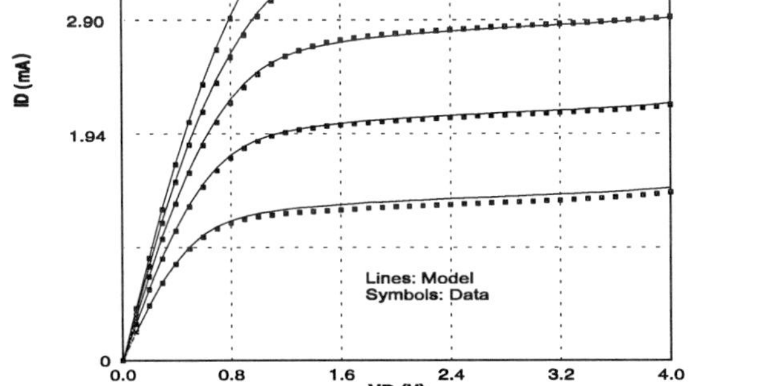

Fig. 12.3.7 Measured and modeled I_{ds}-V_{ds} characteristics of device with W_{drawn}/L_{drawn}=10μm/0.4μm at different gate voltages. After Cheng et al. [12.12].

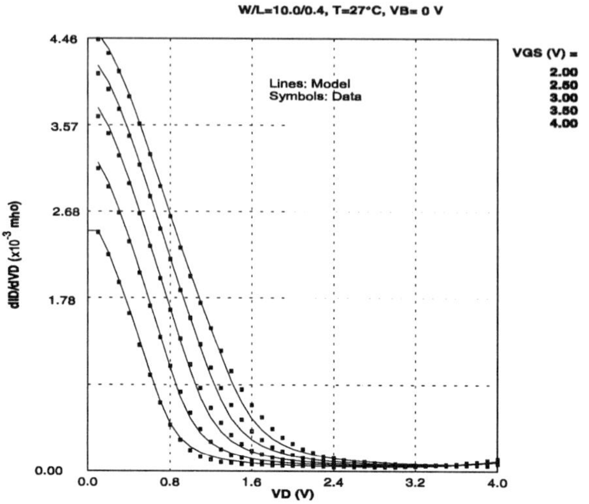

Fig. 12.3.8 Measured and modeled g_{ds}-V_{ds} characteristics of a device with W_{drawn}/ L_{drawn}=10μm/0.4μm at different gate voltages. After Cheng et al. [12.12].

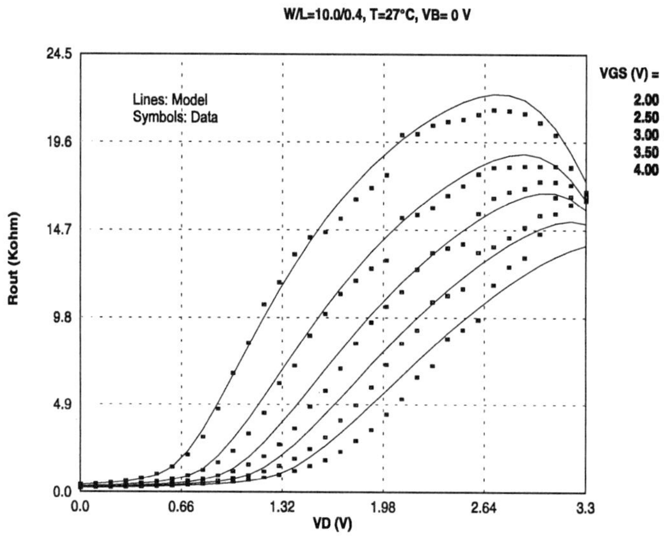

Fig. 12.3.9 Measured and modeled R_{out}-V_{ds} characteristics of a device with W_{drawn}/ L_{drawn}=10μm/0.4μm at different gate voltages. After Cheng et al. [12.12].

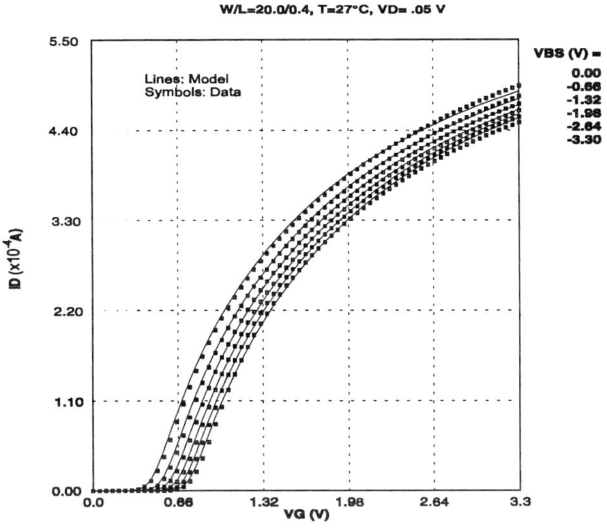

Fig. 12.3.10 Measured and modeled I_{ds}-V_{gs} characteristics (in linear scale) of a device with W_{drawn}/L_{drawn}=20μm/0.4μm at V_{ds}=50mV and different body biases, V_{bs}. After Cheng et al. [12.12].

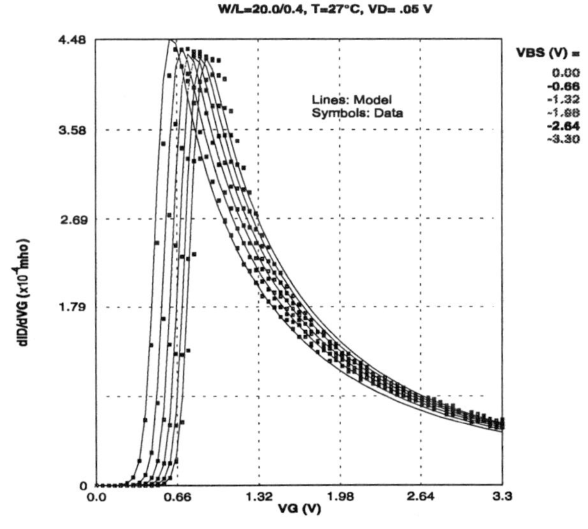

Fig. 12.3.11 Measured and simulated transconductance g_m versus the gate bias, V_{gs}. W_{drawn}/L_{drawn}=20μm/0.4μm at V_{ds}=50mV and different body biases, V_{bs}. After Cheng et al. [12.12].

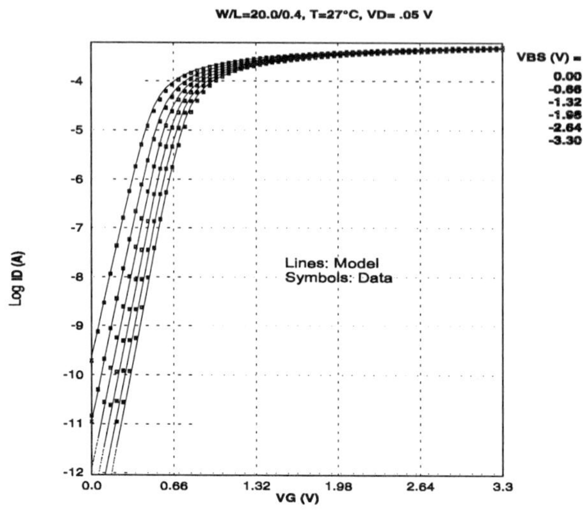

Fig. 12.3.12 Measured and modeled I_{ds}-V_{gs} characteristics (in logarithmic scale). W_{drawn}/L_{drawn}=20μm/0.4μm at V_{ds}=50mV and different body biases, V_{bs}. After Cheng et al. [12.12].

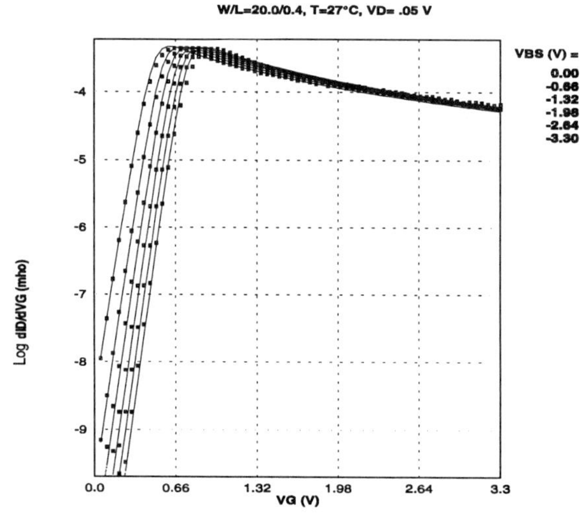

Fig. 12.3.13 Measured (symbols) and simulated (lines) transconductance g_m versus the gate bias V_{gs} in logarithmic scale. W_{drawn}/L_{drawn}=20μm/0.4μm at V_{ds}=50mV and several body biases V_{bs}. After Cheng et al. [12.12].

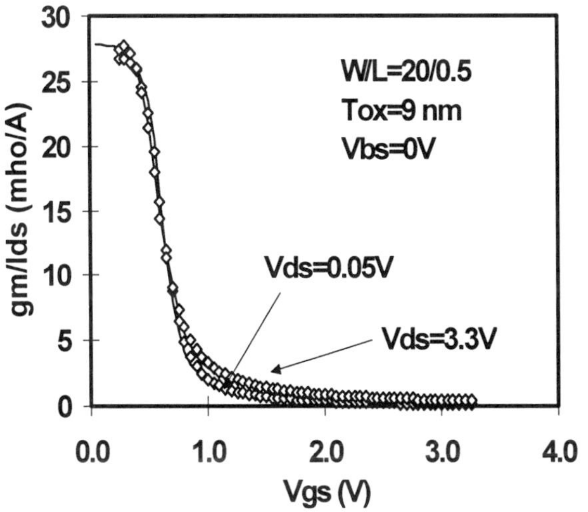

Fig. 12.3.14 Measured (symbols) and modeled (lines) g_m/I_{ds}-V_{gs} characteristics of a device with W_{drawn}/L_{drawn}=20μm/0.5μm at different drain voltages. After Cheng et al. [12.12].

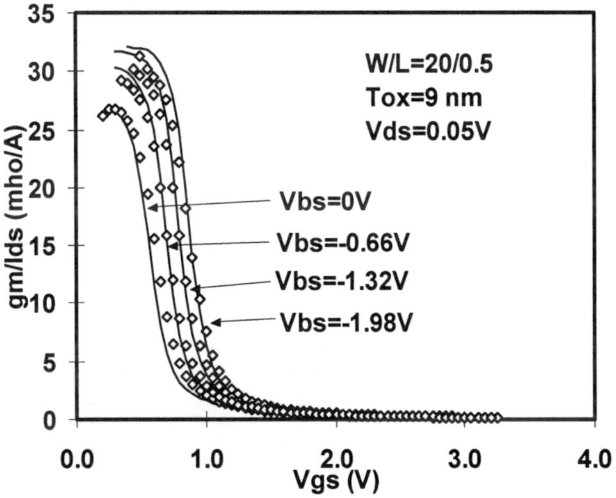

Fig. 12.3.15 Measured (symbols) and modeled (lines) g_m/I_{ds}-V_{gs} characteristics of a device with W_{drawn}/L_{drawn}=20μm/0.5μm at several body bias voltages. After Cheng et al. [12.12].

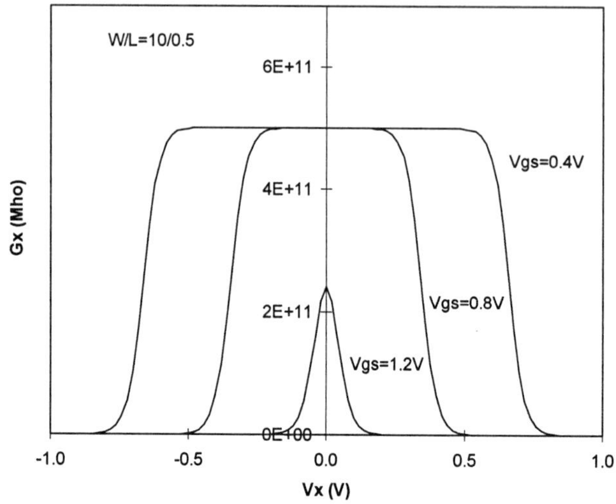

Fig. 12.3.16 Results of the Gummel symmetry test for a device with W/L=10μm/0.5μm in the regions from subthreshold to strong inversion.

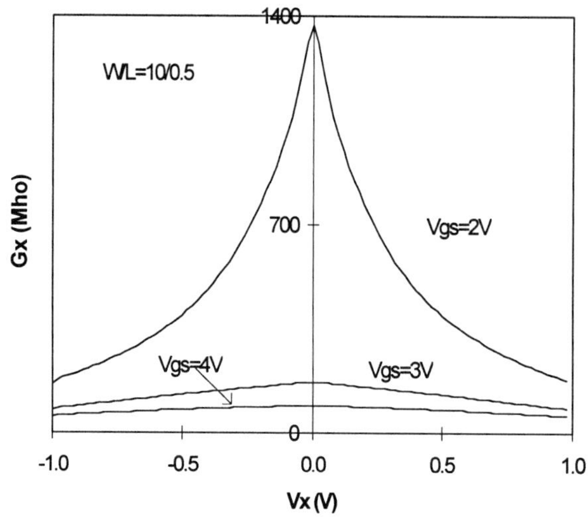

Fig. 12.3.17 Results of the Gummel symmetry test for a device with W/L=10μm/0.5μm in strong inversion region and V_{gs} varied from 2V to 4V.

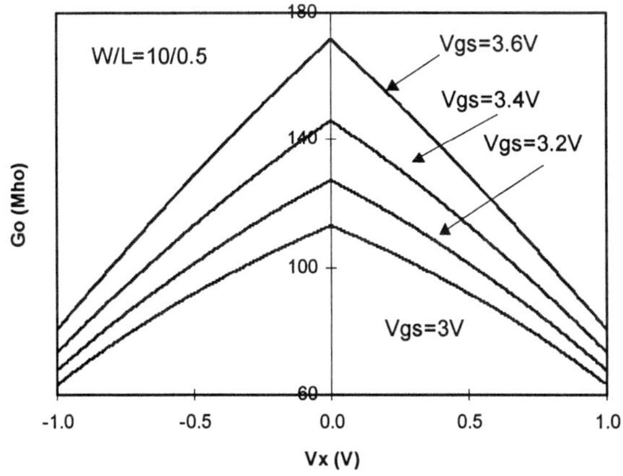

Fig. 12.3.18 Results of the Gummel symmetry test for a device with $W/L=10\mu m/0.5\mu m$ in the strong inversion region with V_{gs} varying from 3V to 3.6V.

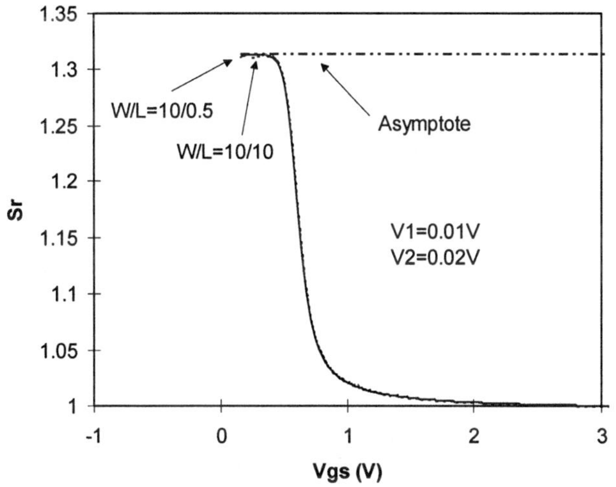

Fig. 12.3.19 Results of the Gummel slope ratio test for devices with $W/L=10\mu m/0.5\mu m$ and $W/L=10\mu m/10\mu m$.

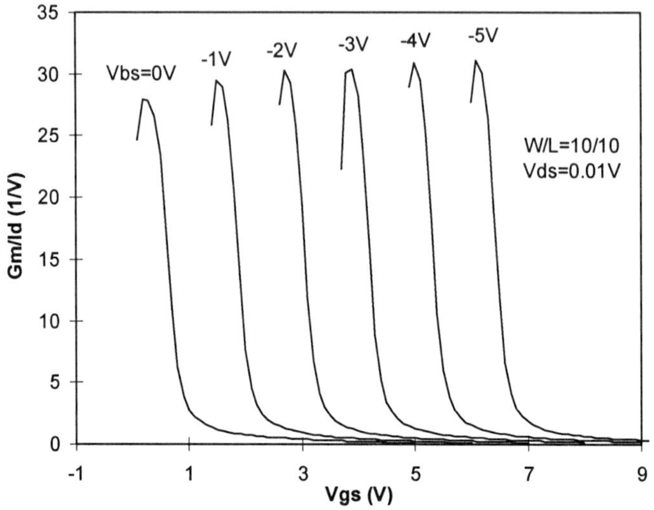

Fig. 12.3.20 (a) Results of the Gummel treetop curve test for a device with $W/L=10\mu m/10\mu m$.

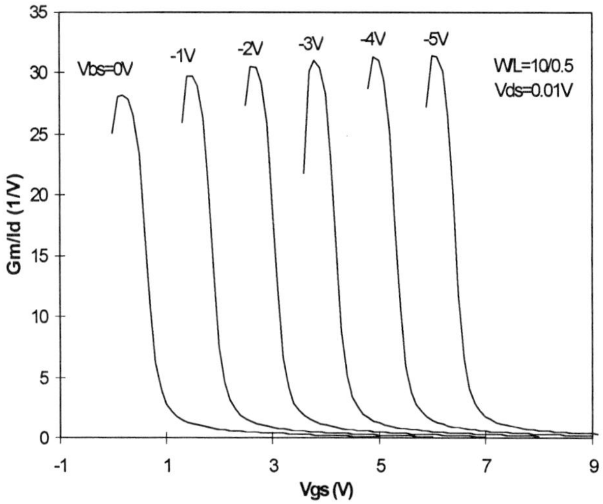

Fig. 12.3.20 (b) Results of the Gummel treetop curve test for a device with $W/L=10\mu m/0.5\mu m$.

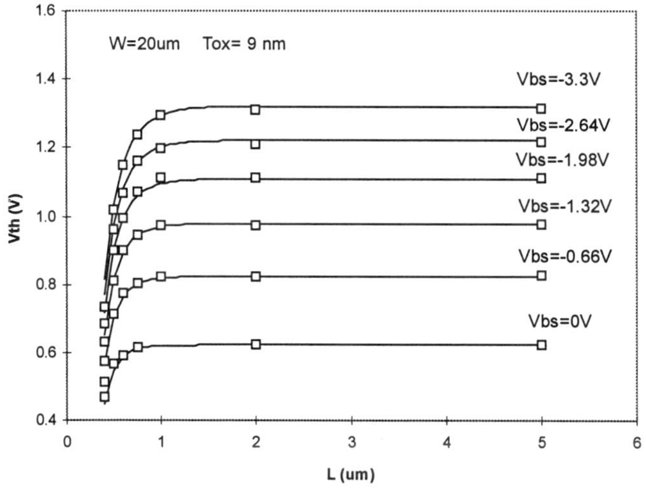

Fig. 12.3.21 Measured (symbols) and modeled (lines) threshold voltage characteristics of devices with different channel lengths. After Cheng et al. [12.12].

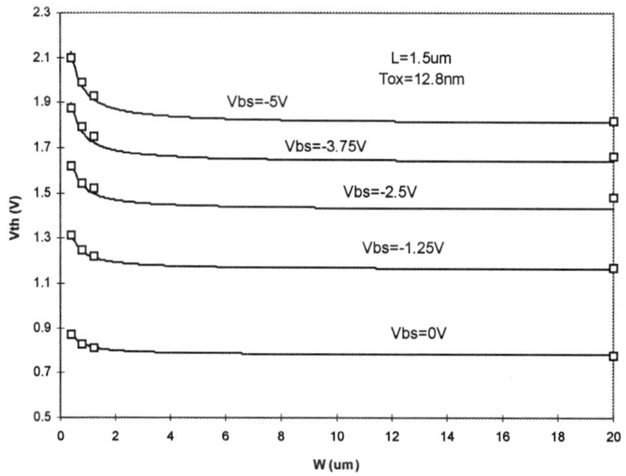

Fig. 12.3.22 Measured (symbols) and (lines) modeled threshold voltage of devices with different channel widths. After Cheng et al. [12.12].

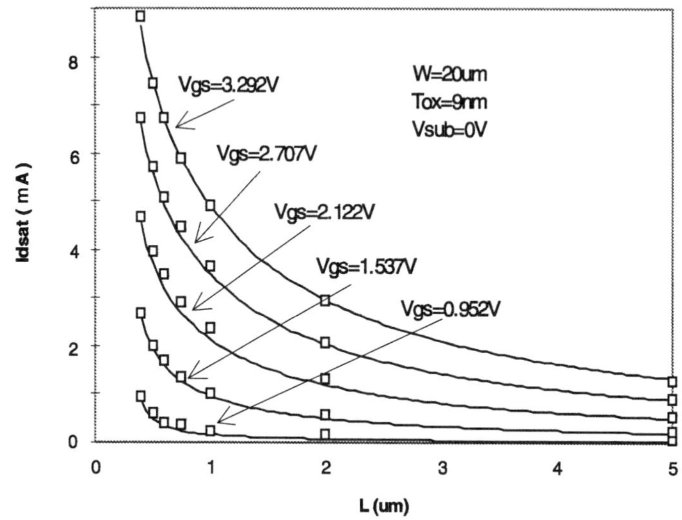

Fig. 12.3.23 Measured (symbols) and modeled (lines) saturation drain current of devices with different channel lengths. After Cheng et al. [12.12].

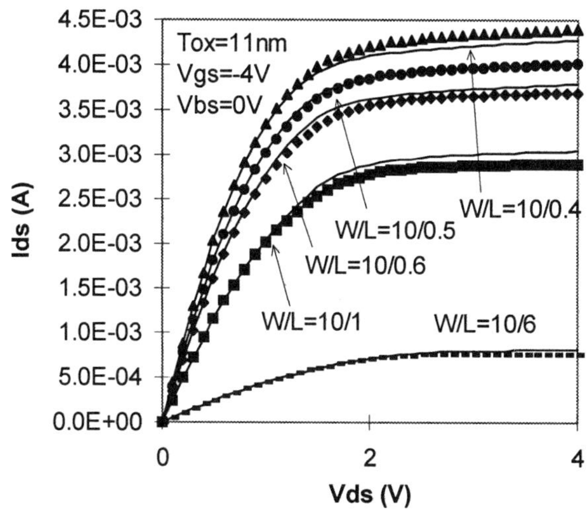

Fig. 12.3.24 Measured (symbols)and simulated (lines) I_{ds} versus V_{ds} curves of devices of different W/L. This is a scalability test. After Cheng et al. [12.12].

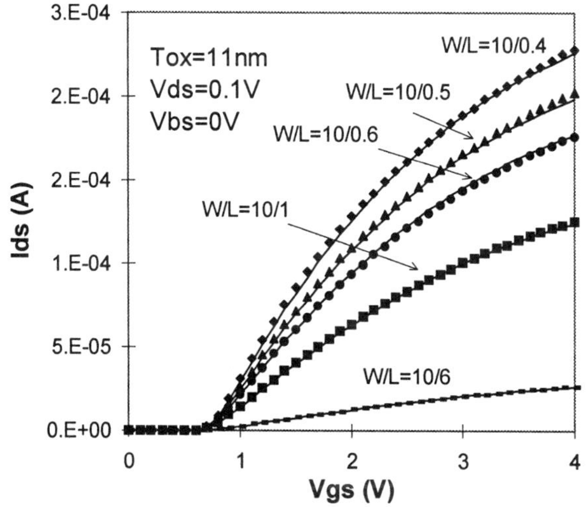

Fig. 12.3.25 Measured (symbols) and simulated (lines) I_{ds} versus V_{gs} curves of devices of different *W*/*L*. After Cheng et al. [12.12].

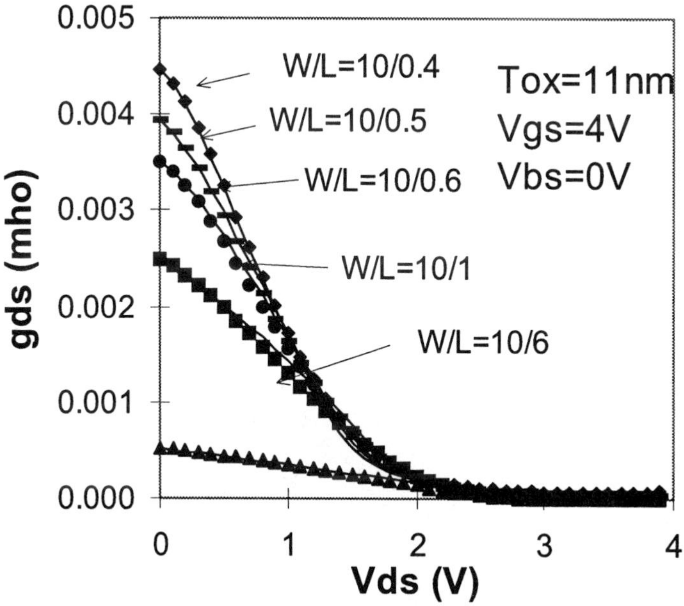

Fig. 12.3.26 Measured (symbols) and simulated (lines) g_{ds} versus V_{ds} curves of devices of different *W*/*L*. This is a scalability test. After Cheng et al. [12.12].

12.4 Helpful Hints

1. Summary of the BSIM3v3 model test results.

(1) The BSIM3v3 *I-V* model passes the qualitative tests discussed above and shows smooth transitions from the triode to saturation regions, and from the subthreshold to strong inversion regions. No negative conductance, kinks, glitches or discontinuities are observed.

(2) It passes the treetop curve and the Gummel slope ratio tests, and demonstrates model symmetry at the first derivative level. These results validate the physics basis of the model in both the strong inversion and subthreshold regions.

(3) It models the current and transconductances of the devices accurately, and has good scalability over a wide device geometry range.

(4) It has been tested with the measured characteristics of devices from different sources [12.20, 12.21, 12.22].

(5) The model shows good temperature dependence up to 125^oC as demonstrated in Chapter 9.

2. Understanding some limitations and shortcomings of the present BSIM3v3 model

As discussed above, the BSIM3v3 model is good for both digital and analog applications. However, it still can be improved to meet even more strict requirements. We discuss some shortcomings of the present model to make users aware of its limitations.

(1) The *I-V* model ensures the continuity and symmetry at V_{ds}=0 at the first derivative level, but fails at the second derivative level [12.27].

(2) The *C-V* model is not symmetric at V_{ds}=0 as we have shown in Chapter 5.

(3) The bias-dependent source/drain series resistance is treated as a virtual parameter to derive the analytical model by assuming that the device is symmetric. This makes it very difficult or impossible to simulate an asymmetric device. Separate source and drain series resistances should be used in the model.

(4) It has been found that the velocity saturation and hot carrier effects can influence the thermal noise characteristics significantly in short channel devices. The present model accounts for velocity saturation only in an empirical manner and should be enhanced to improve the accuracy of the thermal noise model.

(5) The temperature dependence of impact ionization is not included in the present model.

(6) The present model needs to be improved to simulate the high frequency behaviors for RF applications because it does not include the influence of some parasitics such as the gate resistance and the substrate resistances.

3. Additional benchmark tests to validate a model

The benchmark tests discussed in this chapter can be considered the basic tests for validating a model. They check some salient properties of the model. However, more benchmark tests need to be developed. For example, benchmark tests to examine the harmonic distortion behavior of the model are needed.

References

[12.1] Y. Tsividis and K. Suyama, "MOSFET modeling for analog circuit CAD: Problems and prospects," *Tech. Dig.* CICC-93, pp14.1.1-14.1.6, 1993.

[12.2] Marc McSwain and Colin McAndrew, *Compact Model Workshop*, Sunnyvale, CA, Aug., 1995.

[12.3] *Compact Model Workshop*, Dallas, TX, March, 1995.

[12.4] *Compact Model Workshop*, Austin, TX, June, 1995.

[12.5] *Compact Model Council* (http://www.eia.org/eig/CMC).

[12.6] *Compact Model Workshop*, Washington D. C., Dec, 1995.

[12.7] *Compact Model Workshop*, Austin, Taxis, Mar, 1996.

[12.8] *Compact Model Workshop*, Burlington, Vermont, Aug, 1996.

[12.9] B. J. Sheu, D. L. Scharfetter, P. K. Ko, and M. C. Jeng, "BSIM: berkeley short -channel IGFET model for MOS transistors," *IEEE J. Solid-State Circuits*, vol. SC-22, pp.558-565, 1987.

[12.10] M. C. Jeng, *Design and modeling of deep-submicrometer MOSFETs*, ERL memorandum ERL M90/90, University of California, Berkeley, 1990.

[12.11] J. H. Huang et al., *BSIM3 Manual (Version 2.0)*, University of California, Berkeley, March 1994.

[12.12] Y. Cheng et al., “A physical and scalable BSIM3v3 I-V model for analog/digital circuit simulation”, *IEEE Trans. Electron Devices*, Vol. 44, pp.277-287, Feb. 1997.

[12.13] R. M. D. A. Velghe, D. B. M. Klassen, and F. M. Klassen, “Compact MOS modeling for analog circuit simulation”, *IEEE IEDM 93, Tech. Dig.*, pp.485-488, Dec. 1993.

[12.14] J. A. Power and W.A. Lane, “An enhanced SPICE MOSFET model suitable for analog applications”, *IEEE Trans. Computer-Aided Design* vol.CAD-11 pp.1418-1425, 1992.

[12.15] N. D. Arora, R. Rios, C. L. Huang and K. Raol, “PCIM: a physically based continuous short-channel IGFET model for circuit simulation,” *IEEE Trans. Electron Devices*, vol.41, pp. 988-997, 1994.

[12.16] C. C. Enz, F. Krummenacher and E. A. Vittoz, “An analytical MOS transistor model valid in all regions of operation and dedicated to low voltage and low-current applications”, *J. Analog Integrated Circuit and Signal Processing*, Vol. 8, pp.83-114, 1995.

[12.17] M. Shur, T. A. Fjeldly, T. Ytterdal, and K. Lee, “An unified MOSFET model,” *Solid-State Electronics*, 35, pp.1795-1802, 1992.

[12.18] Y. Cheng et al., “A unified BSIM *I-V* mode for circuit simulation”, *1995 International semiconductor device research symposium,* Charlottesville, Dec. 1995.

[12.19] Y. Cheng et al., “An investigation on the robustness, accuracy and simulation performance of a physics-based deep-submicronmeter BSIM model for analog/digital circuit simulation”, *CICC'96*, pp. 321-324, May 1996.

[12.20] C. Lyons and S. Power, *Compact Model Workshop*, Washington DC, Dec. 1995.

[12.21] A. Dognis, *Compact Model Workshop,* Washington DC, Dec. 1995.

[12.22] M. C. Jeng and Z. H. Liu, *Compact Model Workshop,* Washington DC, Dec. 1995.

[12.23] *BSIMpro Manual*, BTA Inc., Old Ironsides Drive, Santa Clara, CA, 1996 (http://www.btat.com).

[12.24] J. H. Huang et al., “A physical model for MOSFET output resistance”, *IEDM, Technical Digest*, Dec. of 1992.

[12.25] Y. Cheng et al., *BSIM3 version 3.0 User's Manual*, University of California, Berkeley, 1995.

[12.26] Y. Cheng et al., *BSIM3 version 3.1 User's Manual*, University of California, Berkeley, Memorandum No. UCB/ERL M97/2, 1997.

[12.27] K. Joardar et al., “An improved MOSFET model for circuit simulation,” *IEEE Trans. Electron Devices*, vol. 45, pp. 134-148, 1998.

CHAPTER 13

Model Parameter Extraction

Parameter extraction is an important part of device modeling. Many different extraction methods have been developed [13.1, 13.2]. The appropriate methodology depends on the model and on the way the model will be used. In the following sections, we discuss the parameter extraction approaches that are often used in the semiconductor industry.

13.1 Overview of Model Parameter Extraction

There are two different optimization strategies which can be used for parameter extraction: global optimization and local optimization [13.3, 13.4]. Global optimization lets the computer find one set of parameters which best fit all the available experimental data. Global optimization may minimize the error between the simulation results and the available experimental data. However, any particular parameter extracted by global optimization may not have a close resemblance to its actual physical value. In local optimization each parameter is extracted in a certain operation region where its corresponding device's behavior is dominant. Parameters optimized locally may not perfectly fit the experimental data in all the operating regions, but they are closely related to the physical processes at work.

Also, there are two different strategies for extracting model parameters [13.5, 13.6]: the single device extraction strategy and the group extraction strategy. In the single device extraction strategy, one uses the experimental data from a single device to extract a complete set of model parameters. This strategy can fit one device very well, but may not fit other devices with different geometries. If only one channel length and width is used, parameters which are related to channel length and channel width dependencies cannot be determined. In the group extraction method, parameters are extracted using the experimental data from multiple devices having different *W*'s and/or *L*'s. This strategy may not fit one device extremely well, but can fit many devices with different geometries reasonably well.

Ideally, one set of model parameters should cover the whole range of device geometries in the circuit design. However, a parameter extraction approach called 'binning' has been used to improve the model accuracy for devices with wide geometry variation. In the binning approach, the interested range of device geometry is divided into many geometry bins, as shown in Fig. 13.1.1. One set of model parameters is used only in each bin. Thus, depending on if users adopt the binning approach, the model parameter extraction can be further divided into a single-bin (or scalable) approach and a multi-bin approach.

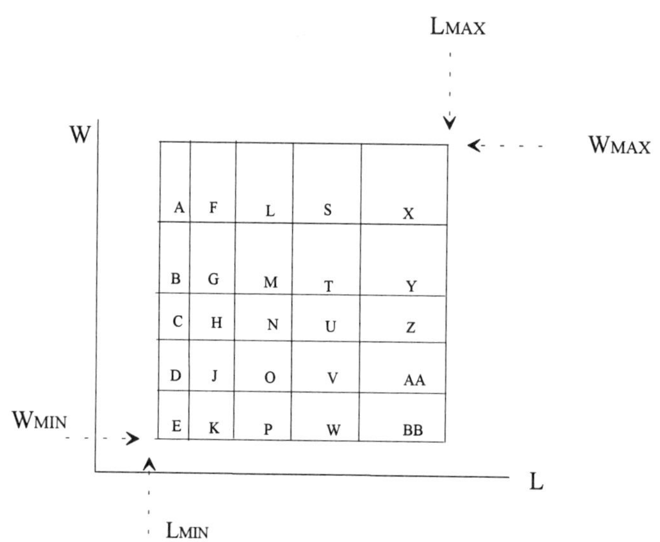

Fig. 13.1.1 *W* and *L* of devices are divided into different bins according to the requirements of the model accuracy. Ideally, only a single bin is used.

13.2 Parameter Extraction for BSIM3v3

13.2.1 Optimization and extraction strategy

Based on the properties of the BSIM3 model, a combination of local optimization and the group device extraction strategy is the best for obtaining the preliminary or initial parameters [13.7, 13.8]. A global optimization may then be used to further improve the overall agreement between the model and the measured data if necessary.

As we have discussed in the previous chapters, BSIM3v3 is a scalable model that can cover a wide geometry range with one set of model parameters [13.9, 13.10]. It will lose this advantage if the single device extraction strategy is used.

13.2.2 Extraction routines

1. Devices and measurements needed for parameter extraction.

For scalable model parameter extraction without using a binning approach [13.11, 13.12], three sets of different size devices are needed to extract the model parameters, as shown in Fig. 13.2.1. One set of devices has a fixed channel width and different channel lengths. One set of devices has a fixed long channel length and different channel widths. One set of devices have a fixed shortest channel length and different channel widths.

The large size device ($W \geq 10\mu m$, $L \geq 10\mu m$) is used to extract such parameters as $\boldsymbol{U_A}$, $\boldsymbol{U_B}$, and $\boldsymbol{U_C}$ for mobility, the long-channel device threshold voltage $\boldsymbol{V_{TH0}}$, and the body effect coefficients $\boldsymbol{K_1}$ and $\boldsymbol{K_2}$ which depend on the vertical doping profile. One set of devices with a fixed large channel width and different channel lengths is used to extract parameters which are related to the short channel effects. The devices having a fixed long channel length and different channel widths are used to extract parameters which are related to narrow width effects. The other devices with the fixed minimum channel length and different channel widths are used to extract the model parameters for the small size effects.

For parameter extraction using the binning approach [13.11, 13.12], more devices will be needed as discussed later in this chapter.

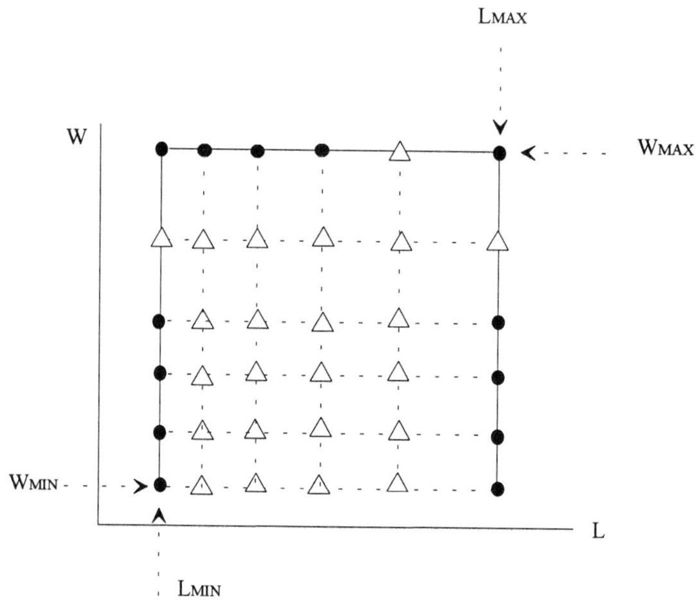

Fig. 13.2.1 Devices used for parameter extraction. •: devices needed for scalable model parameter extraction; Δ: optional devices for parameter checking or for binning model parameter extraction.

Five sets of data are recommended to be measured for each device and saved into different files for DC parameter extraction [13.9]:

(a) I_{ds} vs. V_{gs} at different V_{bs} and V_{ds}=0.05V (linear region measurement). The measurement configuration is shown in Fig. 13.2.2.

(b) I_{ds} vs. V_{ds} at different V_{gs} and V_{bs}=0V (linear & saturation region measurements). The measurement configuration is shown in Fig. 13.2.2.

(c) I_{ds} vs. V_{gs} at different V_{bs} and V_{ds} = Maximum V_{ds} (saturation region measurement). The measurement configuration is shown in Fig. 13.2.2.

(d) I_{ds} vs. V_{gs} at different V_{gs} and V_{bs} = V_{bb} (linear & saturation region measurements, and $|V_{bb}|$ is the maximum body bias). The measurement configuration is shown in Fig. 13.2.2.

(e) I_{sub} vs. V_{gs} at different V_{ds} and V_{bs} =0 and V_{bb} (substrate current measurements). The measurement configuration is shown in Fig. 13.2.3.

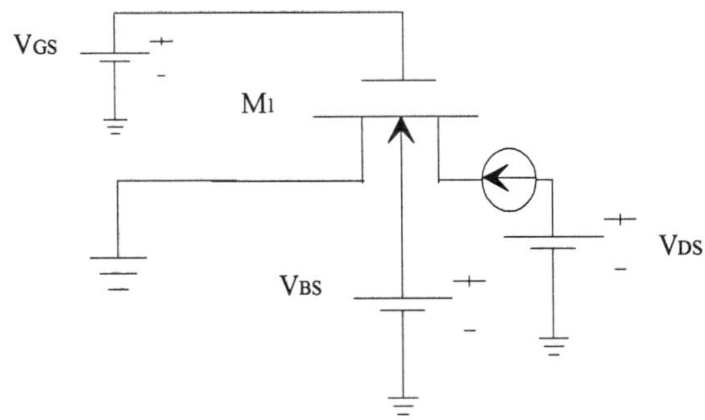

Fig. 13.2.2 Measurement setup for I_{ds}-V_{ds} and I_{ds}-V_{gs} characteristics of n-MOSFETs.

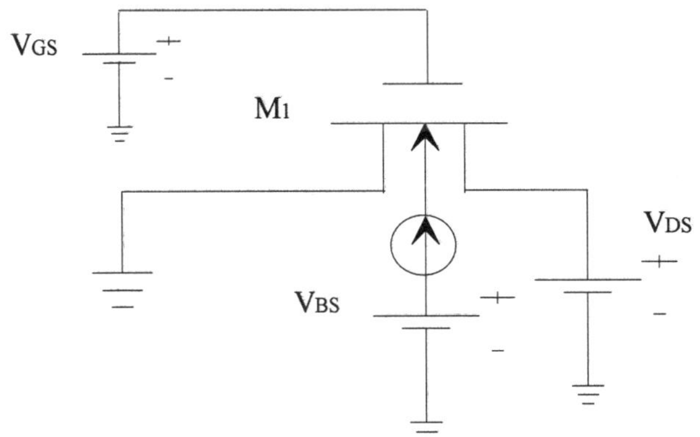

Fig. 13.2.3 Substrate current measurement setup for n-MOSFETs.

2. Optimization method

The optimization process recommended for BSIM3v3 is a combination of Newton-Raphson iteration and a linear-least-square fit with either one, two, or three variables. The flow chart of the optimization process is shown in Fig. 13.2.4 [13.8, 13.9]. The model equation is arranged in a form suitable for Newton-Raphson's iteration as shown in Eq. (13.2.1) [13.13]:

$$f_{\exp}(P_{10}, P_{20}, P_{30}) - f_{sim}(P_1^{(m)}, P_2^{(m)}, P_3^{(m)}) = \frac{\partial f_{sim}}{\partial P_1}\Delta P_1^m + \frac{\partial f_{sim}}{\partial P_2}\Delta P_2^m + \frac{\partial f_{sim}}{\partial P_3}\Delta P_3^m \tag{13.2.1}$$

where *fsim()* is the function to be optimized and *fexp()* is the experimental data. P_{10}, P_{20}, and P_{30} stand for the true parameter values which we are seeking. $P_1^{(m)}$, $P_2^{(m)}$ and $P_3^{(m)}$ represent the parameter values after the m*th* iteration. We change Eq. (13.2.1) into a form that the linear least-square fit routine can use (a form of $y = a + bx1 + cx2$), by dividing both sides of the Eq. (13.2.1) by $\partial fsim / \partial P_1$. After putting the experimental data into Eq. (13.2.1), we find the increments of each parameter for the next iteration, $\Delta P_i^{(m)}$. The parameter values for the *(m+1)th* iteration are given by

$$P_i^{(m+1)} = P_i^{(m)} + \Delta P_i^{(m)} \quad i = 1, 2, 3 \tag{13.2.2}$$

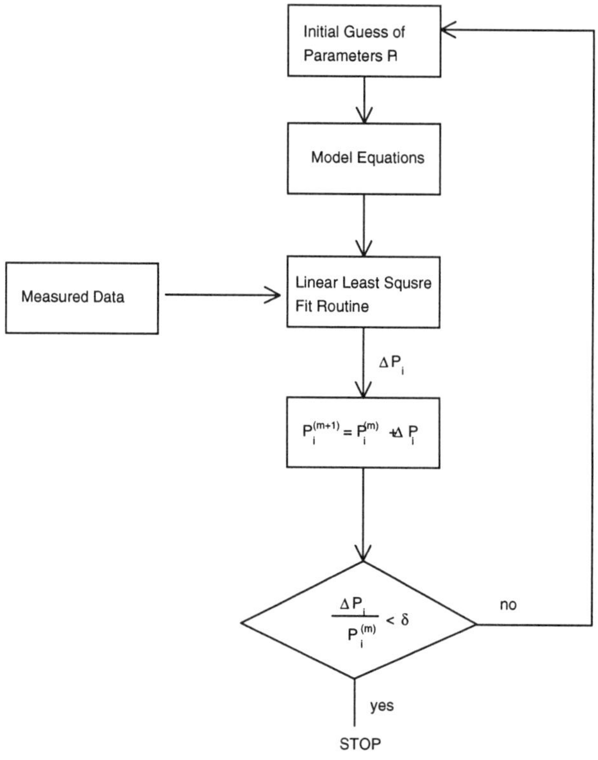

Fig. 13.2.4 Optimization flow for parameter extraction. After Huang et al. [13.7].

The new parameter values will be used for the next iteration until the increments are smaller than some pre-determined values. At this point, we have solved for our parameter values [13.7, 13.10].

3. Parameter extraction procedures

The recommended extraction routine is discussed here. In order to extract the model parameters, some process parameters have to be provided by the users before starting the parameter extraction. They are listed in Table 13.2.1:

Table 13.2.1 Parameters needed for the parameter extraction

Input Parameter Names	Physical Meaning of the Input Parameter
T_{OX}	Gate oxide thickness
N_{CH}	Doping concentration in the channel
T	Temperature at which the data is taken
L_{drawn}	Designated channel length
W_{drawn}	Designated channel width
X_j	Junction Depth

DC model parameters are extracted in the following procedures as shown in Fig. 13.2.5. The procedures are developed based on a understanding of the model and based on the local optimization principle.

Step 1

Extracted Parameters & Fitting Target Data[a]	Device & Experimental Data
VTH0, K1, K2 Fitting Target Data: *Vth(Vbs)*	Large Size Device (Large W & L). *Ids* vs. *Vgs* @ *Vds* = *0.05V* & Different *Vbs* Extracted Experimental Data, *Vth(Vbs)*

a. *Fitting Target Data* is the experimental data that the model wants to match by adjusting parameters.

Step 2

Extracted Parameters & Fitting Target Data	Devices & Experimental Data
$\mu 0, U_A, U_B, U_C$ Fitting Target Data: Strong Inversion region *Ids(Vgs, Vbs)*	Large Size Device (Large *W* & *L*). *Ids* vs. *Vgs* @ *Vds* = *0.05V* & Different *Vbs*

Step 3

Extracted Parameters & Fitting Target Data	Devices & Experimental Data
$\boldsymbol{L_{INT}}$, $R_{ds}(\boldsymbol{R_{DSW}}, W, Vbs)$ Fitting Target Data: Strong Inversion region $I_{ds}(V_{gs}, V_{bs})$	One Set of Devices (Large and Fixed *W* & Different *L*). I_{ds} vs. V_{gs} @ $V_{ds} = 0.05V$ & Different V_{bs}

Step 4

Extracted Parameters & Fitting Target Data	Devices & Experimental Data
$\boldsymbol{W_{INT}}$, $R_{ds}(\boldsymbol{R_{DSW}}, W, V_{bs})$ Fitting Target Data: Strong Inversion region $I_{ds}(V_{gs}, V_{bs})$	One Set of Devices (Large and Fixed *L* & Different *W*). I_{ds} vs. V_{gs}, @ $V_{ds} = 0.05V$ & Different V_{bs}

Step 5

Extracted Parameters & Fitting Target Data	Devices & Experimental Data
$\boldsymbol{R_{DSW}}$, $\boldsymbol{P_{RWB}}$, $\boldsymbol{W_R}$ Fitting Target Data: Strong Inversion region $R_{ds}(\boldsymbol{R_{DSW}}, W, V_{bs})$	One Set of Devices (Large and Fixed *L* & Different *W*). I_{ds} vs. V_{gs}, @ $V_{ds} = 0.05V$ & Different V_{bs}. $R_{ds}(\boldsymbol{R_{DSW}}, W, V_{bs})$

Step 6

Extracted Parameters & Fitting Target Data	Devices & Experimental Data
$\boldsymbol{D_{VT0}}$, $\boldsymbol{D_{VT1}}$, $\boldsymbol{D_{VT2}}$, $\boldsymbol{N_{LX}}$ Fitting Target Data: $V_{th}(V_{bs}, L, W)$	One Set of Devices (Large and Fixed *W* & Different *L*). $V_{th}(V_{bs}, L, W)$

Step 7

Extracted Parameters & Fitting Target Data	Devices & Experimental Data
$\boldsymbol{K_3}$, $\boldsymbol{K_{3B}}$, $\boldsymbol{W_0}$ Fitting Target Data: $V_{th}(V_{bs}, L, W)$	One Set of Devices (Large and Fixed *L* & Different *W*). $V_{th}(V_{bs}, L, W)$

Step 8

Extracted Parameters & Fitting Target Data	Devices & Experimental Data
$\boldsymbol{D_{VT0W}}$, $\boldsymbol{D_{VTW1}}$, $\boldsymbol{D_{VTW2}}$ Fitting Target Data: $V_{th}(V_{bs}, L, W)$	One Set of Devices (Small and Fixed *L* & Different *W*). $V_{th}(V_{bs}, L, W)$

Step 9

Extracted Parameters & Fitting Target Data	Devices & Experimental Data
V_{OFF}, N_{FACTOR}, C_{DSC}, C_{DSCB} Fitting Target Data: Subthreshold region $I_{ds}(V_{gs}, V_{bs})$	One Set of Devices (Large and Fixed W & Different L). I_{ds} vs. V_{gs} @ $V_{ds}= 0.05V$ & Different V_{bs}

Step 10

Extracted Parameters & Fitting Target Data	Devices & Experimental Data
C_{DSCD} Fitting Target Data: Subthreshold region $I_{ds}(V_{gs}, V_{ds})$	One Set of Devices (Large and Fixed W & Different L). I_{ds} vs. V_{gs} @ $V_{bs} = V_{bb}$ & Different V_{ds}

Step 11

Extracted Parameters & Fitting Target Data	Devices & Experimental Data
D_{WB} Fitting Target Data: Strong Inversion region $I_{ds}(V_{gs}, V_{bs})$	One Set of Devices (Large and Fixed L & Different W). I_{ds} vs. V_{gs} @ $V_{ds} = 0.05V$ & Different V_{bs}

Step 12

Extracted Parameters & Fitting Target Data	Devices & Experimental Data
v_{SAT}, A_0, A_{GS} Fitting Target Data: $I_{sat}(V_{gs}, V_{bs})/W$ A_1, A_2 (PMOS Only) Fitting Target Data $V_{Asat}(V_{gs})$	One Set of Devices (Large and Fixed W & Different L). I_{ds} vs. V_{ds} @ $V_{bs} = 0V$ & Different V_{gs}

Step 13

Extracted Parameters & Fitting Target Data	Devices & Experimental Data
$B0$, $B1$ Fitting Target Data: $I_{dsat}(V_{gs}, V_{bs})/W$	One Set of Devices (Large and Fixed L & Different W). I_{ds} vs. V_{ds} @ $V_{bs} = 0V$ & Different V_{gs}

Step 14

Extracted Parameters & Fitting Target Data	Devices & Experimental Data
$\boldsymbol{D_{WG}}$ Fitting Target Data: Strong Inversion region $I_{ds}(V_{gs}, V_{bs})$	One Set of Devices (Large and Fixed L & Different W). I_{ds} vs. V_{gs} @ $V_{bs} = V_{bb}$ & Different V_{ds}

Step 15

Extracted Parameters & Fitting Target Data	Devices & Experimental Data
$\boldsymbol{P_{SCBE1}}$, $\boldsymbol{P_{SCBE2}}$ Fitting Target Data: $R_{out}(V_{gs}, V_{ds})$	One Set of Devices (Large and Fixed W & Different L). I_{ds} vs. V_{ds} @ $V_{bs} = 0V$ & Different V_{gs}

Step 16

Extracted Parameters & Fitting Target Data	Devices & Experimental Data
$\boldsymbol{P_{CLM}}$, $\theta(\boldsymbol{D_{ROUT}}, \boldsymbol{P_{DIBLC1}}, \boldsymbol{P_{DIBLC2}}, L)$, $\boldsymbol{P_{VAG}}$ Fitting Target Data: $R_{out}(V_{gs}, V_{ds})$	One Set of Devices (Large and Fixed W & Different L). I_{ds} vs. V_{ds} @ $V_{bs} = 0V$ & Different V_{gs}

Step 17

Extracted Parameters & Fitting Target Data	Devices & Experimental Data
$\boldsymbol{D_{ROUT}}$, $\boldsymbol{P_{DIBLC1}}$, $\boldsymbol{P_{DIBLC2}}$ Fitting Target Data: $\theta(\boldsymbol{D_{ROUT}}, \boldsymbol{P_{DIBLC1}}, \boldsymbol{P_{DIBLC2}}, L)$	One Set of Devices (Large and Fixed W & Different L). $\theta(\boldsymbol{D_{ROUT}}, \boldsymbol{P_{DIBLC1}}, \boldsymbol{P_{DIBLC2}}, L)$

Step 18

Extracted Parameters & Fitting Target Data	Devices & Experimental Data
$\boldsymbol{P_{DIBLCB}}$ Fitting Target Data: $\theta(\boldsymbol{D_{ROUT}}, \boldsymbol{P_{DIBLC1}}, \boldsymbol{P_{DIBLC2}}, L, V_{bs})$	One Set of Devices (Large and Fixed W & Different L). I_{ds} vs. V_{ds} @ fixed V_{gs} & Different V_{bs}

Step 19

Extracted Parameters & Fitting Target Data	Devices & Experimental Data
E_{TA0}, E_{TAB}, D_{SUB} Fitting Target Data: *Subthreshold region* $I_{ds}(V_{gs}, V_{bs})$	One Set of Devices (Large and Fixed W & Different L). I_{ds} vs. V_{gs} @ $V_{ds} = V_{dd}$ & Different V_{bs}

Step 20

Extracted Parameters & Fitting Target Data	Devices & Experimental Data
K_{ETA} Fitting Target Data: $I_{sat}(V_{gs}, V_{bs})/W$	One Set of Devices (Large and Fixed W & Different L). I_{ds} vs. V_{ds} @ $V_{bs} = V_{bb}$ & Different V_{gs}

Step 21

Extracted Parameters & Fitting Target Data	Devices & Experimental Data
α_0, β_0 Fitting Target Data: $I_{sub}(V_{gs}, V_{bs})/W$	One Set of Devices (Large and Fixed W & Different L). I_{sub} vs. V_{gs} @ $V_{bs} = V_{bb}$ & Different V_{ds}

Fig. 13.2.5 Flow-chart of the DC model parameter extraction procedure. After Cheng et al. [13.10].

To obtain the DC model parameters, the standard transistor test structures are sufficient for the parameter extraction. However, to extract the model parameters for the overlap and intrinsic capacitances as well as the diode model parameters, special device structures are needed to get the required data. For example, to measure the perimeter and area capacitances of the source/drain junctions, a device with a large perimeter but a small area and a device with a large area and a small perimeter are needed. To measure the overlap and intrinsic capacitances, MOSFETs with very wide channels are used to facilitate the capacitance measurement.

The measurements required to extract the capacitance model parameters are the following.

(a) Total gate capacitance versus gate bias from accumulation through strong inversion regions. The measurement configuration is shown in Fig. 13.2.6.

(b) C_{gc} capacitance versus gate voltage from strong inversion to depletion regions. The measurement configuration is shown in Fig. 13.2.7.

(c) C_{gs}/C_{gd} versus gate voltage from accumulation through depletion to strong inversion regions. The measurement configuration is shown in Fig. 13.2.8.

(d) C_{gs}/C_{gd} capacitance versus drain voltage from linear through saturation regions. The measurement configuration is shown in Fig. 13.2.9.

(e) C_{gb} capacitance versus gate voltage at different substrate voltages. The measurement configuration is shown in Fig. 13.2.10.

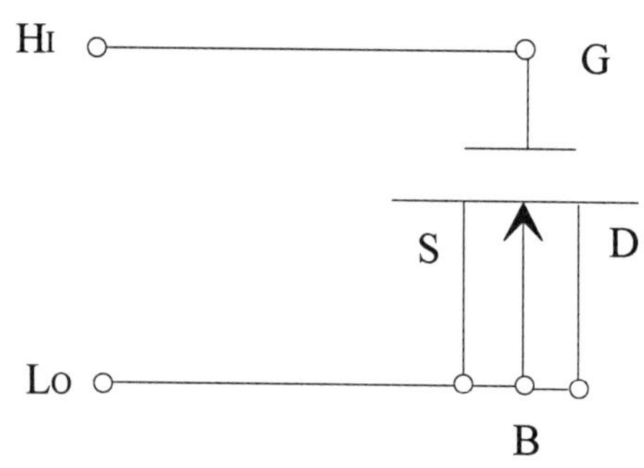

Fig. 13.2.6 Measurement configuration for the gate capacitance C_{gg} vs. gate bias. Hi: the port where AC and DC biases are applied; Lo: the port where the LCR collects the AC current.

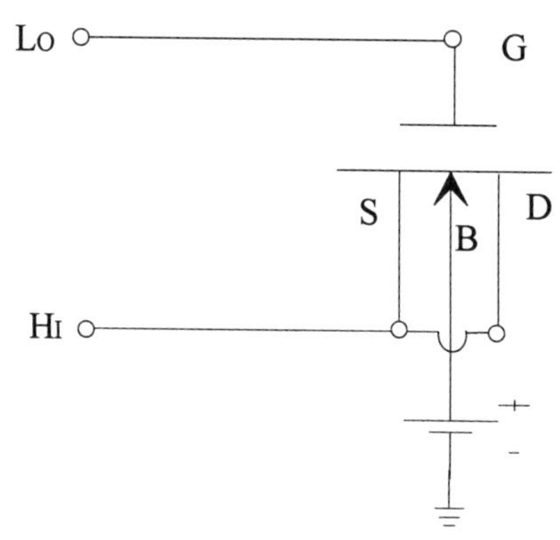

Fig. 13.2.7 Measurement configuration for the gate-to-channel capacitance, C_{gc}, vs. V_{gs}.

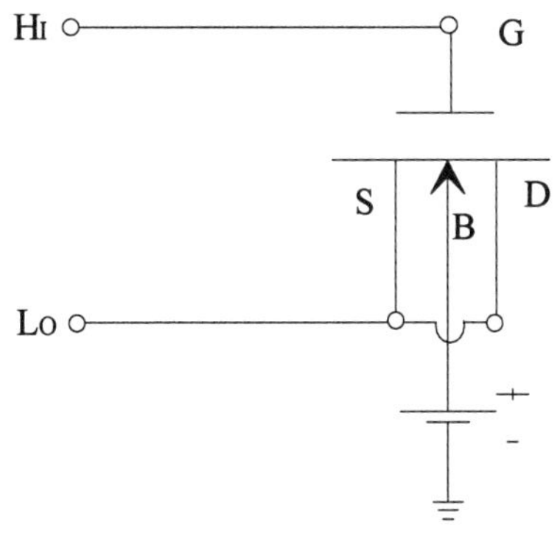

Fig. 13.2.8 Measurement configuration for the gate/source and gate/drain overlap capacitances vs. gate bias.

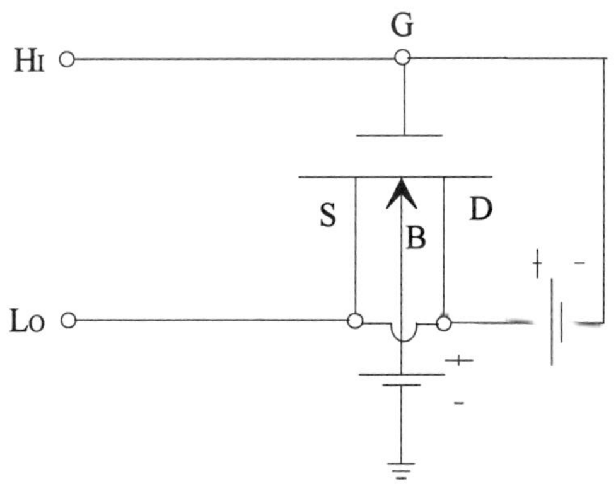

Fig. 13.2.9 Measurement configuration for the gate/source and gate/drain overlap capacitances vs. drain bias.

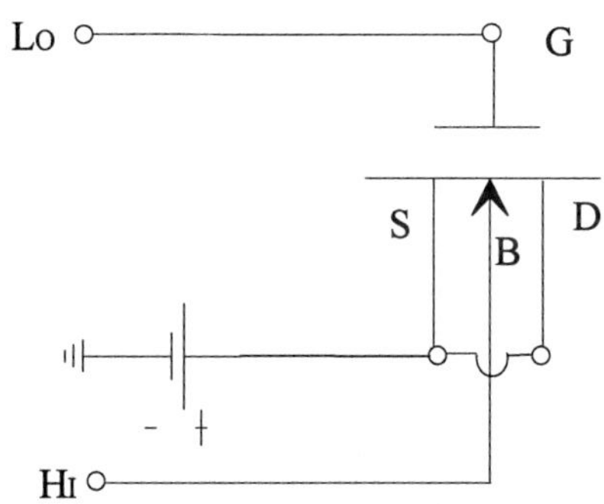

Fig. 13.2.10 Measurement configuration for C_{gb} vs. gate bias.

A flow-chart to extract the overlap and intrinsic capacitance model parameters is given in Fig. 13.2.11.

Step 1

Extracted Parameters & Fitting Target Data	Devices & Experimental Data
T_{OX} Fitting Target Data: C_{gg} versus V_{gs}	One device with large W and L;

Step 2

Extracted Parameters & Fitting Target Data	Devices & Experimental Data
D_{LC}, C_F, C_{GSO}, C_{GDO}, C_{GDL}, C_{GSL}, C_{KAPPA} Fitting Target Data: Overlap C_{gs}/C_{gd} versus V_{gs}	One set of devices with a wide channel width and several channel lengths;

Step 3

Extracted Parameters & Fitting Target Data	Devices & Experimental Data
C_{LE}, C_{LC} Fitting Target Data: C_{gs}/C_{gd} vs V_{ds}	One set of devices with a wide channel width and several channel lengths;

Step 4

Extracted Parameters & Fitting Target Data	Devices & Experimental Data
C_J, M_J, P_B, C_{JSW}, M_{JSW}, P_{BSW}, C_{JSWG}, M_{JSWG}, P_{BSWG} Fitting Target Data: source/drain junction capacitance characteristics.	One set of devices designed to separate area and perimeter junction capacitances;

Fig. 13.2.11 Flow-chart for the extraction of the capacitance model parameters, including the junction capacitances.

To extract the model parameters for the temperature dependencies, the measurements discussed above need to be performed at several different temperatures of interest. The parameter extraction for the temperature dependence parameters is a tedious but straightforward process involving the extraction of the temperature dependent parameters with the equations given in Chapter 8.

13.3 Binning Methodology

The parameter binning approach is supported in BSIM3v3 as an option for users to achieve the ultimate model accuracy. The basic idea of the binning methodology is the following. To improve the model accuracy, the device W and L are divided into many bins as shown in Fig. 13.1.1. The minimum/maximum channel length L_{min}/L_{max} and the minimum/maximum channel width W_{min}/W_{max}, as well as the bin assignment, are determined by the parameter extraction engineer according to the anticipated applications. A different set of model parameters is used in each bin. The numbers of bins are selected according to the required model accuracy. Generally, more bins are assigned to the region of short channel lengths and narrow channel widths, and less bins are needed for the region of large channel lengths and wide channel widths.

Most and perhaps all MOS technologies can be well modeled within the bias and geometry range of interest without using the binning approach (with the help of a high-quality parameter extraction tool and an experienced tool user). It is recommended that the binning approach not be used unless it is absolutely necessary. A single bin model is superior for statistical modeling and for device performance optimization.

The geometry dependence given in Eq. (13.3.1) is used for the model parameters that can be binned. A complete list of the model parameters that can be binned is given in Appendix A.

$$P_i = P_{I0} + P_{IL} / L_{eff} + P_{IW} / W_{eff}' + P_{IP} / (W_{eff}' L_{eff}) \qquad (13.3.1)$$

P_{I0} is the zero-order term of the parameter P_i, P_{IL} accounts for the length dependence, P_{IW} accounts for the width dependence, and P_{IP} account for the cross term dependence of both the length and width product.

To use the binning approach, more devices are needed compared with the scalable model parameter extraction shown in Fig. 13.2.1 P_{IL}, P_{IW} and P_{IP} are chosen to guarantee that the model parameters are continuous at the boundaries of the bins.

A general procedure to generate a library of model parameters for the binning approach is the following:

(a) Extract some important model parameters which cannot be binned, such as oxide thickness and effective channel length and width.

(b) Extract a complete set of model parameters for each bin without using P_{IL}, P_{IW}, and P_{IP}.

(c) Find P_{I0}, P_{IL}, P_{IW}, and P_{IP} for each parameter to ensure the continuity of the parameter at all bin boundaries.

13.4 Recommended Value Range of the Model Parameters

According to the physical basis of the model, each model parameter has a range of reasonable values. Reasonable values of parameters can be used to initialize the parameter extraction, and can greatly help to obtain the parameter set quickly, especially in the case of global optimization. Table 13.4.1 gives the recommended ranges of the BSIM3v3 *I-V* model parameters.

Table 13.4.1 Recommended ranges of BSIM3v3 Model Parameters

Symbols in equation	Symbols in source code	Description	Default	Unit	Recommended range of parameters
V_{TH0}	V_{th0}	Threshold voltage @ V_{bs}=0 for large L. Typically V_{th0}>0 for NMOSFET and V_{th0}<0 for PMOSFET	0.7 for nMOS -0.7 for PMOS	V	-2 ~ 2
V_{FB}	*vfb*	Flat band voltage	calculated	V	-2~2
K_1	*k1*	First-order body effect coefficient	0.53	$V^{1/2}$	0~1
K_2	*k2*	Second-order body effect coefficient	-0.0186	none	-0.05~ 0
K_3	*k3*	Narrow width coefficient	80.0	none	10^{-3}~10^2
K_{3B}	*k3b*	Body effect coefficient of K_3	0.0	1/V	-10~10

W_0	*w0*	Narrow width parameter	2.5x10^{-6}	m	10^{-6}~10^{-5}
N_{LX}	*nlx*	Lateral non-uniform doping coefficient	1.74x10^{-7}	m	10^{-8}~10^{-6}
D_{VT0W}	*dvt0w*	First coefficient of narrow width effect on V_{th} at small L	0	none	0~10
D_{VT1W}	*dvt1w*	Second coefficient of narrow width effect on V_{th} at small L	5.3x10^{6}	1/m	0~1/Leff
D_{VT2W}	*dvt2w*	Body-bias coefficient of narrow width effect on V_{th} at small L	-0.032	1/V	-0.05~0
D_{VT0}	*dvt0*	First coefficient of short-channel effect on V_{th}	2.2	none	0~10
D_{VT1}	*dvt1*	Second coefficient of short-channel effect on V_{th}	0.53	none	0~1
D_{VT2}	*dvt2*	Body-bias coefficient of short-channel effect on V_{th}	-0.032	1/V	-0.05~0
V_{BM}	*vbm*	Maximum applied body bias in V_{th} calculation	-3	V	-3 ~ -10
μ_0	uo	Mobility at $T = T_{NOM}$ NMOSFET PMOSFET	 670.0 250.0	cm^2/V/sec	100~1000
U_A	ua	First-order mobility degradation coefficient	2.25x10^{-9}	m/V	10^{-10}~10^{-8}
U_B	ub	Second-order mobility degradation coefficient	5.87x10^{-19}	(m/V)^2	10^{-21}~10^{-18}

U_C	uc	Body-effect of mobility degradation coefficient	***mobMod***=1,2: -4.65×10^{-11} ***mobMod***=3: -0.0465	m/V^2 1/V	Mobmod=1,2: -10^{-11}~-10^{-8} Mobmod=3: -10^{-3}~0
v_{SAT}	vsat	Saturation velocity at $T = T_{NOM}$	8.0×10^4	m/sec	10^4~10^5
A_0	a0	Bulk charge effect coefficient for channel length	1.0	none	0~2
A_{GS}	ags	Gate bias coefficient of the bulk charge effect	0.0	1/V	-1~1
B_0	b0	Bulk charge effect coefficient for channel width	0.0	m	0~10^{-5}
B_1	b1	Bulk charge effect width offset	0.0	m	0~10^{-7}
K_{ETA}	keta	Body-bias coefficient of the bulk charge effect	-0.047	1/V	-10^{-3}~0
A_1	a1	First non-saturation parameter	0.0	1/V	0~0.1
A_2	a2	Second non-saturation parameter	1.0	none	0.4~1
R_{DSW}	rdsw	Parasitic resistance per unit width	0.0	Ω-μm^{Wr}	10^2~10^{-3}
P_{RWG}	prwg	Gate bias effect coefficient of R_{ds}	0	V^{-1}	-10^{-3}~0
P_{RWB}	prwb	Body bias effect coefficient of R_{ds}	0	$V^{-1/2}$	-10^{-3}~0
W_R	wr	Width offset from W_{eff} for R_{ds} calculation	1.0	none	1~5
W_{INT}	wint	Width offset fitting parameter without bias effect	0.0	m	0~3×10^{-7}
L_{INT}	lint	Length offset fitting parameter without bias effect	0.0	m	0~3×10^{-7}

D_{WG}	dwg	Coefficient of W_{eff}'s gate dependence	0.0	m/V	$0\sim10^{-7}$
D_{WB}	dwb	Coefficient of W_{eff}'s body bias dependence	0.0	$m/V^{1/2}$	$0\sim10^{-7}$
V_{OFF}	voff	Offset voltage in the subthreshold region at large *W* and *L*	-0.08	V	-0.15~0
N_{FACTOR}	nfactor	Subthreshold swing factor	1.0	none	0~2
E_{TA0}	eta0	*DIBL* coefficient in subthreshold region	0.08	none	0~1
E_{TAB}	etab	Body-bias coefficient for the subthreshold *DIBL* effect	-0.07	1/V	$-10^{-3}\sim0$
P_{CLM}	pclm	Channel length modulation parameter	1.3	none	0.1~10
P_{DIBLC1}	pdiblc1	First output resistance *DIBL* effect correction parameter	0.39	none	0~1
P_{DIBLC2}	pdiblc2	Second output resistance *DIBL* effect correction parameter	0.0086	none	$10^{-5}\sim10^{-2}$
P_{DIBLCB}	pdiblcb	Body effect coefficient of *DIBL* correction parameters	0	1/V	$-10^{-3}\sim0$
D_{ROUT}	drout	*L* dependence coefficient of the *DIBL* correction parameter in R_{out}	0.56	none	0~1
P_{SCBE1}	pscbe1	First substrate current induced body-effect parameter	4.24×10^{8}	V/m	$10^{8}\sim8\times10^{8}$

P_{SCBE2}	pscbe2	Second substrate current induced body-effect parameter	1.0x10^{-5}	m/V	10^{-9}~10^{-4}
P_{VAG}	pvag	Gate dependence of Early voltage	0.0	none	-10~10
δ	delta	Effective V_{ds} parameter	0.01	V	10^{-3}~0.03
N_{GATE}	ngate	Poly gate doping concentration	0	cm^{-3}	2x10^{18}~ 9x10^{24}
D_{SUB}	dsub	*DIBL* coefficient exponent in sub-threshold region	D_{ROUT}	none	0~1
C_{IT}	cit	Interface trap capacitance	0.0	F/m^2	-10^{-4}~10^{-3}
C_{DSC}	cdsc	Drain/Source to channel coupling capacitance	2.4x10^{-4}	F/m^2	0~10^{-3}
C_{DSCD}	cdscd	Drain-bias sensitivity of C_{DSC}	0.0	F/Vm^2	0~10^{-3}
C_{DSCB}	cdscb	Body-bias sensitivity of C_{DSC}	0.0	F/Vm^2	-10^{-4}~0

13.5 Automated Parameter Extraction Tool

The task of parameter extraction can be greatly simplified with the help of an automated software tool. A high quality tool provides the local and global optimization routines and supports the single device and group device approaches, as well as the single-bin and multi-bin methodologies. Such a tool can usually extract the parameter for many popular MOSFET compact models as well as bipolar transistor models. Convenient user interfaces including curve plotting functions are among the features that can be expected from these tools.

All the example model files used in this book are extracted using the parameter extraction software - BSIMPro [13.14].

References

[13.1] D. E. Ward and K. Doganis, “Optimized extraction of MOS model parameters,” *IEEE Trans. Computer-aided Design*, CAD-1, pp.163-168, 1982.

[13.2] P. R. Karlsson and K. O. Jeppson, “An efficient parameter extraction algorithm for MOS transistor models,” *IEEE Trans. Electron Devices*, vol. 39, No. 9, pp. 2070-2076, 1992.

[13.3] L. C. W. Dixon and G. P. Szego (Eds.), *Towards Global Optimization*, North-Holland, Amsterdam, 1979.

[13.4] M. F. Hamer, “First-order parameter extraction on enhancement silicon MOS transistors,” *IEE Proc.* vol. 133, Pt. I, pp.49-54, 1986.

[13.5] S. J. Wamg, J. Y. Lee, and C. Y. Chang, “An efficient and reliable approach for semiconductor device parameter extraction,” *IEEE Trans. Computer-aided Design,* CAD-6, pp. 170-178, 1986.

[13.6] B. Ankele et al., “Enhanced MOS parameter extraction and SPICE modeling,” *Proc. IEEE Int. Conf. on Microelectronic Test Structures*, vol.2, pp. 73-78, 1989.

[13.7] J. H. Huang et al., *BSIM3 Manual (Version 2.0)*, University of California, Berkeley, March 1994.

[13.8] Y. Cheng et al., *BSIM3 version 3.0 User's Manual*, University of California, Berkeley, 1995.

[13.9] Y. Cheng et al., “An investigation on the robustness, accuracy and simulation performance of a physics-based deep-submicrometer BSIM model for analog/digital circuit simulation,” *CICC’96*, pp. 321-324, May 1996.

[13.10] Y. Cheng et al., *BSIM3 version 3.1 User's Manual*, University of California, Berkeley, Memorandum No. UCB/ERL M97/2, 1997.

[13.11] B. J. Sheu et al., "BSIM: Berkeley short -channel IGFET model for MOS transistors," *IEEE J. solid-state Circuits*, vol. SC-22, pp.558-565, 1987.

[13.12] J. S. Duster et al., *User's guide for BSIM2 parameter extraction program and the SPICE3 with BSIM implementation*, University of California, Berkeley, 1988.

[13.13] M. C. Jeng et al., *Theory, algorithm, and users's guide for BSIM and SCALP*, Memorandum No. UCB/ERL M87/35, University of California, Berkeley, 1987.

[13.14] *BSIMpro Manual*, BTA Inc., Old Ironsides Drive, Santa Clara, CA, 1996 (http://www.btat.com).

CHAPTER 14

RF and Other Compact Model Applications

Since BSIM3v3 was selected to start the industry standardization of MOSFET compact models in 1995, it has been widely used for digital and analog circuit design. In this chapter, we discuss some examples of new applications using BSIM3v3 in RF modeling, statistical modeling, and technology prediction.

14.1 RF Modeling

With the advent of submicron technologies, GHz RF circuits can now be realized in a standard CMOS process [14.1]. A major barrier to the realization of commercial CMOS RF components is the lack of adequate models which accurately predict MOSFET device behavior at high frequencies. The conventional microwave table-look-up approach requires a large database obtained from numerous device measurements. This method becomes prohibitively complex when used to simulate highly integrated CMOS communication systems. Furthermore, in this measurement-to-table approach, there is no built-in check and correction for the considerable errors of the measurement data. Hence, a compact model, valid for a broad range of bias conditions, device sizes, and operating frequencies is desirable.

A complete compact RF model does not yet exist in commercial circuit simulators. A common modeling approach for RF applications is to build sub-circuits based on MOSFET models that are suitable for analog/digital applications. BSIM3v3-based subcircuit models have been reported and tested against measured high frequency data [14.2, 14.3, 14.4, 14.5, 14.6]. In the sub-circuit, parasitic elements around gate, source, drain, and substrate are added to improve the model accuracy at high frequency.

An example of a subcircuit model for RF applications is given in Fig. 14.1.1 [14.3]. Similar subcircuit models with simpler or more complex substrate RC networks have also been reported [14.5, 14.6, 14.7]. An important part of RF modeling is to establish physical and scalable model equations for the parasitic elements at the source, drain, gate and substrate. The scalability of the intrinsic device is ensured by the core model, for example, BSIM3v3.

We will first discuss the modeling of the gate and substrate resistances. Then we will give an example of using the BSIM3v3-based RF MOSFET model in GHz circuit design.

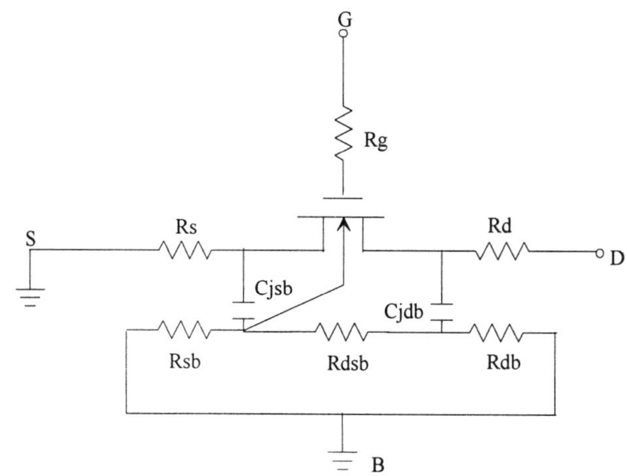

Fig. 14.1.1 A subcircuit with parasitic elements added to an intrinsic MOSFET model for RF circuit simulation. After Cheng et al. [14.3].

14.1.1 Modeling of the gate resistance

Gate resistance of a MOSFET at low frequency can be calculated with the following simple equation:

$$Rg = \frac{W_{eff}}{L_{eff}} R_{gsquare} \tag{14.1.1}$$

where $R_{gsquare}$ is the gate sheet resistance per square, W_{eff} is the effective channel width, and L_{eff} is the effective channel length. The typical sheet resistance for a polysilicon gate ranges between 20-40 Ω/square, and can be reduced by a factor of 10 with a silicide process, and even more with a metal stack process.

The correct modeling of the gate resistance at high frequency is more complex because of the distributed transmission-line effect. The lumped equivalent gate resistance is 1/3 of the end-to-end resistance [14.8]

$$Rg = \frac{W_{eff}}{3L_{eff}} R_{gsquare} \tag{14.1.2}$$

The factor 3 is introduced to account for the distributed RC effects when the gate electrode is contacted at only one end. The factor is 12 when the electrode is contacted on both ends. The derivation of Eq. (14.1.2) can be found in the literature, for example, [14.9].

However, it has been found that not only the distributed RC effect of the gate but also the non-quasi-static effect or the distributed RC effect of the channel must be accounted for in modeling MOSFET high frequency behavior. It is convenient to add an additional component to the gate resistance to represent the channel distributed RC effect. A physical effective gate resistance model incorporating the first-order non-quasi-static effect and the distributed gate resistance has been developed [14.10], and is discussed next.

When a MOSFET operates at high frequency, the contribution to the effective gate resistance is not only from the physical gate electrode resistance but also from the distributed channel resistance which can be "seen" by the signal applied to the gate. Thus, the effective gate resistance consists of two parts: the distributed gate electrode resistance (R_{geltd}) and the distributed channel resistance seen from the gate (R_{gch}), as shown in Fig. 14.4.2 [14.10].

$$Rg = R_{geltd} + R_{gch} \tag{14.1.3}$$

Since R_{geltd} is insensitive to bias and frequency, its value can be obtained from the gate electrode sheet resistance (R_{eltd}),

$$R_{geltd} = R_{eltd}(\alpha W / L + \beta) \tag{14.1.4}$$

where α is 1/3 when the gate terminal is brought out from one side, and 1/12 when connected on both sides. β models the external gate resistance.

There are two origins of R_{gch}: one is the static channel resistance (R_{st}), which accounts for the DC channel resistance. The other is the excess-diffusion channel resistance (R_{ed}) due to the change of channel charge distribution by the AC excitation of the gate voltage. R_{st} and R_{ed} together determine the time constant of the non-quasi-static effect. R_{st} is modeled by integrating the resistance along the channel under the quasi-static assumption,

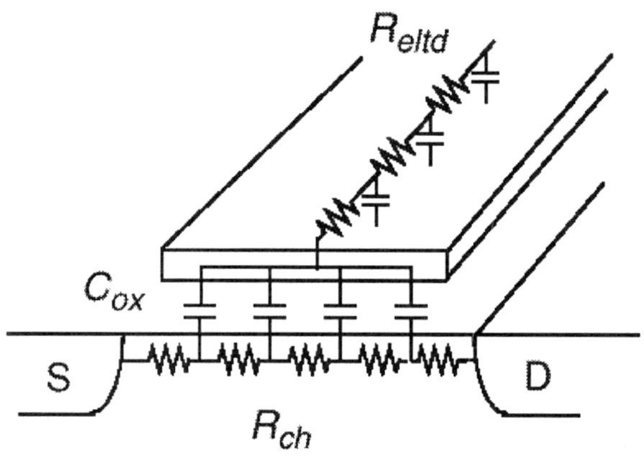

Fig. 14.1.2 Illustration of the distributed nature of gate electrode resistance R_{eltd}, channel resistance R_{ch}, and gate capacitance C_{ox}. After Jin et al. [14.10].

$$R_{st} = \int dR = \int dV / I_d$$

$$= V_{ds} / I_{ds} \text{ in triode region; or} \tag{14.1.5}$$

$$= V_{dsat} / I_{ds} \text{ in saturation region,} \tag{14.1.6}$$

where V_{dsat} is the saturation drain voltage.

R_{ed} can be derived from the diffusion current as

$$R_{ed} = \frac{qL_{eff}}{\eta W_{eff} \mu C_{ox} K_B T} \tag{14.1.7}$$

where η is a technology-dependent constant.

The overall channel resistance seen from the gate is

$$\frac{1}{R_{gch}} = \gamma\left(\frac{1}{R_{st}} + \frac{1}{R_{ed}}\right) \tag{14.1.8}$$

where γ is a parameter accounting for the distributed nature of the channel resistance and C_{ox} (see Fig. 14.1.2). γ is 12 if the resistance is uniformly distributed along the channel [14.10]. Since this assumption is not valid in the saturation region, γ is left as a fitting parameter.

To extract R_g, two-port S-parameters are converted to Y-parameters and the input resistance is

$$R_{in} = real(1 / Y_{11}) \tag{14.1.9}$$

where the gate is connected to port 1 and the drain to port 2. R_{in} includes the influence of R_g and the source/drain resistance (R_s and R_d). Since R_s and R_d are known from DC measurement, the value of R_g can be extracted from R_{in}.

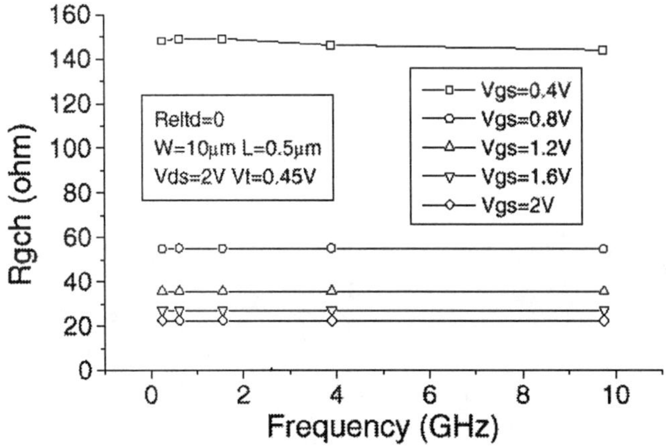

Fig. 14.1.3 2-D simulation of the effective channel resistance R_{gch} extracted from Y_{11}. Gate electrode sheet resistance R_{eltd} is set to zero. After Jin et al. [14.10].

In order to verify this physical model of R_{gch}, 2-D simulation of Y_{11} was carried out with R_{eltd} set to zero. Fig. 14.1.3 shows the simulation results of a 10μm/0.5μm NMOS at various biases and frequencies. Essentially R_{gch} is independent of frequency but sensitive to bias. The results are rearranged in Fig. 14.1.4 to emphasize the bias dependency of R_{gch}. Good agreement between 2-D simulation and the proposed model is observed with $\eta=1$ and $\gamma=14$. Note the model will deviate from data at low V_{gs} if R_{ed} is ignored. Fig. 14.1.5 shows the good agreement in both the triode and saturation regions.

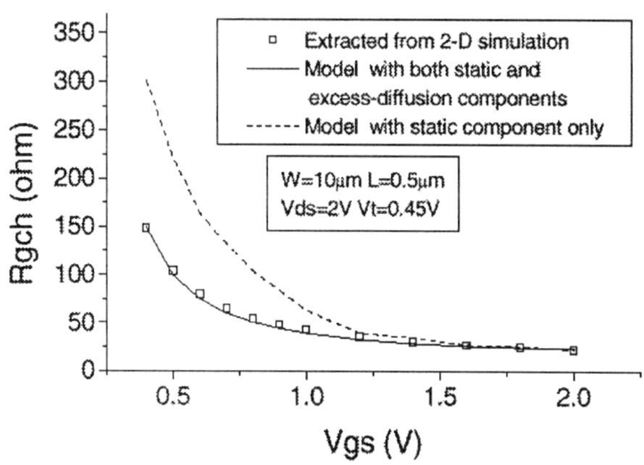

Fig. 14.1.4 The effective R_g model agrees well with 2-D simulation. After Jin et al. [14.10].

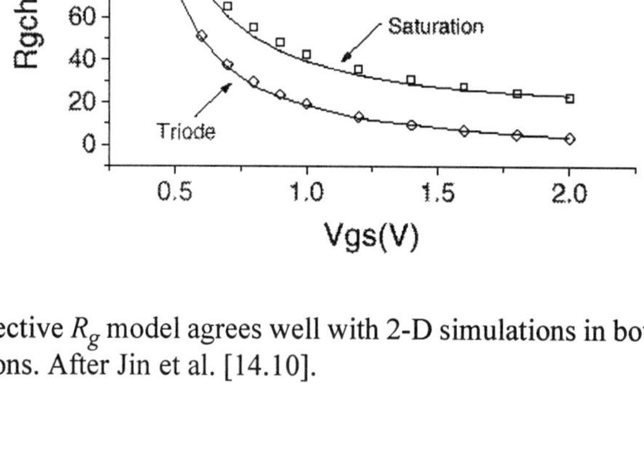

Fig. 14.1.5 Effective R_g model agrees well with 2-D simulations in both the triode and saturation regions. After Jin et al. [14.10].

RF device test patterns were designed and fabricated in a 0.35μm process. Fig. 14.1.6 shows the measured R_g at various biases and frequencies. Fig. 14.1.7 shows an excellent agreement between the data and the model. In this case R_{geltd} contributes less than 1Ω of the total effective R_g. This R_g model is accurate up to $f_T/3$ of the MOSFET [14.10].

The gate resistance plays a very significant role in RF noise modeling because R_g not only affects the input impedance but also contributes to the thermal noise. The overall thermal noise consists of three parts: one is the noise from R_g; another is the noise from the physical resistance R_d, R_s, R_{db}, R_{sb}, and R_{dsb} (as shown in Fig. 14.1.1). The third component is the channel thermal noise current which can be described by its power spectral density as [14.11]

$$Sid = \frac{4KB}{LeffIds} \int_0^{Vds} Te(V)(\mu_{eff}W_{eff}Q_{inv}(V) - \frac{Ids}{Ec})^2 \qquad (14.1.10)$$

where T_e is the electron temperature accounting for the hot electron effect, μ_{eff} is the effective carrier mobility, Q_{inv} is the inversion layer charge density, and E_c is the critical electrical field when carriers reach the saturation velocity.

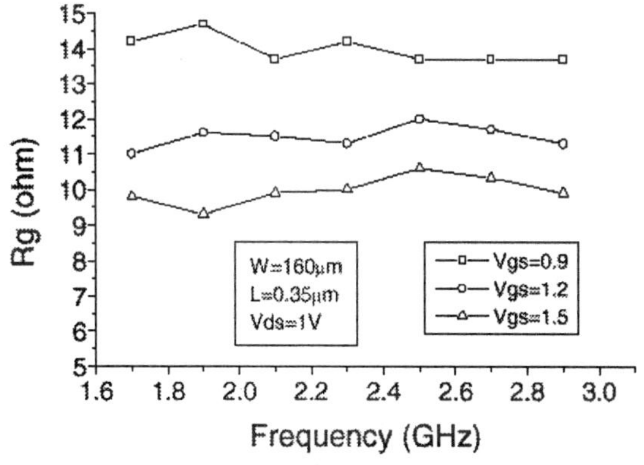

Fig. 14.1.6 R_g extracted from s-parameter data of a 16-finger n-MOSFET. After Jin et al. [14.10].

To obtain the overall noise of a device, the direct calculation method [14.12] is employed. Fig. 14.1.8 shows good agreement between the measured and modeled minimum noise figure (N_{Fmin}) at various biases and frequencies. Fig. 14.1.9 shows N_{Fmin} vs. bias current with a fixed V_{ds}. A minimum N_{Fmin} is observed at a particular bias condition, an important point for RF circuits such as low-noise amplifiers (LNA). These comparisons show that the proposed effective gate resistance model accurately predicts the device noise behavior in addition to the input impedance and other AC parameters of CMOS devices.

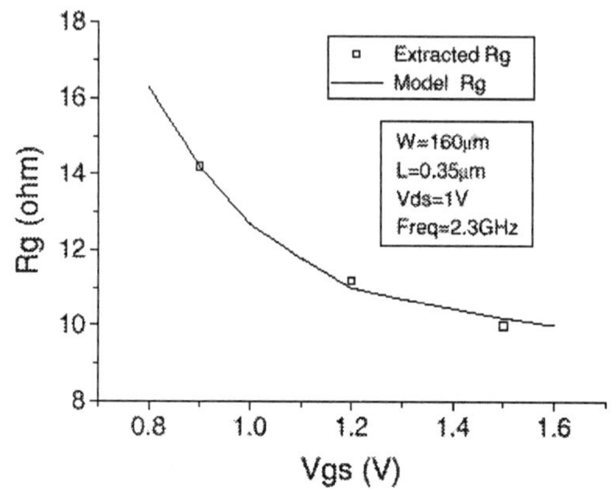

Fig. 14.1.7 Effective R_g model agrees well with measured data at various V_{gs}. The gate electrode resistance R_{geltd} contributes less than 1Ω of the overall R_g. After Jin et al. [14.10].

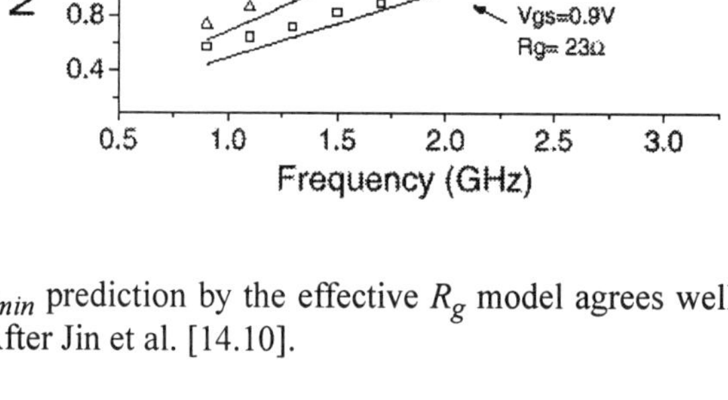

Fig. 14.1.8 N_{Fmin} prediction by the effective R_g model agrees well with data at two different V_{gs}. After Jin et al. [14.10].

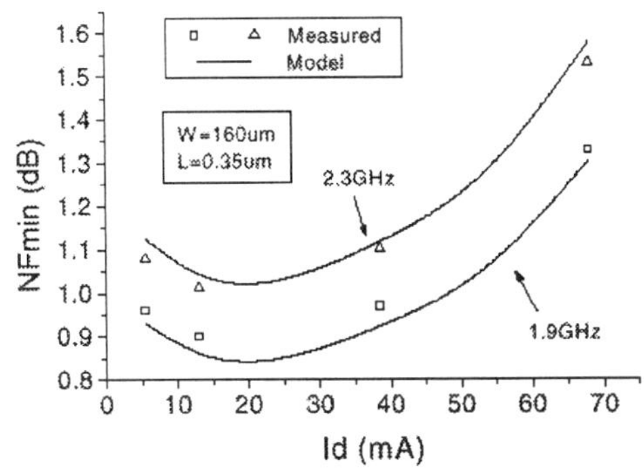

Fig. 14.1.9 Comparison of N_{Fmin} between the measured data and the model. A minimum N_{Fmin} is observed at a particular bias. After Jin et al. [14.10].

14.1.2 Modeling the substrate network

Modeling of the substrate parasitic elements is not necessary in low-frequency digital/analog applications. However, it has been found that the substrate network is very important in RF applications [14.3, 14.22]. To obtain the desired scalable RF model, a scalable model for the substrate components is critical.

An equivalent circuit has been proposed to describe the high frequency contribution of the substrate parasitic elements, as shown in Fig. 14.1.10 [14.3].

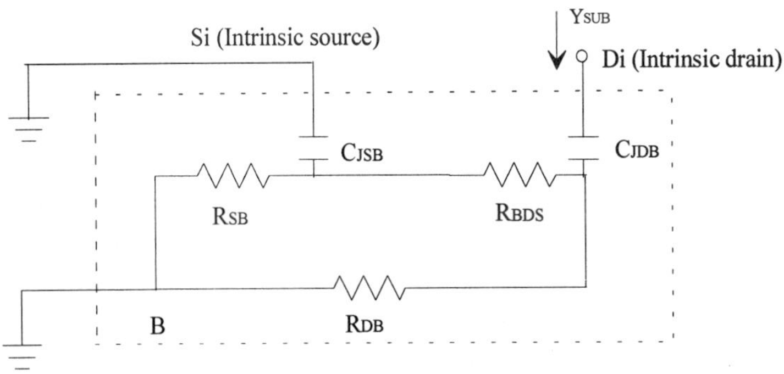

Fig. 14.1.10 An equivalent circuit for the substrate network. C_{jsb} and C_{jdb} are capacitances of source/bulk and drain/bulk junctions. After Cheng et al. [14.3].

As an example, Fig. 14.1.11 gives a comparison of the Y parameters from the two port substrate network and device simulation [14.3]. The RC network model is accurate up to at least 10GHz for devices with 0.35μm channel length. The above substrate network has been implemented in a complete sub-circuit model for a RF MOSFET, and has been verified with measured data for its accuracy [14.3, 14.22]. Again, good match of Y_{22} between the model and measured data shows that this substrate network is accurate up to 10GHz, as shown in Fig. 14.1.12 for a device with W/L of 120μm/0.36μm (10 fingers).

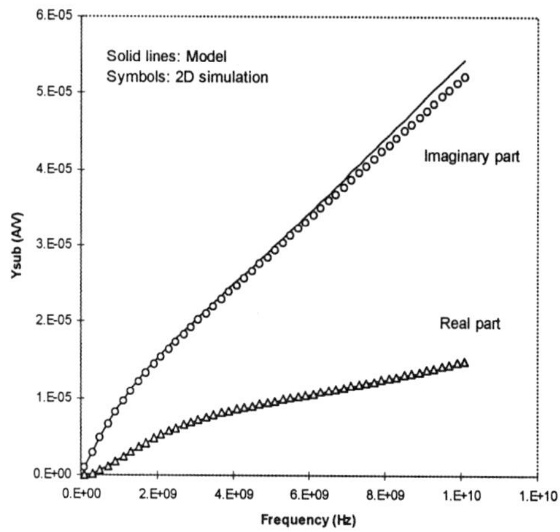

Fig. 14.1.11 Comparison of two port Y_{sub} parameters (see Fig. 14.1.10) between the model and 2-D device simulation for a 3 finger device with 0.35μm channel length at different V_{ds}. After Cheng et al. [14.3].

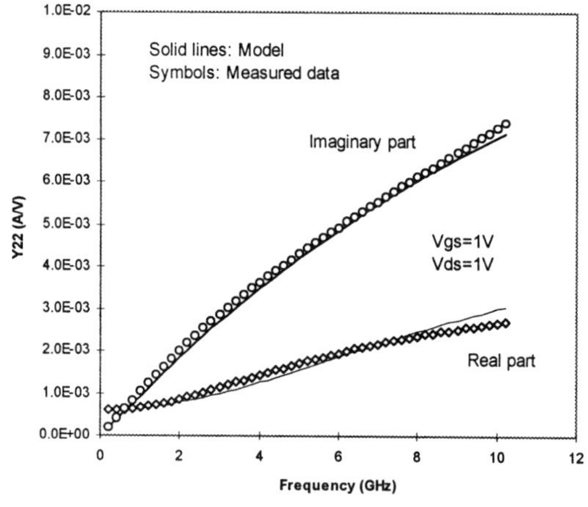

Fig. 14.1.12 Comparison of modeled and measured Y_{22} vs. frequency at V_{gs}=V_{ds}=1V for a device with 0.36μm channel length. After Cheng et al. [14.3].

14.1.3 A RF MOSFET model based on BSIM3v3 for GHz communication IC's

In this section, we present a unified device model realized with a lumped resistance network suitable for simulations of both RF and baseband analog circuits. This model has been verified with measured data on both the device and circuit levels [14.6].

The RF MOSFET model is realized with the addition of three resistors R_g, R_{subd}, and R_{subs} to the existing BSIM3v3.1 model (shown in Fig. 14.1.14). R_g models both the physical gate resistance as well as the non-quasi-static (NQS) effect. R_{subd} and R_{subs} are the lumped substrate resistances between the source/drain junctions and the substrate contacts. The values of R_{subd} and R_{subs} may not be equal as they are functions of the transistor layout (illustrated in Fig. 14.1.15).

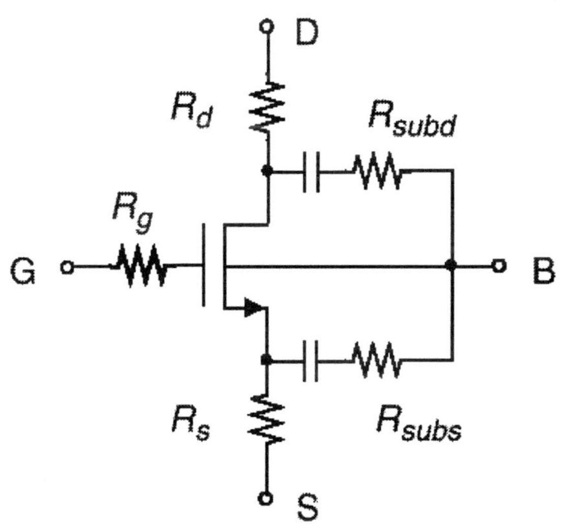

Fig. 14.1.14 A MOSFET RF model based on BSIM3v3. After Ou et al. [14.6].

To demonstrate the accuracy of the model, S-parameters of the BSIM3v3 RF model and measured data of a 0.35μm NMOS device are plotted in Fig. 14.1.16. The BSIM3v3 model without the subcircuit components is also given in the plot. The improvement can be clearly seen in S_{22} but hardly in S_{11}. A better picture with more physical insight may be obtained by separating the terminal impedance into the real and imaginary parts with the following six parameters,

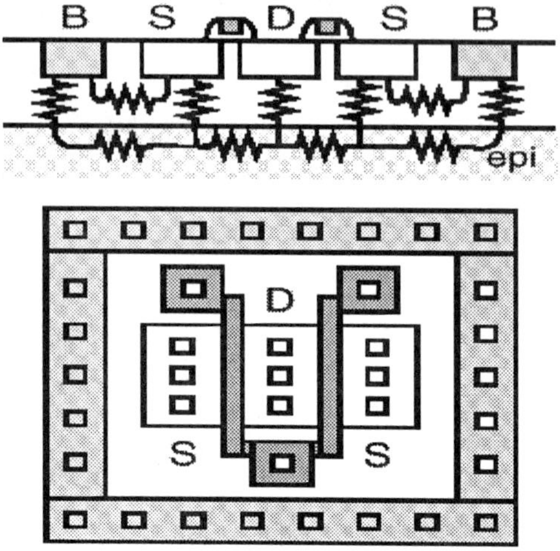

Fig. 14.1.15 Cross-sectional view and top view of a typical transistor cell layout. After Ou et al. [14.6].

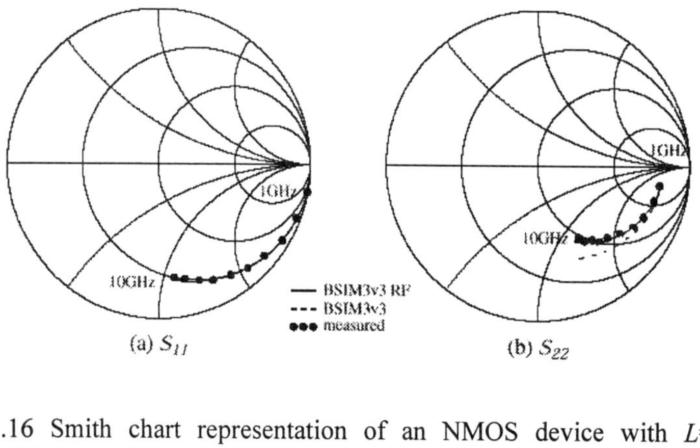

Fig. 14.1.16 Smith chart representation of an NMOS device with L=0.35μm, W=160μm, W_{finger}=10μm, V_{gs}=2V, and V_{ds}=2V. After Ou et al. [14.6].

$$R_{in} = \text{real}(1/Y_{11}) \qquad (14.1.11)$$

$$C_{in} = -1/\text{imag}(1/Y_{11})/\omega \qquad (14.1.12)$$

$$R_{out} = 1/\text{real}(Y_{22}) \qquad (14.1.13)$$

$$C_{out} = \text{imag}(Y_{22})/\omega \qquad (14.1.14)$$

$$g_m = \text{real}(Y_{21}) \qquad (14.1.15)$$

$$C_{fb} = -\text{imag}(Y_{12})/\omega \qquad (14.1.16)$$

where ω is the frequency in rad/s (gate is port1, drain is port2, and body is shorted to the source).

As seen in Fig. 14.1.17, excellent agreement up to 10GHz has been achieved. In particular, the proposed model significantly improves the accuracy of the model for R_{in} and R_{out} over a wide frequency range.

A unique problem in modeling CMOS is the body bias effect. To study this effect at high frequencies, a 2-D device simulator was used to generate both dc and S-parameter data. The results show that the body bias mainly affects the device dc characteristics but not the high frequency behavior. Fig. 14.1.18 shows the simulated R_{out} and C_{out} of a 0.5μm NMOS device with various body biases. Good agreement has been achieved between the 2-D simulation and the proposed model.

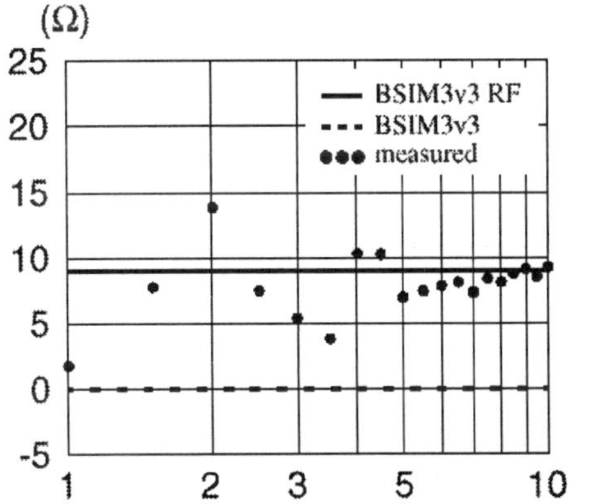

(a) R_{in}

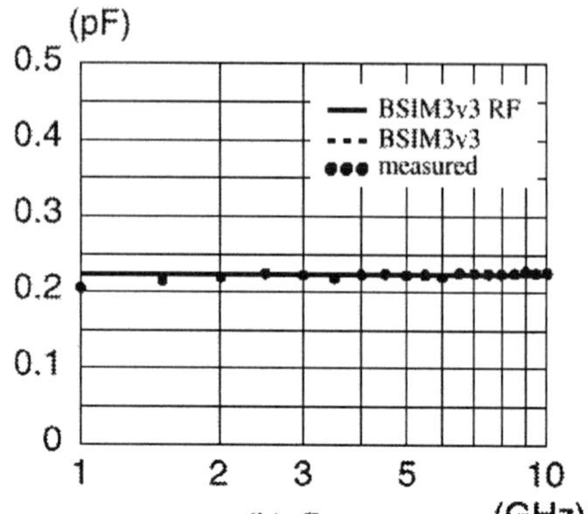

(b) C_{in}

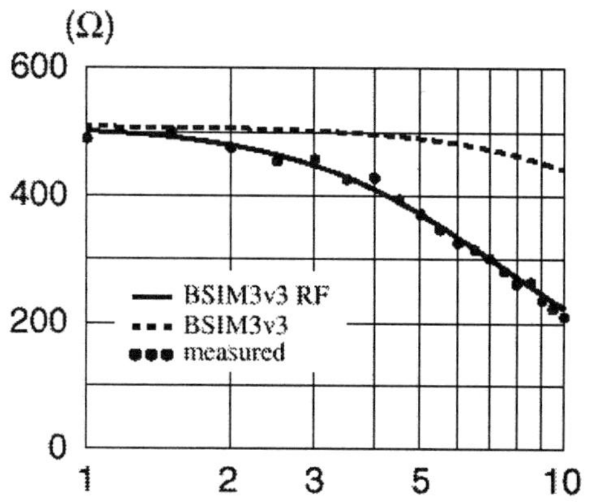

(c) R_{out}

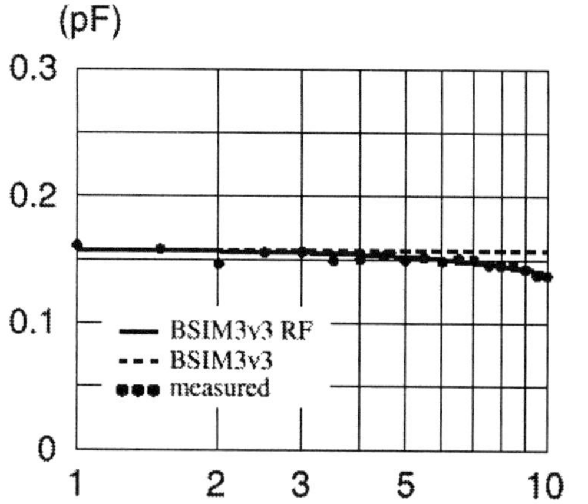

(d) C_{out}

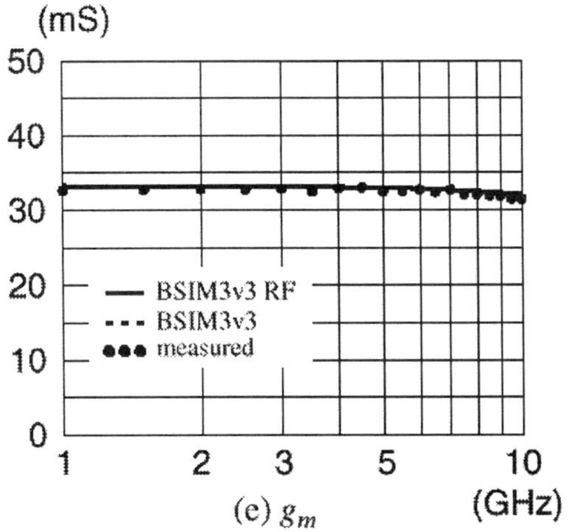

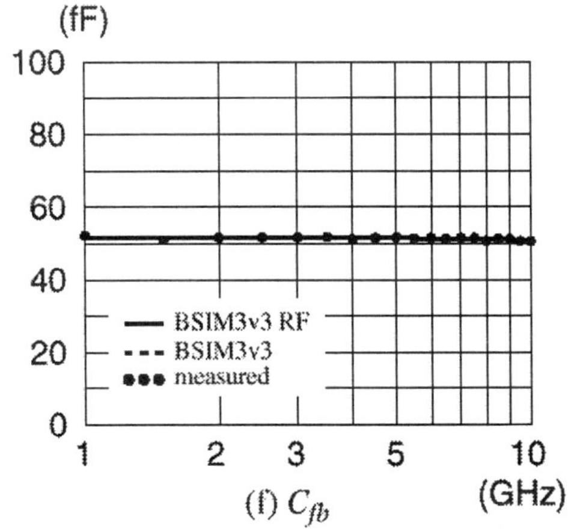

Fig. 14.1.17 Terminal impedance illustration of an NMOS device with L=0.35μm, W=160μm, W_{finger} =10μm, V_{gs}=2V and V_{ds}=2V. R_g =9Ω and R_{subd} =90Ω are extracted in this case. After Ou et al. [14.6].

R_g can be extracted in part from the gate sheet resistance. With the NQS effect, a lumped R_g may be obtained from the measured R_{in}. For a fixed cell layout, R_{subd} can be extracted from R_{out} by connecting the drain as port 2. Similarly R_{subs} is found by using the source as port 2. It is recommended to adjust the low frequency source/drain junction capacitance to fit the S-parameter data to account for distributed RC effects and any measurement inaccuracies.

To test the robustness of the BSIM3v3-based RF model, a circuit level evaluation was performed using two different approaches. As the first example, a 5GHz single-ended low-noise amplifier (LNA) using a 0.35μm device was simulated, as shown in Fig. 14.1.19 (a). The table-lookup method was employed to compute the overall circuit performance from the measured device data and the results were compared with SPICE simulation using the compact model shown in Fig. 14.1.14. Fig. 14.1.19(b) shows good agreement between the two methods for S_{21}.

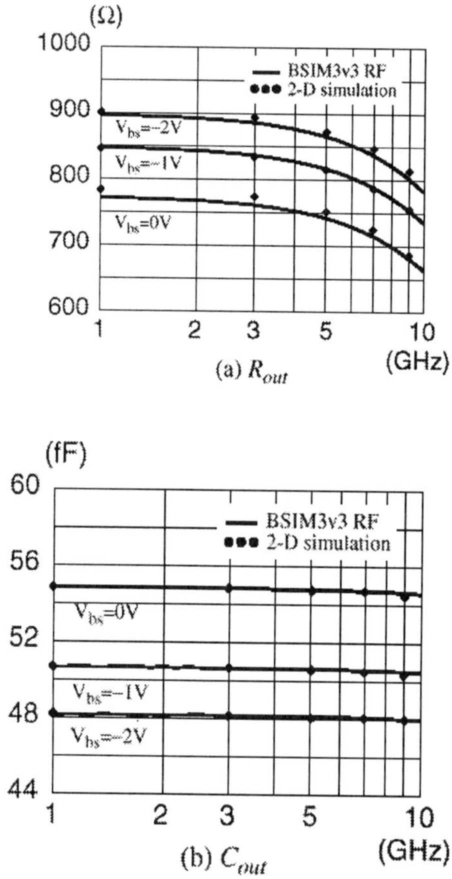

Fig. 14.1.18 2-D simulation result vs. BSIM3v3 RF model at various body biases. L=0.5μm, W=100μm. After Ou et al. [14.6].

In the second example, a 2GHz differential LNA was designed and fabricated in a 0.6μm CMOS process, as shown in Fig. 14.1.20(a). Fig.14.1.20(b) shows the measured voltage gain compared to the simulated results. Clearly the low frequency BSIM3v3 model overestimates the peak voltage gain by 2dB, while the RF compact model accurately predicts the circuit performance within the frequency range of interest.

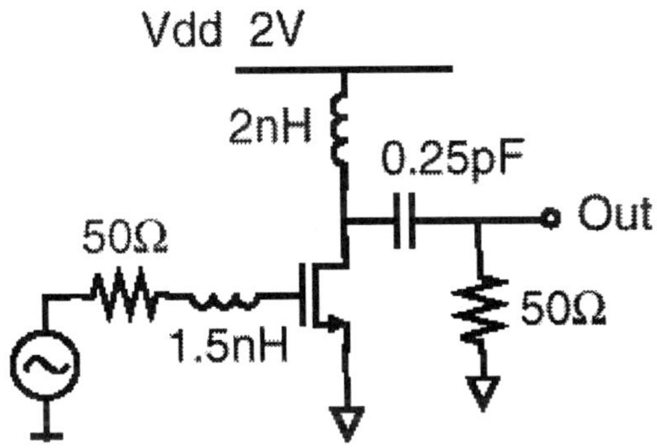

(a) circuit schematic

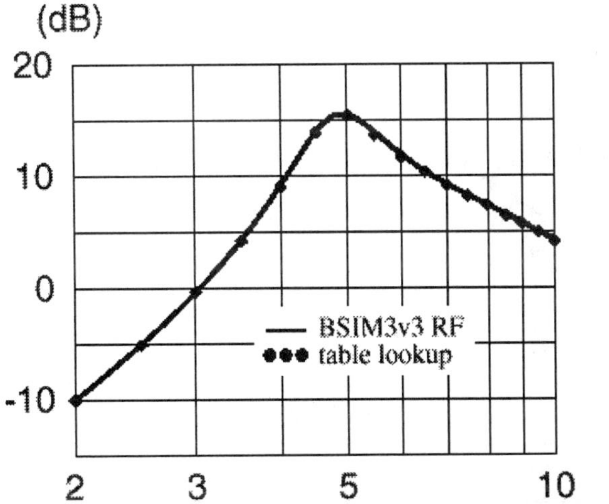

(b) $|S_{21}|$

Fig. 14.1.19 A single-ended 5GHz LNA using a $160\mu m/0.35\mu m$ NMOS device. After Ou et al. [14.6].

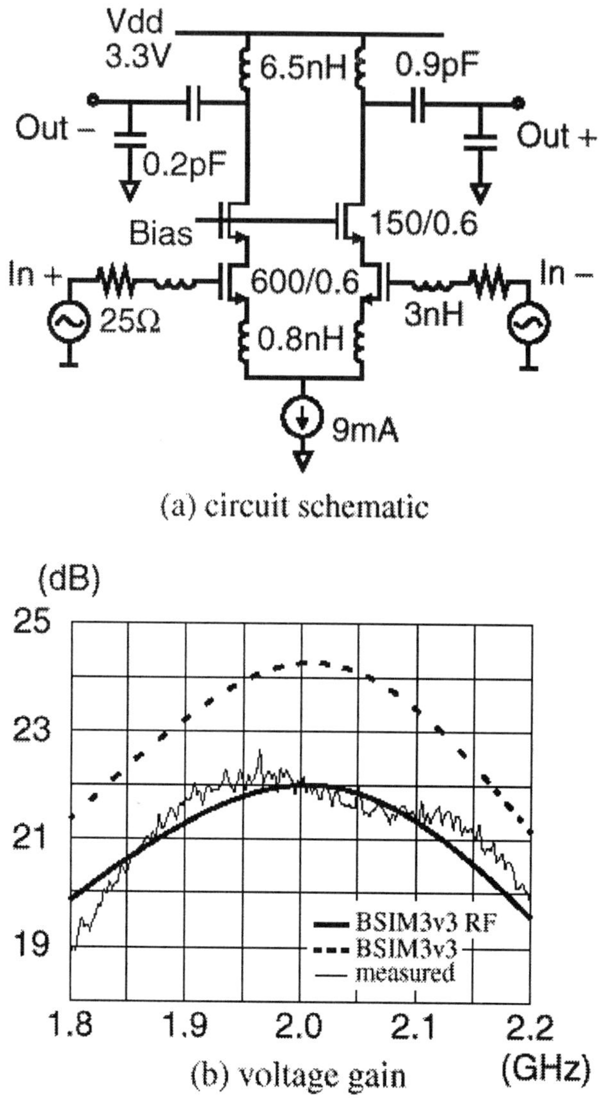

Fig. 14.1.20 A 2GHz differential LNA designed and fabricated in a 0.6μm CMOS process. After Ou et al. [14.6].

14.2 Statistical Modeling

Because the statistical variation in device and process parameters such as channel length, channel width, threshold voltage, channel/substrate doping, and oxide thickness do not always scale with the parameters themselves, device mismatch and performance variation tend to increase as the device feature size is reduced. To improve circuit reliability and yield, statistical design techniques based on physical statistical modeling will be highly desirable [14.13, 14.14, 14.15].

In order to meet the above requirements, a physical statistical device model with built-in physical model parameters and clear extraction methodology is needed. Recently, some statistical modeling work based on BSIM3v3 has been reported [14.16, 14.17]. These studies developed statistical parameter extraction methodologies that translate actual process variation into SPICE model parameter variation and generate the worst-case models from electrical-test (E-T) data.

Fig.14.2.1 illustrates the basic idea of Electrical-Test Based Statistical Modeling (EBSM). From a nominal die, one SPICE model file is extracted for a group of devices. This step involves the traditional use of a SPICE model parameter extractor. The extracted parameters are then grouped into two parts: those which are assumed to vary between different die locations and those which do not (fixed at extracted nominal values). For the former SPICE parameters, E-T data is either used as direct replacements or as inputs into a calculator that produces the corresponding values. This procedure is repeated for all dies of interest. Using E-T data as direct replacements for SPICE model parameters is relatively straightforward. Many BSIM3v3 parameters have direct counterparts in the E-test data (i.e. $\boldsymbol{T_{OX}}$, dL, dW, $\boldsymbol{N_{CH}}$, and R_{ds}) [14.18]. However, to ensure the methodology actually produces parameters that reflect the I-V characteristics accurately, other electrical parameters from the E-T database such as I_{dsat} (saturation current) can be used to calculate other SPICE model parameters. This calculation amounts to solving the BSIM3v3 current equation. For example, a measured I_{dsat} value can be inserted on the left hand side of the current expression. The BSIM3v3 current expression with the specific model parameter of unknown value is on the right hand side. Once a solution is found, the calculated SPICE model parameter value will guarantee that the simulated I_{dsat} value at that bias condition will be equal to the measured value. This idea can be extended to calculate other SPICE model parameters by solving several equations simultaneously. There is flexibility in deciding which electrically measured E-T parameters are used in the calculation of other SPICE model parameters. However, special attention should be paid to the fact that a SPICE model parameter may be more sensitive to a particular E-T parameter at certain bias points than others.

E-T Based Statistical Modeling was used to model device performance variations (Fig. 14.2.2) for an experimental 0.5μm process. Five E-T parameters ($\boldsymbol{T_{OX}}$, dL, dW, $\boldsymbol{K_1}$, R_{ds}) were used as direct replacements for BSIM3v3 model parameters. In additional, 9 electrical E-T data were chosen to ensure accurate modeling of threshold voltage and saturation current. These are listed in Table 14.2.1 and were used to calculate 9 NMOS BSIM3v3 parameters. PMOS parameters were similarly calculated. This process was repeated for 220 dies from 5 wafers and 2 lots. The average time spent was 30 seconds per die. Fig. 14.2.3 and Fig. 14.2.4 show a comparison between simulated and E-T measured threshold voltages as functions of L_{drawn} and W_{drawn}, respectively [14.16].

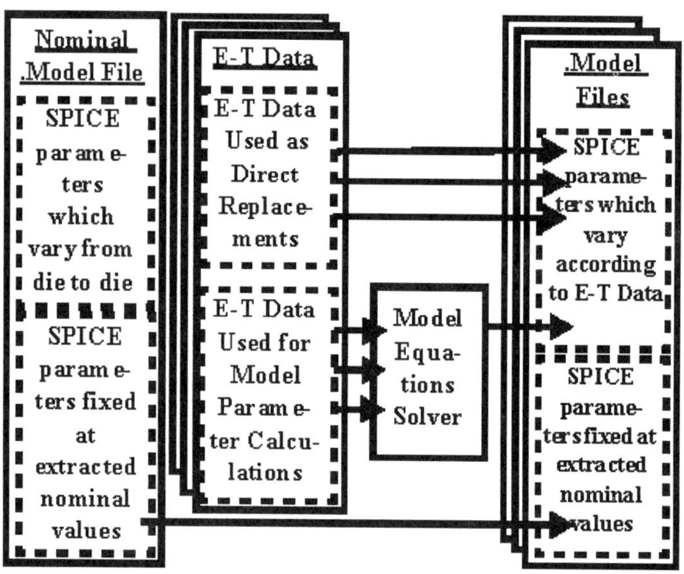

Fig. 14.2.1 Illustration of E-T based statistical modeling methodology. After Chen et al. [14.16].

Fig. 14.2.5 shows simulated and measured drain current characteristics for the 95th percentile (or the highest 5%) current characteristics for a W/L=20μm/0.5μm NMOS transistor. The RMS error is 4.76%. Excellent agreement was also observed (RMS error of 4.31%) for the 5th percentile (or the lowest 5%) drain current characteristics. These observations illustrate the ability of E-T

Based Statistical Modeling methodology to model a wide variety of process induced device performance variations.

The distribution of one calculated BSIM3v3 parameter is shown in Fig. 14.2.6. To assess overall *I-V* curve fit, the RMS error percentage is calculated between measured and simulated I_d -V_d curves for all dies. A distributive nature is observed (Fig. 14.2.7). The results are summarized in Table 14.2.2 and display excellent fit for all device geometries and both MOS channel types.

Table 14.2.1 NMOS E-T data used to calculate BSIM3v3 model parameters [14.16].

				BSIM3v3 SPICE Parameters for the Following Speed Percentiles						
Calculated NMOS BSIM3v3 Parameter	NMOS E-T Data Used for Calculation	Mean	Sigma	(fast) 2%	10%	25%	50%	75%	90%	(slow) 98%
Vth0	Vth (20μm/20μm)	0.5839	0.00848	0.5472	0.5712	0.5803	0.5859	0.5894	0.5826	0.5870
Nlx (x10^{-7})	Vth (20μm/1μm)	1.958	0.1193	1.817	2.110	1.923	1.966	2.068	1.913	1.965
Dvt1	Vth (20μm/0.6μm)	0.3526	0.03425	0.4441	0.3638	0.3812	0.3320	0.3311	0.2971	0.3416
Dvt0	Vth (20μm/0.5μm)	3.528	0.7705	6.515	3.875	4.295	2.895	2.795	2.325	3.545
K3	Vth (5μm/0.5μm)	25.47	15.05	111.756	44.636	36.056	21.356	6.586	23.996	26.666
Dvt0w	Vth (1μm/0.5μm)	0.4423	0.4976	2.409	1.371	0.965	0.244	-0.404	0.317	0.490
U0	Idsat (20μm/20μm)	340.6	4.279	340.2	349.7	350.5	342.7	337.1	337.2	339.3
Vsat	Idsat (20μm/0.5μm)	142100	12460	148570	121090	127960	145280	159930	145590	143590
Dwg (x10^{-9})	Idsat (1μm/0.5μm)	-8.006	6.383	1.119	-0.991	-1.472	-0.442	-0.766	-0.597	-0.901

Note: Similar calculators were performed for PMOS transistors. Percentile columns represent 7 SPICE files which can be used for statistical circuit simulations.

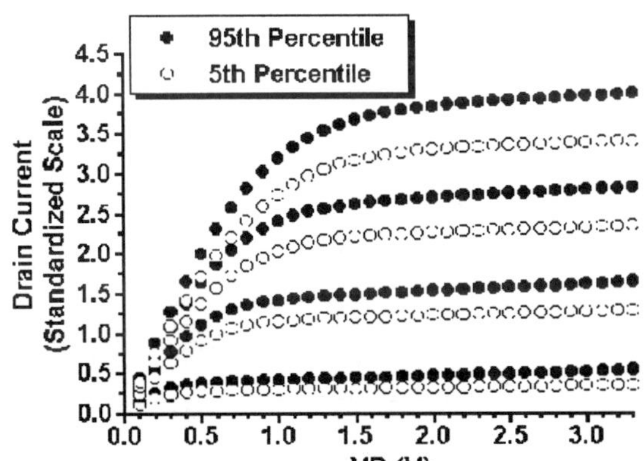

Fig. 14.2.2. Comparison of the 95th percentile and 5th percentile measured current characteristics for W/L=20μm/0.5μm. After Chen et al. [14.16].

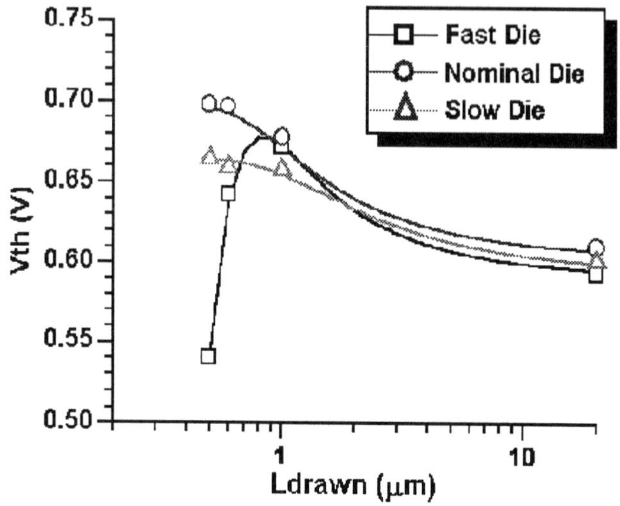

Fig. 14.2.3. Comparison between simulated (solid line) and measured (markers) threshold voltage characteristics for three different dies as a function of L_{drawn}. After Chen et al. [14.16].

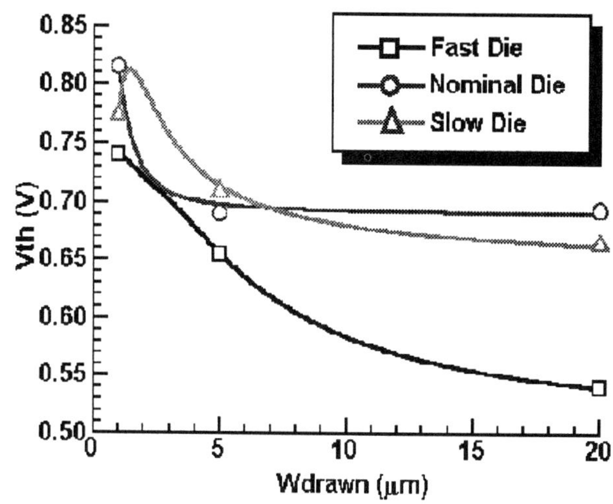

Fig.14.2.4. Comparison between simulated (solid line) and measured (markers) threshold voltage characteristics for three different dies as a function of W_{drawn}. After Chen et al. [14.16].

The total number of extracted SPICE files is only limited by the number of available E-T data sets. These extracted SPICE files can then be used immediately (or after Principal Component transformation) for Compact Statistical Circuit Simulation (CSCS) to be described below.

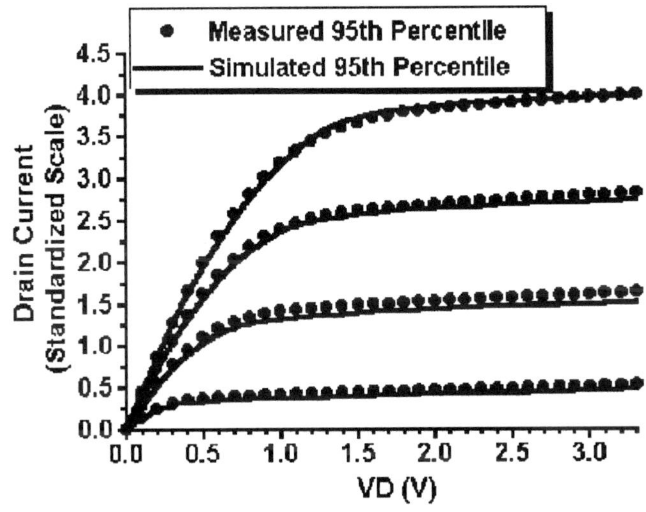

Fig. 14.2.5. Comparison between simulated (solid line) and measured (markers) 95th percentile drain current characteristics. The RMS error is 4.76%. After Chen et al. [14.16].

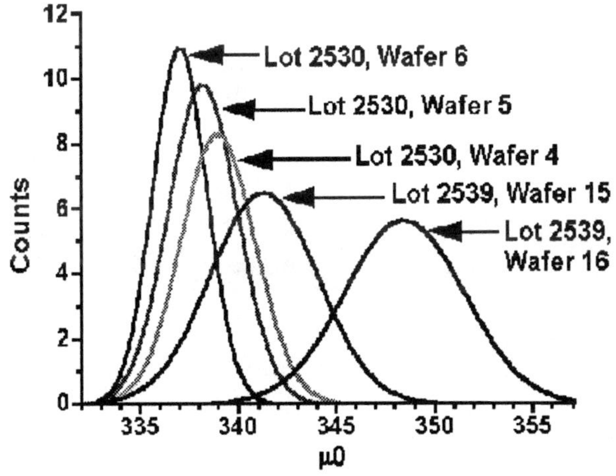

Fig. 14.2.6. Results from E-T Based Modeling calculation of one BSIM3v3 model parameter, μ_0. After Chen et al. [14.16].

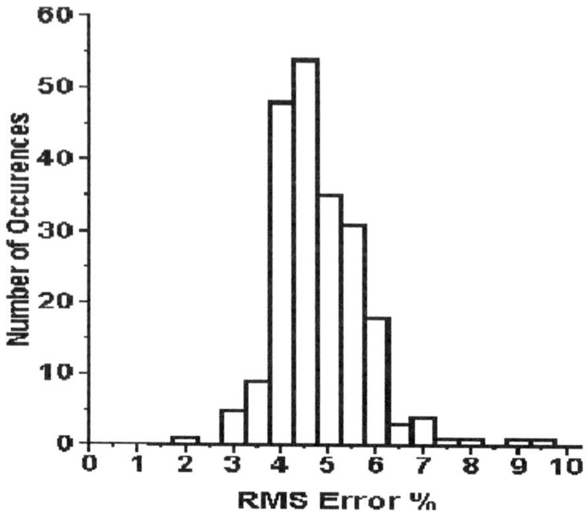

Fig. 14.2.7. Distribution of Id-V_d RMS error percentages for NMOS W/L=20μm / 0.5μm transistors. After Chen et al. [14.16].

Table 14.2.2. Summary of I_d-V_d RMS Error percent-ages for various devices. Low values for the mean and sigma are observed for all device geometries [14.16].

Device	Mean (%)	Sigma (%)
NMOS: W/L=20μm/20μm	1.6813	0.1311
NMOS: W/L=20μm/0.5μm	4.8086	0.9804
NMOS: W/L=1μm/0.5μm	4.4581	2.1461
PMOS: W/L=20μm/20μm	4.9612	1.7780
PMOS: W/L=20μm/0.5μm	4.4868	2.3991
PMOS: W/L=1μm/0.5μm	6.4220	2.6335

Wafer to wafer and lot to lot variation can be readily seen. In order to avoid time consuming repetitive simulation of large circuits, representative smaller digital circuits are simulated for all 220 extracted SPICE files. The basis for this efficient methodology is that all digital gates (e.g. inverter, NAND, NOR, etc.) speeds are highly correlated with one another. Since these smaller circuit blocks comprise the larger circuit; "worst", "best", or other percentile case files corresponding to these smaller circuits can be said to correspond to similar performance of the larger circuits. As an example, an inverter was simulated 220 times and the SPICE model files corresponding to seven specific delay percentiles were then retained. These were then used for seven simulations of a 4-bit adder. The results are compared (Fig. 14.2.8) against actual 4-bit adder simulations from all 220 extracted SPICE files. The agreement is good and shows that accurate speed distributions can be efficiently determined.

The methodologies using Electrical Test data to directly generate SPICE model parameter sets and using CSCS to find "tail" models is an efficient simulation strategy. CSCS circumvents repetitive simulation of large circuits to produce accurate distribution of circuit speeds.

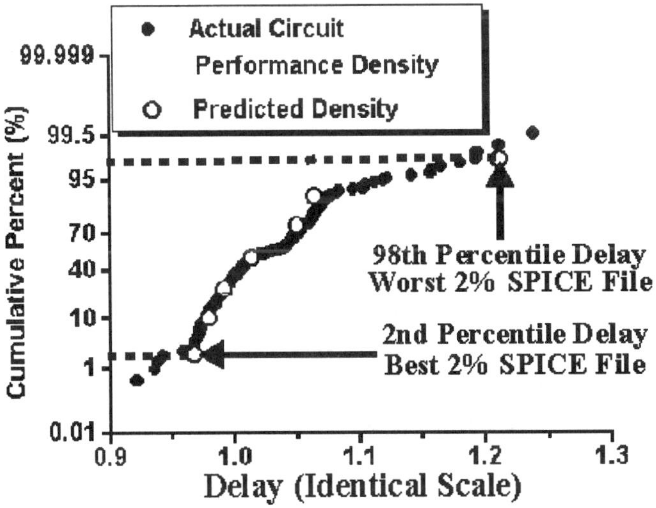

Fig. 14.2.8. Actual simulated delay density (220 simulations) of a 4-bit adder is compared with that predicted from just seven simulations. The latter used SPICE files selected for seven specific delay percentiles of small logic circuits. Excellent quantitative agreement is observed. After Chen et al. [14.16].

14.3 Technology Extrapolation and Prediction Using BSIM3 Model

As discussed in previous chapters, BSIM3 employs physical models for mobility, short channel effects, etc. Important process parameters such as

T_{OX} and N_{CH} explicitly appear in BSIM3 elements such as threshold voltage V_{th}, saturation voltage V_{dsat}, output resistance R_{out}, and substrate current I_{sub} etc. With built-in geometry and process parameters [14.19], BSIM3 can be used to predict device characteristics for future technologies. This concept is called Technology Extrapolation and Prediction [14.20, 14.21]. Technology Extrapolation (TE) starts with an existing parameter set that has been extracted from devices fabricated by an existing technology. The major parameters (usually size and process parameters) such as effective channel length (L_{eff}) and width (W_{eff}), parasitic series resistances (R_{ds}), long channel device threshold voltage (V_{TH0}), gate oxide thickness (T_{OX}), and doping concentration (N_{CH}), are slightly altered. This alteration immediately generates a new set of model parameter for a new or extrapolated technology. TE is useful for exploring the effect of, say, modestly reducing T_{OX} or L on the performance of a product. Technology Prediction (TP) further extends the model's capability beyond TE by allowing the size and process parameters to take quite different values that may be appropriate for one or more generations of technology ahead. In the extreme case, one may use default values for all unknown model parameters.

The BSIM3 model has been studied for TE and TP of N-channel MOSFETs with L_{eff} as small as $0.25\mu m$ using 8 different technologies (see Table 14.3.1) [14.21]. T_{OX} ranges from 5.5 to $15.6nm$. The needed model parameters were extracted using BSIMPro [14.22].

Table 14.3.1 Process parameters of technologies used in the study of Technology Extrapolation and Prediction [14.21]

Technology	T_{OX} (nm)	N_{CH} (cm^{-3})
a	10	2.910^{17}
b	9	2.510^{17}
c	11	$8.9x10^{16}$
d	13	$1.3x10^{17}$
e	12	$1.8x10^{17}$
f	5.5	$1.0x10^{17}$
g	7	$1.0x10^{17}$
h	15.6	$1.0x10^{17}$

The TE and TP capabilities of BSIM3 are demonstrated in Figs. 14.3.1 to 14.3.5, where all devices were fabricated by the same process except for different T_{OX} ($5.5nm$, $7nm$, and $15.6nm$). The parameters were extracted from devices with $T_{OX} = 7nm$. The model predicts the drain current within 10%

error for all T_{OX}. This results show that BSIM3 is very useful for technology developers to estimate the device performance for the next generation. Figs. 14.3.6 and 14.3.7 show the TE results of I_{dsat}, obtained with BSIM3v3 based on the parameters from technology b, for the next generation technology (2.5V power supply). It can be seen in Fig. 14.3.6 that, compared with the device with a 3.3V power supply, the same driving capability (I_{dsat}) can be achieved for the device with same channel length but 2.5V power supply if T_{OX} is properly reduced. In Fig. 14.3.6, higher performance can be obtained for the 5*nm* T_{OX} and 0.15μm L device at 2.5V power supply, compared with that for 9*nm* T_{OX} and 0.25μm L device at 3.3V power supply. In Fig. 14.3.6, a similar V_{th} was used, that is, $V_{th}(L)$ follows the same relationship for the short channel effects when channel length L decreases. In Fig. 14.3.7, the TP results of I_{dsat}/I_{dsato} for different V_{th} values are given. It can be seen that the reduction of V_{th} can improve I_{dsat}, but not very significantly. For the device of 0.25μm channel length and 5.5*nm* T_{OX} at 2.5V power supply, only about a 3% increase of I_{dsat} can be obtained if V_{th} decreases from 0.5V to 0.3V. Because the decrease of V_{th} will increase the off-state leakage current, it should be done very carefully if one wants to reduce the V_{th} to improve the current drive. The benefit of reducing V_{th} is of course larger at smaller V_{dd}'s.

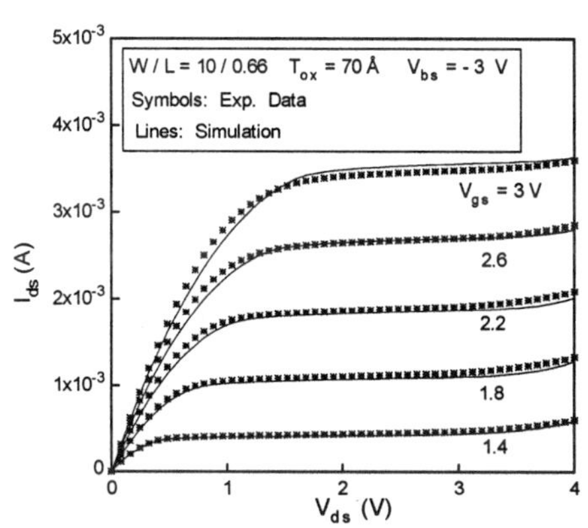

Fig. 14.3.1 Modeled and measured I_d-V_d curves for device h, from which BSIM3 parameters were extracted. After Huang et al. [14.20].

TP, using default parameters, is a much bigger challenge for BSIM3v3 as many different technologies exist in the world. Thus, it is very important to choose suitable default values for the model parameters so that we can use the set of default parameters in TP for many technologies. Using the default parameters of BSIM3v3 (except for T_{OX}, N_{CH}, R_{ds}, and V_{TH0}) the TP results are shown in Fig. 14.3.8 to Fig. 14.3.10. The maximum error in these plots is less than 14%.

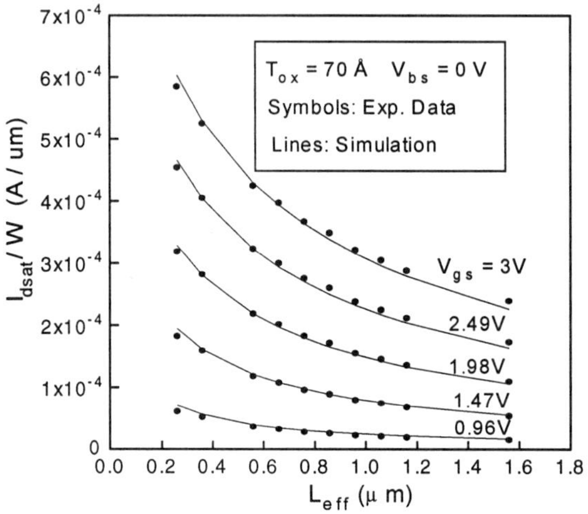

Fig.14.3.2 Modeled and measured I_{dsat} of device *h*. After Huang et al. [14.20].

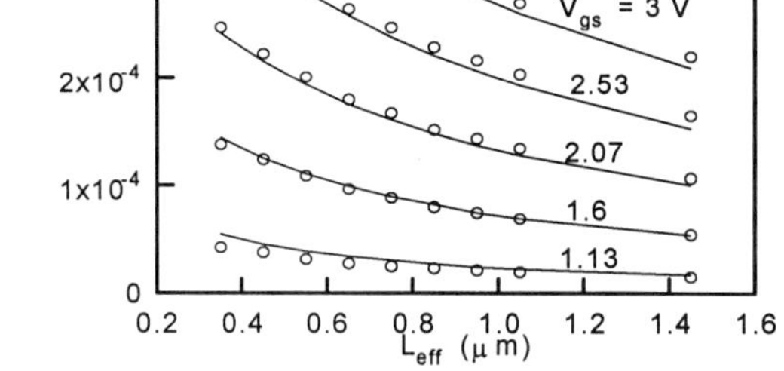

Fig. 14.3.3 TE of I_{dsat} for devices of T_{OX}=55Å, using the same BSIM3 parameters as

those for Fig. 14.3.1 except for T_{OX} and N_{CH}. After Huang et al. [14.20].

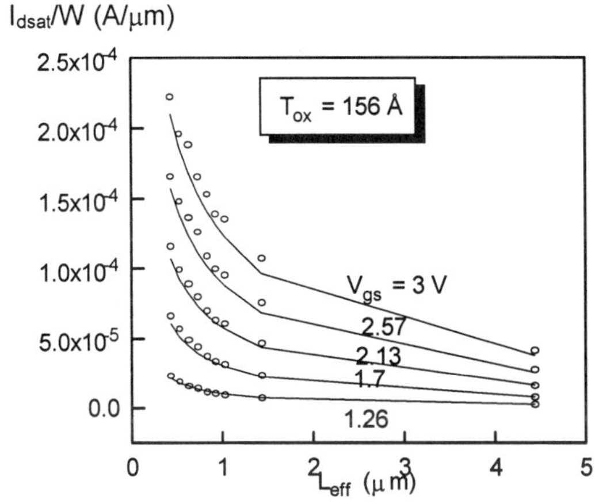

Fig. 14.3.4 Saturation currents for devices with T_{OX} = *156Å* using the same BSIM3 parameters as those in Fig. 14.3.1 except for T_{OX} and N_{CH}. After Huang et al. [14.20].

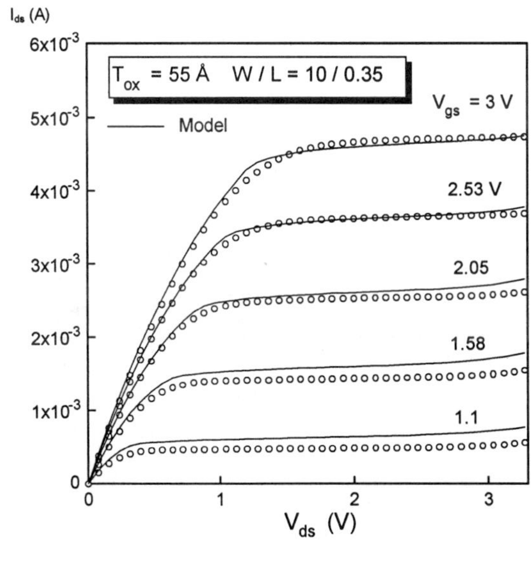

Fig. 14.3.5 TE of drain current with T_{OX} = $55\AA$ using the same BSIM3 parameters as those in Fig. 14.3.1 except for T_{OX} and N_{CH}. After Huang et al. [14.20].

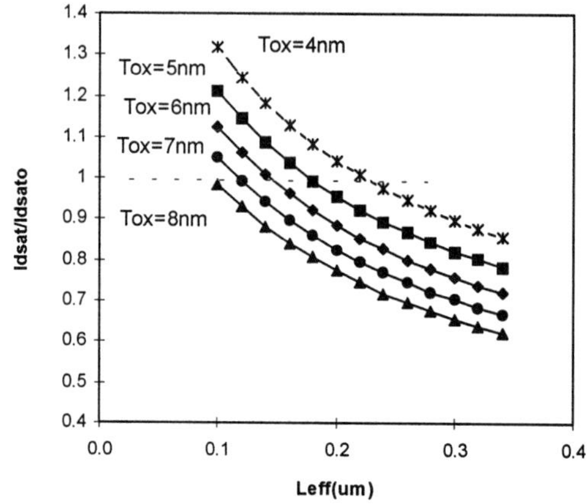

Fig. 14.3.6 TP of I_{dsat} for different T_{OX}. I_{dsat} is the saturation current at V_{dd}=2.5V. I_{dsato} is for device b with L_{eff}=0.25μm and V_{th} =0.63V at V_{dd} =3.3V. After Cheng et al. [14.21].

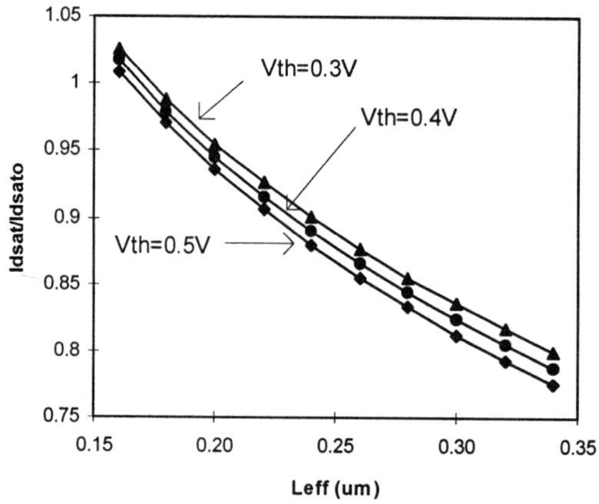

Fig. 14.3.7 TP of I_{dsat} for different V_{th}. I_{dsat} is the saturation current for V_{dd} =2.5V and T_{OX}=5.5nm. I_{dsato} is defined in Fig 14.3.6. After Cheng et al. [14.21].

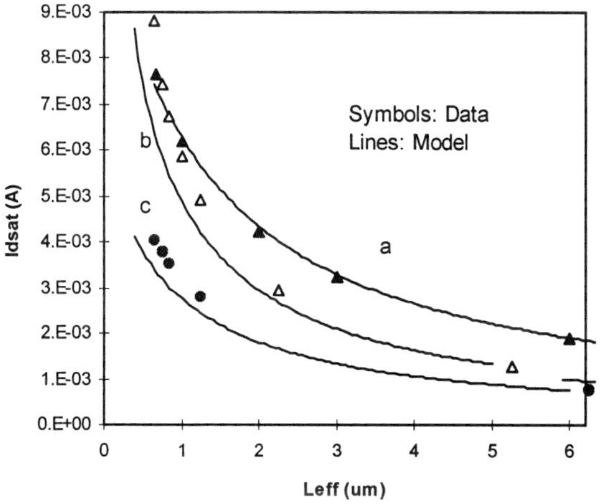

Fig. 14.3.8 Technology Prediction of saturation currents of NMOSFETs for different technologies using BSIM3v3 default parameters except for T_{OX}, N_{CH}, R_{ds}, and V_{TH0}. After Cheng et al. [14.21].

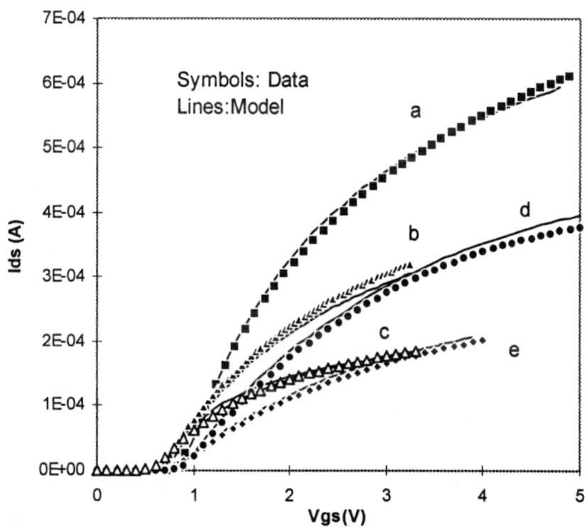

Fig. 14.3.9 Technology Prediction of I_{ds}- V_{gs} characteristics of NMOSFETs (at fixed V_{ds} and V_{bs}) for several technologies using BSIM3v3 default parameters except for T_{OX}, N_{CH}, R_{ds}, and V_{TH0}. After Cheng et al. [14.21].

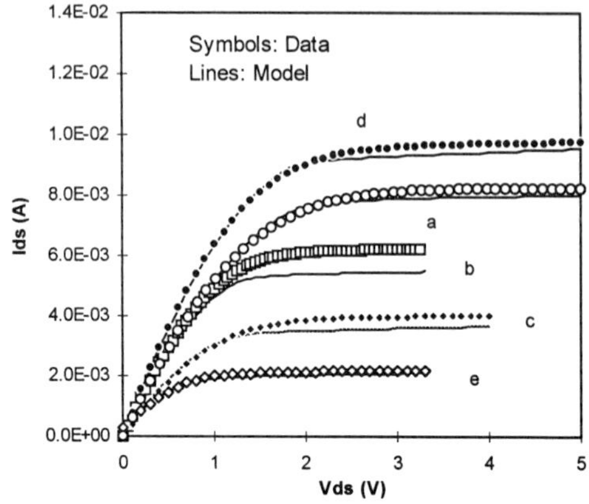

Fig. 14.3.10 Technology Prediction of I_{ds}- V_{ds} characteristics of NMOSFETs (at fixed V_{gs} and V_{bs}) for several technologies using BSIM3v3 default parameters except for $\boldsymbol{T_{OX}}$, $\boldsymbol{N_{CH}}$, R_{ds}, and V_{TH0}. After Cheng et al. [14.21].

References

[14.1] A. A. Abidi et al., "The future of CMOS wireless transceivers," *ISSCC Dig. Tech. Papers*, pp.118-119, 1997.

[14.2] W. Liu et al., "RF MOSFET modeling accounting for distributed sub-state and channel resistance with emphasis on the BSIM3v3 SPICE model," in *IEDM Tech. Dig.*, pp. 309-312, Dec. 1997.

[14.3] Y. Cheng et al., "RF modeling issues of deep-submicron MOSFETs for circuit design," *1998 International Conference of Solid-state and Integrated Circuit Technology*, pp.416-419, 1998.

[14.4] M. C Ho et al., "Scalable RF Si MOSFET distributed lumped element model based on BSIM3v3," *Electronics Letters*, vol.33. No. 23, pp. 1992-1993, 1997.

[14.5] D. R. Pehlke et al., "High frequency application of MOS compact model and their development for scalable RF model libraries," *Proc. of CICC*, pp.219-222, May 1998.

[14.6] J.-J. Ou et al., "CMOS RF modeling for GHz communication IC's," *VLSI Symp. on Tech., Dig. of Tech. Papers*, pp. 94-95, June 1998.

[14.7] S. H. Jen et al., "Accurate modeling and parameter extraction for MOS transistor valid up to 10-GHz", *ESDERC'98*, Sept. 1998.

[14.8] R. Goyal, *High-frequency analog integrated circuit design*, John Willy & Sons, Inc., New York, 1994.

[14.9] W. Liu and M. C. Chang, "Transistor transient studies including transcapacitive current and distributive gate resistance for inverter circuits," *IEEE Trans. On Circuit and Systems-I: Fundamental Theory and Applications*, vol. 45, no.4, 1998.

[14.10] X. Jin et al., "An effective gate resistance model for CMOS RF and noise modeling," *IEDM Tech Dig. 1998.*

[14.11] A. van der Ziel, *Noise in Solid State Devices and Circuits*, New York, Wiley, 1986.

[14.12] C.-H. Chen et al.,"Direct calculation of the MOSFET high frequency noise parameters," *Proc. of the 14th International Conference on Noise and Physical Systems and 1/f Fluctuations*, pp. 488-491, July 1997.

[14.13] R. Rios et al., "A Physical compact MOSFET model, including quantum mechanical effects for statistical circuit design applications", *IEDM Tech Dig*, pp. 937-940, 1995.

[14.14] J. Power et al., "Relating statistical MOSFET model parameter variabilities to IC manufacturing process fluctuations enabling realistic worst case design", *IEEE Trans on Semiconductor Manufacturing*, Aug. 1994, pp. 306-318.

[14.15] M.J. van Dort and D.B. Klassen, "Circuit sensitivity analysis in terms of process parameters," *IEDM Tech Dig*, pp. 941-944. 1995.

[14.16] J. C. Chen et al., "E-T based statistical modeling and compact statistical circuit simulation methodologies," *IEDM Tech Dig.* pp. 635-638, 1996.

[14.17] J. A. Power et al., "Statistical modeling for a 0.6μm BiCMOS technology," *IEEE BCTM Tech. Dig.*, pp. 24-27, 1997.

[14.18] Y. Cheng et al., *BSIM3 version 3.0 User's Manual*, University of California, Berkeley, 1995.

[14.19] Y. Cheng et al., *BSIM3 version 3.1 User's Manual*, University of California, Berkeley, Memorandum No. UCB/ERL M97/2, 1997.

[14.20] J. H. Huang et al., *BSIM3 Manual (Version 2.0)*, University of California, Berkeley, March 1994.

[14.21] Y. Cheng et al., "A Study of deep-submicon MOSFET technology prediction and scaling with BSIM3", *Techcon'97*, 1996.

[14.22] C. Enz and Y. Cheng, "MOS transistor modeling issues for RF circuit design", *Workshop of Advances in Analog Circuit Design*, France, March, 1999.

APPENDIX A

BSIM3v3 Parameter Table

A.1 Model control parameters

Symbols in equation/ book	Symbols in source code	Description	Default	Unit	Can be binned?
Level	level	BSIM3v3 model selector in the simulator	8 (In Berkeley Spice3)	none	No
Version	version	BSIM3v3 version selector	3.2	none	No
mobMod	mobmod	Mobility model selector	1	none	No
capMod	capmod	Capacitance model selector	3 (in BSIM3v3.2)	none	No
nqsMod	nqsmod	NQS model selector	0	none	No
noiMod	noimod	Noise model selector	1	none	No
paramChk	paramchk	Flag parameter for parameter checking	1	none	No
binUnit	binunit	Flag parameter for the units of size parameters in binning approach	1	none	No

A.2 Process parameters

Symbols in equation	Symbols in source code	Description	Default	Unit	Can be binned?
T_{OX}	*tox*	Gate oxide thickness	$1.5\text{x}10^{-8}$	m	No
T_{OXM}	*toxm*	Nominal T_{ox} at which parameter are extracted	T_{ox}	m	No
X_J	*xj*	Junction Depth	$1.5\text{x}10^{-7}$	m	Yes
N_{CH}	*nch*	Channel doping concentration	$1.7\text{x}10^{17}$	$1/\text{cm}^3$	Yes
N_{SUB}	*nsub*	Substrate doping concentration	$6.0\text{x}10^{16}$	$1/\text{cm}^3$	Yes
γ_1	gamma1	Body-effect coefficient near the interface	Calculated	$\text{V}^{1/2}$	Yes
γ_2	gamma2	Body-effect coefficient in the bulk	Calculated	$\text{V}^{1/2}$	Yes
X_T	xt	Doping depth	$1.55\text{x}10^{-7}$	m	Yes
V_{BX}	vbx	V_{bs} at which the depletion width equals X_T	Calculated	V	Yes

A.3 Parameters for V_{th} model

Symbols in equation	Symbols in source code	Description	Default	Recommended value range	Unit	Can be binned ?
V_{TH0}	V_{th0}	Threshold voltage @ V_{bs} =0 for large L. Typically V_{th0}>0 for NMOSFET and V_{th0}<0 for PMOSFET	0.7 for nMOS -0.7 for PMOS	-2~2	V	Yes
V_{FB}	*vfb*	Flat band voltage	calculated	-2~2	V	No
K_1	*k1*	First-order body effect coefficient	0.53	0~1	$\text{V}^{1/2}$	Yes
K_2	*k2*	Second-order body effect coefficient	-0.0186	-0.05~ 0	none	Yes
K_3	*k3*	Narrow width coefficient	80.0	10^{-3}~10^2	none	Yes

K_{3B}	*k3b*	Body effect coefficient of K_3	0.0	-10~10	1/V	Yes
W_0	*w0*	Narrow width parameter	2.5x10^{-6}	10^{-6}~10^{-5}	m	Yes
N_{LX}	*nlx*	Lateral non-uniform doping coefficient	1.74x10^{-7}	10^{-8}~10^{-6}	m	Yes
D_{VT0W}	*dvt0w*	First coefficient of narrow width effect on V_{th} at small L	0	0~10	none	Yes
D_{VT1W}	*dvt1w*	Second coefficient of narrow width effect on V_{th} at small L	5.3x10^{6}	0~1/Leff	1/m	Yes
D_{VT2W}	*dvt2w*	Body-bias coefficient of narrow width effect on V_{th} at small L	-0.032	-0.05~0	1/V	Yes
D_{VT0}	*dvt0*	First coefficient of short-channel effect on V_{th}	2.2	0~10	none	Yes
D_{VT1}	*dvt1*	Second coefficient of short-channel effect on V_{th}	0.53	0~1	none	Yes
D_{VT2}	*dvt2*	Body-bias coefficient of short-channel effect on V_{th}	-0.032	-0.05~0	1/V	Yes
V_{BM}	*vbm*	Maximum applied body bias in V_{th} calculation	-3	-3 ~ -10	V	Yes

A.4 Parameters for *I-V* model

Symbols in equation	Symbols in source code	Description	Default	Recommended value range	Unit	Can be binned ?
μ_0	uo	Mobility at $T = T_{NOM}$ NMOSFET PMOSFET	670.0 250.0	100~1000	cm^2/V/sec	Yes
U_A	ua	First-order mobility degradation coefficient	2.25x10^{-9}	10^{-10}~10^{-8}	m/V	Yes
U_B	ub	Second-order mobility degradation coefficient	5.87x10^{-19}	10^{-21}~10^{-18}	(m/V)^2	Yes

U_C	uc	Body-effect of mobility degradation coefficient	***mobMod***= 1,2: -4.65x10^{-11} ***mobMod***= 3: -0.0465	Mobmod=1,2: -10^{-11}~-10^{-8} Mobmod=3: -10^{-3}~0	m/V^2 1/V	Yes
v_{SAT}	vsat	Saturation velocity at $T = T_{NOM}$	8.0x10^4	10^4~10^5	m/sec	Yes
A_0	a0	Bulk charge effect coefficient for channel length	1.0	0~2	none	Yes
A_{GS}	ags	Gate bias coefficient of the bulk charge effect	0.0	-1~1	1/V	Yes
B_0	b0	Bulk charge effect coefficient for channel width	0.0	0~10^{-5}	m	Yes
B_1	b1	Bulk charge effect width offset	0.0	0~10^{-7}	m	Yes
K_{ETA}	keta	Body-bias coefficient of the bulk charge effect	-0.047	-10^{-3}~0	1/V	Yes
A_1	a1	First non-saturation parameter	0.0	0~0.1	1/V	Yes
A_2	a2	Second non-saturation parameter	1.0	0.4~1	none	Yes
R_{DSW}	rdsw	Parasitic resistance per unit width	0.0	10^2~10^{-3}	Ω-μm^{Wr}	Yes
P_{RWG}	prwg	Gate bias effect coefficient of R_{ds}	0	-10^{-3}~0	V^{-1}	Yes
P_{RWB}	prwb	Body bias effect coefficient of R_{ds}	0	-10^{-3}~0	V$^{-1/2}$	Yes
W_R	wr	Width offset from W_{eff} for R_{ds} calculation	1.0	1~5	none	Yes
W_{INT}	wint	Width offset fitting parameter without bias effect	0.0	0~3x10^{-7}	m	No

L_{INT}	lint	Length offset fitting parameter without bias effect	0.0	0~3x10^{-7}	m	No
D_{WG}	dwg	Coefficient of W_{eff}'s gate dependence	0.0	0~10^{-7}	m/V	Yes
D_{WB}	dwb	Coefficient of W_{eff}'s body bias dependence	0.0	0~10^{-7}	m/$V^{1/2}$	Yes
V_{OFF}	voff	Offset voltage in the subthreshold region at large *W* and *L*	-0.08	-0.15~0	V	Yes
N_{FACTOR}	nfactor	Subthreshold swing factor	1.0	0~2	none	Yes
E_{TA0}	eta0	*DIBL* coefficient in subthreshold region	0.08	0~1	none	Yes
E_{TAB}	etab	Body-bias coefficient for the subthreshold *DIBL* effect	-0.07	-10^{-3}~0	1/V	Yes
P_{CLM}	pclm	Channel length modulation parameter	1.3	0.1~10	none	Yes
P_{DIBLC1}	pdiblc1	First output resistance *DIBL* effect correction parameter	0.39	0~1	none	Yes
P_{DIBLC2}	pdiblc2	Second output resistance *DIBL* effect correction parameter	0.0086	10^{-5}~10^{-2}	none	Yes
P_{DIBLCB}	pdiblcb	Body effect coefficient of *DIBL* correction parameters	0	-10^{-3}~0	1/V	Yes
D_{ROUT}	drout	*L* dependence coefficient of the *DIBL* correction parameter in R_{out}	0.56	0~1	none	Yes

P_{SCBE1}	pscbe1	First substrate current induced body-effect parameter	4.24x10^8	10^8~8x10^8	V/m	Yes
P_{SCBE2}	pscbe2	Second substrate current induced body-effect parameter	1.0x10^{-5}	10^{-9}~10^{-4}	m/V	Yes
P_{VAG}	pvag	Gate dependence of Early voltage	0.0	-10~10	none	Yes
δ	delta	Effective V_{ds} parameter	0.01	10^{-3}~0.03	V	Yes
N_{GATE}	ngate	Poly gate doping concentration	0	2x10^{18}~ 9x10^{24}	cm^{-3}	Yes
D_{SUB}	dsub	*DIBL* coefficient exponent in sub-threshold region	D_{ROUT}	0~1	none	Yes
C_{IT}	cit	Interface trap capacitance	0.0	-10^{-4}~10^{-3}	F/m ^{2}	Yes
C_{DSC}	cdsc	Drain/Source to channel coupling capacitance	2.4x10^{-4}	0~10^{-3}	F/m ^{2}	Yes
C_{DSCD}	cdscd	Drain-bias sensitivity of C_{DSC}	0.0	0~10^{-3}	F/Vm ^{2}	Yes
C_{DSCB}	cdscb	Body-bias sensitivity of C_{DSC}	0.0	-10^{-4}~0	F/Vm ^{2}	Yes

A.5 Parameters for capacitance model

Symbols in equation	Symbols in source code	Description	Default	Unit	Can be binned?
X_{PART}	xpart	Charge partitioning flag	0	none	No
C_{GSO}	cgso	Non LDD region source-gate overlap capacitance per channel length	calculated	F/m	No
C_{GDO}	cgdo	Non LDD region drain-gate overlap capacitance per channel length	calculated	F/m	No
C_{GBO}	cgbo	Gate bulk overlap capacitance per unit channel length	0.0	F/m	No

C_{GSL}	cgsl	Light doped source-gate region overlap capacitance	0.0	F/m	Yes
C_{GDL}	cgdl	Light doped drain-gate region overlap capacitance	0.0	F/m	Yes
C_{KAPPA}	ckappa	Coefficient for lightly doped region overlap capacitance	0.6	F/m	Yes
C_F	cf	Fringing field capacitance	calcu-lated	F/m	Yes
C_{LC}	clc	Constant term for the short channel model	1.0×10^{-7}	m	Yes
C_{LE}	cle	Exponential term for the short channel model	0.6	none	Yes
D_{LC}	dlc	Length offset fitting parameter	lint	m	No
D_{WC}	dwc	Width offset fitting parameter	wint	m	No
V_{FBCV}	vfbcv	Flat-band voltage parameter (for *capMod*=0 only)	-1	V	Yes
N_{OFF}	noff	*C-V* parameter for $V_{gsteff,cv}$	1.0	none	Yes
V_{OFFCV}	voffcv	Offset voltage parameter of V_{th} from weak to strong inversion in *C-V* model	0.0	V	Yes
A_{CDE}	acde	Exponential coefficient for the charge thickness in accumulation and depletion regions	1.0	m/V	Yes
M_{OIN}	moin	Coefficient for the gate-bias dependent surface potential	15.0	$V^{1/2}$	Yes

A.6 Parameters for effective channel length/width in *I-V* model

Symbols in equation	Symbols in source code	Description	Default	Unit	Can be binned?
W_{LN}	wln	Power of length dependence of width offset	1.0	none	No
W_L	wl	Coefficient of length dependence for width offset	0.0	m^{Wln}	No

Symbols in equation	Symbols in source code	Description	Default	Unit	Can be binned?
W_{WN}	wwn	Power of width dependence of width offset	1.0	none	No
W_W	ww	Coefficient of width dependence for width offset	0.0	m^{Wwn}	No
W_{WL}	wwl	Coefficient of length and width cross term for width offset	0.0	$m^{Wwn+W\ln}$	No
L_{LN}	lln	Power of length dependence for length offset	1.0	none	No
L_L	ll	Coefficient of length dependence for length offset	0.0	$m^{L\ln}$	No
L_{WN}	lwn	Power of width denpendence for length offset	1.0	none	No
L_W	lw	Coefficient of width dependence for length offset	0.0	m^{Lwn}	No
L_{WL}	lwl	Coefficient of length and width cross term for length offset	0.0	$m^{Lwn+L\ln}$	No

A.7 Parameters for effective channel length/width in *C-V* model

Symbols in equation	Symbols in source code	Description	Default	Unit	Can be binned?
L_{LC}	llc	Coefficient of length dependence for channel length offset in *C-V* models	L_L	m^{LLN}	No
L_{WC}	lwc	Coefficient of width dependence for channel length offset in *C-V* models	L_W	m^{LWN}	No

Symbols in equation	Symbols in source code	Description	Default	Unit	Can be binned?
L_{WLC}	lwlc	Coefficient of length and width dependence for channel length offset in C-V models	L_{WL}	$m^{LWN+LLN}$	No
W_{LC}	wlc	Coefficient of length dependence for channel width offset in C-V models	W_L	m^{WLN}	No
W_{WC}	wwc	Coefficient of width dependence for channel width offset in C-V models	W_W	m^{WWN}	No
W_{WLC}	wwlc	Coefficient of length and width dependence for channel width offset in C-V models	W_{WL}	$m^{WLN+WWN}$	No

A.8 Parameters for substrate current model

Symbols in equation	Symbols in source code	Description	Default	Unit	Can be binned?
α_0	alpha0	The first parameter of substrate current	0	m/V	Yes
α_1	alpha1	The length scaling parameter of substrate current model	0	1/V	Yes
β_0	beta0	The second parameter of substrate current	30	V	Yes

A.9 Parameters for noise models

Symbols in equation	Symbols in source code	Description	Default	Unit	Can be binned?
A_F	af	Flicker noise exponent	1	none	No
E_F	ef	Flicker Frequency exponent	1	none	No

E_M	em	Saturation electrical field parameter	$4.1\text{x}10^7$	V/m	No
K_F	kf	Flicker noise coefficient	0	$s^{1-Ef}A^{2-Af}F$	No
N_{OIA}	noia	Noise parameter A	(nmos) 10^{20} (pmos) $9.9\text{x}10^{18}$	$s^{1-Ef}m^{-3}ev^{-1}$	No
N_{OIB}	noib	Noise parameter B	(nmos) $5\text{x}10^4$ (pmos) $2.4\text{x}10^3$	$s^{1-Ef}m^{-1}ev^{-1}$	No
N_{OIC}	noic	Noise parameter C	(nmos) $-1.4\text{x}10^{-12}$ (pmos) $1.4\text{x}10^{-12}$	$s^{1-Ef}mev^{-1}$	No

A.10 Parameters for models of parasitic components

Symbols in equation	Symbols in source code	Description	Default	Unit	Can be binned?
R_{SH}	rsh	Sheet resistance in source/drain regions	0	Ω/square	No
A_S	as	Area of the source region	0	m^2	No
A_D	ad	Area of the drain region	0	m^2	No
P_S	ps	Perimeter of the source region	0	m	No
P_D	pd	Perimeter of the drain region	0	m	No
N_{RS}	nrs	Numbers of the squares in the source region	1	none	No
N_{RD}	nrd	Numbers of the squares in the drain region	1	none	No
J_{S0}	js	Saturation current density of bottom junction diode	10^{-4}	A/m^2	No
J_{S0SW}	jssw	Saturation current density of sidewall junction diode	0	A/m	No
N_J	nj	Emission coefficient of source/drain junctions	1	none	Yes
X_{TI}	xti	Temperature exponent coefficient of junction current	3.0	none	Yes
I_{JTH}	ijth	Diode limiting current	0.1	A	No

C_J	cj	Source/drain (S/D) bottom junction capacitance per unit area at zero bias	$5x10^{-4}$	F/m^2	No
M_J	mj	S/D bottom junction capacitance grading coefficient	0.5	none	No
P_B	pb	Bottom junction built-in potential	1.0	V	No
C_{JSW}	cjsw	S/D field oxide sidewall junction capacitance per unit length at zero bias	$5x10^{-10}$	F/m	No
M_{JSW}	mjsw	S/D field oxide sidewall junction capacitance grading coefficient	0.33	none	No
P_{BSW}	pbsw	Source/drain field oxide sidewall junction built-in potential	1.0	V	No
C_{JSWG}	cjswg	S/D gate edge sidewall junction capacitance per unit length at zero bias	Cjsw	F/m	No
M_{JSWG}	mjswg	S/D gate edge sidewall junction capacitance grading coefficient	Mjsw	none	No
P_{BSWG}	pbswg	Built-in potential of the source/drain gate edge sidewall junction	Pbsw	V	No

A.11 Parameters for models of temperature effects

Symbols in equation	Symbols in source code	Description	Default	Unit	Can be binned?
T_{NOM}	tnom	Temperature at which parameters are extracted	27	°C	No
P_{RT}	prt	Temperature coefficient for R_{dsw}	0.0	Ω-μm	Yes
μ_{TE}	ute	Mobility temperature exponent	-1.5	none	Yes
K_{T1}	kt1	Temperature coefficient for threshold voltage	-0.11	V	Yes

K_{T1L}	kt1l	Channel length sensitivity of temperature coefficient for threshold voltage	0.0	Vm	Yes
K_{T2}	kt2	Body-bias coefficient of the V_{th} temperature effect	0.022	none	Yes
U_{A1}	ua1	Temperature coefficient for Ua	4.31×10^{-9}	m/V	Yes
U_{B1}	ub1	Temperature coefficient for Ub	-7.61×10^{-18}	$(m/V)^2$	Yes
U_{C1}	uc1	Temperature coefficient for Uc	***mobMod***=1, 2: -5.6×10^{-11} ***mobMod***=3: -0.056	m/V^2 1/V	Yes
A_T	at	Temperature coefficient for saturation velocity	3.3×10^4	m/sec	Yes
N_J	nj	Emission coefficient of junction	1.0	none	Yes
X_{TI}	xti	Temperature exponent coefficient of junction current	3.0	none	Yes
T_{CJ}	tcj	Temperature coefficient of C_j	0.0	1/K	No
T_{CJSW}	tcjsw	Temperature coefficient of C_{jsw}	0.0	1/K	No
T_{CJSWG}	tcjswg	Temperature coefficient of C_{jswg}	0.0	1/K	No
T_{PB}	tpb	Temperature coefficient of P_b	0.0	V/K	No
T_{PBSW}	tpbsw	Temperature coefficient of P_{bsw}	0.0	V/K	No
T_{PBSWG}	tpbswg	Temperature coefficient of P_{bswg}	0.0	V/K	No

A.12 Parameters for NQS model

Symbols in equation	Symbols in source code	Description	Default	Unit	Can be binned?
E_{LM}	elm	Elmore constant of the channel	5	none	Yes

APPENDIX B

BSIM3v3 Model Equations

B.1 V_{th} equations

$$V_{th} = V_{TH0OX} + K_{1OX}\sqrt{\phi_s - V_{bseff}} - K_{2OX}V_{bseff} + K_{1OX}\left(\sqrt{1+\frac{\boldsymbol{N_{LX}}}{L_{eff}}}-1\right)\sqrt{\phi_s} + (\boldsymbol{K_3} + \boldsymbol{K_{3B}}V_{bseff})\frac{\boldsymbol{T_{OX}}}{W_{eff}'+\boldsymbol{W_0}}\phi_s$$
$$- \boldsymbol{D_{VT0}}\left(\exp(-\boldsymbol{D_{VT1}}\frac{L_{eff}}{2l_t}) + 2\exp(-\boldsymbol{D_{VT1}}\frac{L_{eff}}{l_t})\right)(V_{bi}-\phi_s)$$
$$- \left(\exp(-\boldsymbol{D_{SUB}}\frac{L_{eff}}{2l_{to}}) + 2\exp(-\boldsymbol{D_{SUB}}\frac{L_{eff}}{l_{to}})\right)(\boldsymbol{E_{TA0}} + \boldsymbol{E_{TAB}}V_{bseff})V_{ds}$$
$$- \boldsymbol{D_{VT0w}}\left(\exp(-\boldsymbol{D_{VT1W}}\frac{W_{eff}'L_{eff}}{2l_{tw}}) + 2\exp(-\boldsymbol{D_{VT1W}}\frac{W_{eff}'L_{eff}}{l_{tw}})\right)(V_{bi}-\phi_s)$$

$$V_{th0ox} = \boldsymbol{V_{TH0}} - \boldsymbol{K_1}\sqrt{\phi_s}$$

$$K_{1ox} = \boldsymbol{K_1}\frac{\boldsymbol{T_{OX}}}{\boldsymbol{T_{OXM}}}$$

$$K_{2ox} = \boldsymbol{K_2}\frac{\boldsymbol{T_{OX}}}{\boldsymbol{T_{OXM}}}$$

$$l_t = \sqrt{\varepsilon_{si} X_{dep} / C_{ox}} (1 + \boldsymbol{D_{VT2}} V_{bseff})$$

$$l_{tw} = \sqrt{\varepsilon_{si} X_{dep} / C_{ox}} (1 + \boldsymbol{D_{VT2w}} V_{bseff})$$

$$l_{to} = \sqrt{\varepsilon_{si} X_{dep0} / C_{ox}}$$

$$X_{dep} = \sqrt{\frac{2\varepsilon_{si}(\phi_s - V_{bseff})}{q\boldsymbol{N_{CH}}}}$$

$$X_{dep0} = \sqrt{\frac{2\varepsilon_{si}\phi_s}{q\boldsymbol{N_{CH}}}}$$

$$V_{bseff} = V_{bc} + 0.5[V_{bs} - V_{bc} - \delta_1 + \sqrt{(V_{bs} - V_{bc} - \delta_1)^2 - 4\delta_1 V_{bc}}\ ]\ (\delta_1 = 0.001)$$

$$V_{bc} = 0.9(\phi_s - \frac{\boldsymbol{K_1}^2}{4\boldsymbol{K_2}^2})$$

B.2 Effective V_{gs}-V_{th}

$$V_{gsteff} = \frac{2\,n\,v_t \ln\left[1 + \exp(\frac{V_{gs} - V_{th}}{2\,n\,v_t})\right]}{1 + 2\,n\ C_{OX}\sqrt{\frac{2\phi_s}{q\varepsilon_{si}\boldsymbol{N_{CH}}}}\exp(-\frac{V_{gs} - V_{th} - 2\boldsymbol{V_{OFF}}}{2\,n\,v_t})}$$

$$n = 1 + \boldsymbol{N_{FACTOR}}\frac{C_{dep}}{C_{ox}} + \frac{\boldsymbol{C_{IT}}}{C_{ox}} + \frac{(\boldsymbol{C_{DSC}} + \boldsymbol{C_{DSCD}} V_{ds} + \boldsymbol{C_{DSCB}} V_{bseff})\left(\exp(-\boldsymbol{D_{VT1}}\frac{L_{eff}}{2l_t}) + 2\exp(-\boldsymbol{D_{VT1}}\frac{L_{eff}}{l_t})\right)}{C_{ox}}$$

$$C_{dep} = \frac{\varepsilon_{si}}{X_{dep}}$$

B.3 Mobility

For ***mobMod***=1

$$\mu_{eff} = \frac{\mu_o}{1 + (U_A + U_C V_{bseff})\left(\frac{V_{gsteff} + 2V_{th}}{T_{OX}}\right) + U_B\left(\frac{V_{gsteff} + 2V_{th}}{T_{OX}}\right)^2}$$

For ***mobMod***=2

$$\mu_{eff} = \frac{\mu_o}{1 + (U_A + U_C V_{bseff})\left(\frac{V_{gsteff}}{T_{OX}}\right) + U_B\left(\frac{V_{gsteff}}{T_{OX}}\right)^2}$$

For ***mobMod***=3

$$\mu_{eff} = \frac{\mu_o}{1 + \left[U_A\left(\frac{V_{gsteff} + 2V_{th}}{T_{OX}}\right) + U_B\left(\frac{V_{gsteff} + 2V_{th}}{T_{OX}}\right)^2\right](1 + U_C V_{bseff})}$$

B.4 Drain Saturation Voltage

For $R_{ds}>0$ or $\lambda \neq 1$

$$V_{dsat} = \frac{-b - \sqrt{b^2 - 4ac}}{2a}$$

$$a = A_{bulk}^2 W_{eff} v_{sat} C_{ox} R_{ds} + \left(\frac{1}{\lambda} - 1\right) A_{bulk}$$

$$b = -\left((V_{gsteff} + 2v_t)\left(\frac{2}{\lambda} - 1\right) + A_{bulk} E_{sat} L_{eff} + 3A_{bulk}(V_{gsteff} + 2v_t) W_{eff} v_{sat} C_{ox} R_{ds} \right)$$

$$c = (V_{gsteff} + 2v_t) E_{sat} L_{eff} + 2(V_{gsteff} + 2v_t)^2 W_{eff} v_{sat} C_{ox} R_{ds}$$

$$\lambda = A_1 V_{gsteff} + A_2$$

For $R_{ds}=0$, $\lambda=1$

$$V_{dsat} = \frac{E_{sat} L_{eff} (V_{gsteff} + 2v_t)}{A_{bulk} E_{sat} L_{eff} + V_{gsteff} + 2v_t}$$

$$A_{bulk} = (1 + \frac{K_{1OX}}{2\sqrt{\phi_s - V_{bseff}}} \{ \frac{\boldsymbol{A_0} L_{eff}}{L_{eff} + 2\sqrt{\boldsymbol{X_J} X_{dep}}} [1 - \boldsymbol{A_{GS}} V_{gsteff} (\frac{L_{eff}}{L_{eff} + 2\sqrt{\boldsymbol{X_J} X_{dep}}})^2] + \frac{\boldsymbol{B_0}}{W_{eff}' + \boldsymbol{B_1}} \}) \frac{1}{1 + \boldsymbol{K_{ETA}} V_{bseff}}$$

$$E_{sat} = \frac{2v_{sat}}{\mu_{eff}}$$

B.5 Effective V_{ds}

$$V_{dseff} = V_{dsat} - \frac{1}{2}\left(V_{dsat} - V_{ds} - \boldsymbol{\delta} + \sqrt{(V_{dsat} - V_{ds} - \boldsymbol{\delta})^2 + 4\boldsymbol{\delta} V_{dsat}} \right)$$

B.6 Drain Current Expression

$$I_{ds} = \frac{I_{dso(V_{dseff})}}{1 + \frac{R_{ds} I_{dso(V_{dseff})}}{V_{dseff}}} \left(1 + \frac{V_{ds} - V_{dseff}}{V_A} \right) \left(1 + \frac{V_{ds} - V_{dseff}}{V_{ASCBE}} \right)$$

$$I_{ds0} = \frac{W_{eff} C_{ox} \mu_{eff} V_{gsteff} V_{dseff} (1 - \frac{V_{dseff}}{2V_b})}{L_{eff} (1 + \frac{V_{dseff}}{E_{sat} L_{eff}})}$$

$$V_b = \frac{V_{gsteff} + 2v_t}{A_{bulk}}$$

$$V_A = V_{Asat} + (1 + \frac{\boldsymbol{P_{VAG}} V_{gsteff}}{E_{sat} L_{eff}}) (\frac{1}{V_{ACLM}} + \frac{1}{V_{ADIBLC}})^{-1}$$

$$V_{Asat} = \frac{E_{sat}L_{eff} + V_{dsat} + 2R_{ds}v_{sat}C_{ox}W_{eff}V_{gsteff}[1 - \frac{A_{bulk}V_{dsat}}{2(V_{gsteff} + 2v_t)}]}{2/\lambda - 1 + R_{ds}v_{sat}C_{ox}W_{eff}A_{bulk}}$$

$$V_{ACLM} = \frac{A_{bulk}E_{sat}L_{eff} + V_{gsteff}}{\mathbf{P_{CLM}}A_{bulk}E_{sat}\,l}(V_{ds} - V_{dseff})$$

$$V_{ADIBLC} = \frac{(V_{gsteff} + 2v_t)}{\theta_{rout}(1 + \mathbf{P_{DIBLCB}}V_{bseff})}(1 - \frac{A_{bulk}V_{dsat}}{A_{bulk}V_{dsat} + V_{gsteff} + 2v_t})$$

$$\theta_{rout} = \mathbf{P_{DIBLC1}}\left[\exp(-\mathbf{D_{ROUT}}\frac{L_{eff}}{2l_{t0}}) + 2\exp(-\mathbf{D_{ROUT}}\frac{L_{eff}}{l_{t0}})\right] + \mathbf{P_{DIBLC2}}$$

$$\frac{1}{V_{ASCBE}} = \frac{\mathbf{P_{SCBE2}}}{L_{eff}}\exp\left(\frac{-\mathbf{P_{SCBE1}}\,l}{V_{ds} - V_{dseff}}\right)$$

$$l = \sqrt{\frac{\varepsilon_{si}}{\varepsilon_{ox}}T_{ox}X_j}$$

B.7 Substrate current

$$I_{sub} = (\boldsymbol{\alpha_1} + \frac{\boldsymbol{\alpha_0}}{L_{eff}})(V_{ds} - V_{dseff})\exp(-\frac{\boldsymbol{\beta_0}}{V_{ds} - V_{dseff}})I_{dsa}$$

$$I_{dsa} = \frac{I_{dso}}{1 + \frac{R_{ds}I_{dso}}{V_{dseff}}}\left(1 + \frac{V_{ds} - V_{dseff}}{V_A}\right)$$

B.8 Polysilicon depletion effect

$$V_{gs\_eff} = \boldsymbol{V_{FB}} + \phi_s + \frac{q\varepsilon_{si}\boldsymbol{N_{GATE}T_{OX}^2}}{\varepsilon_{ox}^2}\left(\sqrt{1+\frac{2\varepsilon_{ox}^2(V_{gs}-\boldsymbol{V_{FB}}-\phi_s)}{q\varepsilon_{si}\boldsymbol{N_{GATE}T_{OX}^2}}}-1\right)$$

B.9 Effective channel length and width

$$L_{eff} = L_{drawn} - 2dL$$

$$W_{eff} = W_{drawn} - 2dW$$

$$W_{eff}' = W_{drawn} - 2dW'$$

$$dW = \boldsymbol{W_{INT}} + \boldsymbol{D_{WG}}V_{gsteff} + \boldsymbol{D_{WB}}(\sqrt{\phi_s - V_{bseff}} - \sqrt{\phi_s})$$
$$+ \frac{\boldsymbol{W_L}}{L^{\boldsymbol{W_{LN}}}} + \frac{\boldsymbol{W_W}}{W^{\boldsymbol{W_{WN}}}} + \frac{\boldsymbol{W_{WL}}}{L^{\boldsymbol{W_{LN}}}W^{\boldsymbol{W_{WN}}}}$$

$$dW' = \boldsymbol{W_{INT}} + \frac{\boldsymbol{W_L}}{L^{\boldsymbol{W_{LN}}}} + \frac{\boldsymbol{W_W}}{W^{\boldsymbol{W_{WN}}}} + \frac{\boldsymbol{W_{WL}}}{L^{\boldsymbol{W_{LLN}}}W^{\boldsymbol{W_{WN}}}}$$

$$dL = \boldsymbol{L_{INT}} + \frac{\boldsymbol{L_L}}{L^{\boldsymbol{L_{LN}}}} + \frac{\boldsymbol{L_W}}{W^{\boldsymbol{L_{WN}}}} + \frac{\boldsymbol{L_{WL}}}{L^{\boldsymbol{L_{LN}}}W^{\boldsymbol{L_{WN}}}}$$

B.10 Drain/Source resistance

$$R_{ds} = \frac{\boldsymbol{R_{DSW}}[1 + \boldsymbol{P_{RWG}}V_{gsteff} + \boldsymbol{P_{RWB}}(\sqrt{\phi_s - V_{bseff}} - \sqrt{\phi_s})]}{(10^6 W_{eff}')^{\boldsymbol{W_R}}}$$

B.11 Capacitance model equations

(A) Dimension Dependence

$$L_{active} = L_{drawn} - 2\delta L_{eff}$$

$$W_{active} = W_{drawn} - 2\delta W_{eff}$$

$$\delta W_{eff} = D_{WC} + \frac{W_{LC}}{L^{W_{LN}}} + \frac{W_{WC}}{W^{W_{WN}}} + \frac{W_{WLC}}{L^{W_{LN}} W^{W_{WN}}}$$

$$\delta L_{eff} = D_{LC} + \frac{L_{LC}}{L^{L_{LN}}} + \frac{L_{WC}}{W^{L_{WN}}} + \frac{L_{WLC}}{L^{L_{LN}} W^{L_{WN}}}$$

(B) Overlap Capacitance

(1) For ***capMod***= 0:

(I) Source overlap capacitance

$$\frac{Q_{overlap,s}}{W_{active}} = C_{GSO} V_{gs}$$

(II) Drain overlap capacitance

$$\frac{Q_{overlap,d}}{W_{active}} = C_{GDO} V_{gd}$$

(III) Gate overlap capacitance

$$Q_{overlap,g} = -\left(Q_{overlap,s} + Q_{overlap,d}\right)$$

(2) For ***capMod***= 1:

(I) Source overlap capacitance

If Vgs <0

$$\frac{Q_{overlap,s}}{W_{active}} = C_{GSO} V_{gs} - \frac{C_{KAPPA} C_{GSL}}{2}\left(-1 + \sqrt{1 - \frac{4V_{gs}}{C_{KAPPA}}}\right)$$

Otherwise

$$\frac{Q_{overlap,s}}{W_{active}} = (C\mathit{GSO} + C\mathit{KAPPA}\ C\mathit{GSL})\,\mathrm{V_{gs}}$$

(II) Drain overlap capacitance

If V_{gd}<0

$$\frac{Q_{overlap,d}}{W_{active}} = C\mathit{GDO}V_{gd} - \frac{C\mathit{KAPPA}\ C\mathit{GDL}}{2}\left(-1+\sqrt{1-\frac{4V_{gd}}{C\mathit{KAPPA}}}\right)$$

Otherwise

$$\frac{Q_{overlap,d}}{W_{active}} = (C\mathit{GDO} + C\mathit{KAPPA}\ C\mathit{GDL})\,\mathrm{V_{gd}}$$

(III) Gate overlap capacitance

$$Q_{overlap,g} = -\left(Q_{overlap,s} + Q_{overlap,d}\right)$$

(3) For ***capMod***= 2:

(I) Source overlap capacitance

$$\frac{Q_{overlap,s}}{W_{active}} = C\mathit{GSO}V_{gs} + C\mathit{GSL}[V_{gs} - V_{gs,overlap} - \frac{C\mathit{KAPPA}}{2}(-1+\sqrt{1-\frac{4V_{gs,overlap}}{C\mathit{KAPPA}}})]$$

$$V_{gs,overlap} = \frac{1}{2}[(V_{gs}+\delta_1) - \sqrt{(V_{gs}+\delta_1)^2 + 4\delta_1}]$$

where $\delta_1 = 0.02$.

(II) Drain overlap capacitance

$$\frac{Q_{overlap,d}}{W_{active}} = C\mathit{GD0}V_{gd} + C\mathit{GDL}[V_{gd} - V_{gd,overlap} - \frac{C\mathit{KAPPA}}{2}(-1+\sqrt{1-\frac{4V_{gd,overlap}}{C\mathit{KAPPA}}})]$$

$$V_{gd,overlap} = \frac{1}{2}[(V_{gd} + \delta_2) - \sqrt{(V_{gd} + \delta_2)^2 + 4\delta_2}]$$

where $\delta_2 = 0.02$.

(III) Gate overlap capacitance

$$Q_{overlap,g} = -\left(Q_{overlap,s} + Q_{overlap,d}\right)$$

(C) Intrinsic Charges

(1) For ***capMod***=0

$$Q_G + Q_{INV} + Q_B = 0$$

$$Q_{INV} = Q_S + Q_D$$

$$V_{th} = \boldsymbol{V_{FBCV}} + \phi s + K_{1OX}\sqrt{\phi s - V_{bseff}}$$

(I) If $V_{gs} < \boldsymbol{V_{FBCV}} + V_{bs}$

$$Q_B = -W_{active} L_{active} C_{ox}\left(V_{gs} - V_{bseff} - \boldsymbol{V_{FBCV}}\right)$$

$$Q_G = -Q_B$$

(II) If $\boldsymbol{V_{FBCV}} + V_{bs} < V_{gs} < V_{th}$

$$Q_B = -W_{active} L_{active} C_{ox} \frac{K_{1OX}^2}{2}\left(-1 + \sqrt{1 + \frac{4\left(V_{gs} - \boldsymbol{V_{FBCV}} - V_{bseff}\right)}{K_{1OX}^2}}\right)$$

Q Q

(III) If $V_{gs} > V_{th}$

$$V_{dsat,cv} = \frac{Vgs - Vth}{A_{bulk}'}$$

$$A_{bulk}' = A_{bulk0}\left(1 + \left(\frac{\boldsymbol{CLC}}{L_{active}}\right)^{\boldsymbol{CLE}}\right)$$

$$Abulk0 = \left(1 + \frac{K1OX}{2\sqrt{\phi s - Vbseff}}\left\{\frac{\boldsymbol{A0}\,Leff}{Leff + 2\sqrt{\boldsymbol{XJ}Xdep}} + \frac{\boldsymbol{B0}}{Weff' + \boldsymbol{B1}}\right\}\right)\frac{1}{1 + \boldsymbol{KETA}Vbseff}$$

i. 50/50 charge partition (X_{PART}=0.5)

If $V_{ds} < V_{dsat}$

$$QG = W_{active} L_{active} C_{ox}\left(Vgs - \boldsymbol{VFBCV} - \phi s - \frac{1}{2}V_{ds} + \frac{A_{bulk}'V_{ds}^2}{12\left(Vgs - Vth - \frac{A_{bulk}'}{2}V_{ds}\right)}\right)$$

$$QS = QD = -\frac{W_{active}L_{active}C_{ox}}{2}\left(Vgs - Vth - \frac{A_{bulk}'}{2}V_{ds} + \frac{A_{bulk}'^2 V_{ds}^2}{12\left(Vgs - Vth - \frac{A_{bulk}'}{2}V_{ds}\right)}\right)$$

$$QB = W_{active}L_{active}C_{ox}\left(\boldsymbol{VFBCV} - Vth + \phi s + \frac{1 - A_{bulk}'}{2}V_{ds} - \frac{(1 - A_{bulk}')A_{bulk}'V_{ds}^2}{12\left(Vgs - Vth - \frac{A_{bulk}'}{2}V_{ds}\right)}\right)$$

If $V_{ds} > V_{dsat,cv}$

$$QG = W_{active} L_{active} C_{ox}\left(Vgs - \boldsymbol{VFBCV} - \phi s - \frac{1}{3}Vdsat,cv\right)$$

$$Q_S = Q_D = -\frac{W_{active} L_{active} C_{ox}}{3}\left(V_{gs} - V_{th}\right)$$

$$Q_B = W_{active} L_{active} C_{ox}\left(V_{FBCV} - V_{th} + \phi_s + \frac{1 - A_{bulk}'}{3} V_{dsat,cv}\right)$$

ii. 40/60 charge partition (X_{PART}<0.5)

When V_{ds}<$V_{dsat,cv}$

$$Q_G = W_{active} L_{active} C_{ox}\left(V_{gs} - V_{FBCV} - \phi_s - \frac{1}{2} V_{ds} + \frac{A_{bulk}' V_{ds}^{2}}{12\left(V_{gs} - V_{th} - \frac{A_{bulk}'}{2} V_{ds}\right)}\right)$$

$$Q_B = W_{active} L_{active} C_{ox}\left(V_{FBCV} - V_{th} + \phi_s + \frac{1 - A_{bulk}'}{2} V_{ds} - \frac{(1 - A_{bulk}') A_{bulk}' V_{ds}^{2}}{12\left(V_{gs} - V_{th} - \frac{A_{bulk}'}{2} V_{ds}\right)}\right)$$

$$Q_D = -W_{active} L_{active} C_{ox}[\frac{V_{gs} - V_{th}}{2} - \frac{A_{bulk}'}{2} V_{ds} + \frac{A_{bulk}' V_{ds}[\frac{(V_{gs} - V_{th})^2}{6} - \frac{A_{bulk}'(V_{gs} - V_{th})V_{ds}}{8} + \frac{(A_{bulk}' V_{ds})^2}{40}}{\left(V_{gs} - V_{th} - \frac{A_{bulk}'}{2} V_{ds}\right)^2}]$$

$$Q_S = -(Q_G + Q_B + Q_D)$$

When $V_{ds} \geq V_{dsat,cv}$

$$Q_G = W_{active} L_{active} C_{ox}\left(V_{gs} - V_{FBCV} - \phi_s - \frac{1}{3} V_{dsat,cv}\right)$$

$$Q_D = -\frac{4 W_{active} L_{active} C_{ox}}{15}\left(V_{gs} - V_{th}\right)$$

$$Q_B = W_{active} L_{active} C_{ox} \left(V_{FBCV} - V_{th} + \phi_s + \frac{1 - A_{bulk}'}{3} V_{dsat,cv} \right)$$

$$Q_S = -(Q_G + Q_B + Q_D)$$

iii. 0/100 charge partition (***X_{PART}***>0.5)

When $V_{ds}<V_{dsat,cv}$

$$Q_G = W_{active} L_{active} C_{ox} \left(V_{gs} - V_{FBCV} - \phi_s - \frac{1}{2} V_{ds} + \frac{A_{bulk}' V_{ds}^2}{12 \left(V_{gs} - V_{th} - \frac{A_{bulk}'}{2} V_{ds} \right)} \right)$$

$$Q_B = W_{active} L_{active} C_{ox} \left(V_{FBCV} - V_{th} + \phi_s + \frac{1 - A_{bulk}'}{2} V_{ds} - \frac{(1 - A_{bulk}') A_{bulk}' V_{ds}^2}{12 \left(V_{gs} - V_{th} - \frac{A_{bulk}'}{2} V_{ds} \right)} \right)$$

$$Q_D = -W_{active} L_{active} C_{ox} \left(\frac{V_{gs} - V_{th}}{2} + \frac{A_{bulk}'}{4} V_{ds} - \frac{(A_{bulk}' V_{ds})^2}{24 \left(V_{gs} - V_{th} - \frac{A_{bulk}'}{2} V_{ds} \right)} \right)$$

$$Q_S = -(Q_G + Q_B + Q_D)$$

When $V_{ds} \geq V_{dsat,cv}$

$$Q_G = W_{active} L_{active} C_{ox} \left(V_{gs} - V_{FBCV} - \phi_s - \frac{1}{3} V_{dsat,cv} \right)$$

$$Q_D = 0$$

$$Q_B = W_{active} L_{active} C_{ox} \left(V_{FBCV} - V_{th} + \phi_s + \frac{1 - A_{bulk}'}{3} V_{dsat,cv} \right)$$

$$Q_S = -(Q_G + Q_B + Q_D)$$

(2) For ***capMod***=1

$$Q_G = -(Q_B + Q_{INV})$$

$$Q_B = Q_{DEP} + Q_{ACC}$$

$$Q_{DEP} = Q_{DEP0} + \delta Q_{DEP}$$

$$v_{fb} = V_{th} - \phi_s - K_{1OX}\sqrt{\phi_s - V_{bseff}}$$

$$V_{gsteff,cv} = nv_t \boldsymbol{N_{OFF}} \ln\left(1 + \exp\left(\frac{V_{gs} - V_{th} - \boldsymbol{V_{OFFCV}}}{nv_t \boldsymbol{N_{OFF}}}\right)\right)$$

$$V_{dsat,cv} = \frac{V_{gsteff,cv}}{A_{bulk}'}$$

$$A_{bulk}' = A_{bulk0}\left(1 + \left(\frac{\boldsymbol{C_{LC}}}{L_{active}}\right)^{\boldsymbol{C_{LE}}}\right)$$

(I) When $V_{gs} < v_{fb} + V_{bs} + V_{gsteff,cv}$

$$Q_{G0} = -W_{active} L_{active} C_{ox} \left(V_{gs} - v_{fb} - V_{bseff} - V_{gsteff,cv}\right)$$

$$Q_B = Q_{ACC} = -Q_{G0}$$

(II) When $V_{gs} \geq v_{fb} + V_{bs} + V_{gsteff,cv}$

$$Q_{G0} = W_{active} L_{active} C_{ox} \frac{K_{1OX}^2}{2}\left(-1 + \sqrt{1 + \frac{4\left(V_{gs} - v_{fb} - V_{gsteffcv} - V_{bseff}\right)}{K_{1OX}^2}}\right)$$

$$Q_B = Q_{DEP0} = -Q_{G0}$$

If $0 < V_{ds} \le V_{dsat,cv}$

$$Q_G = Q_{G0} + W_{active} L_{active} C_{ox} \left(\left(V_{gsteff,cv} - \frac{V_{ds}}{2} \right) + \frac{A_{bulk}'^2 V_{ds}^2}{12 \left(V_{gsteff,cv} - \frac{A_{bulk}'}{2} V_{ds} \right)} \right)$$

$$Q_{DEP} = Q_{DEP0} + \delta Q_{DEP}$$

$$\delta Q_{DEP} = W_{active} L_{active} C_{ox} \left(\frac{1 - A_{bulk}'}{2} V_{ds} - \frac{(1 - A_{bulk}') A_{bulk}' V_{ds}^2}{12 \left(V_{gsteff,cv} - \frac{A_{bulk}'}{2} V_{ds} \right)} \right)$$

If $V_{ds} > V_{dsat,cv}$

$$Q_G = Q_{G0} + W_{active} L_{active} C_{ox} \left(V_{gsteff,cv} - \frac{V_{dsat,cv}}{3} \right)$$

$$Q_B = Q_{DEP} = Q_{DEP0} + \delta Q_{DEP}$$

$$\delta Q_{DEP} = -W_{active} L_{active} C_{ox} \frac{V_{gsteff,cv} - V_{dsat,cv}}{3}$$

i. 50/50 charge partition

If $0 < V_{ds} \le V_{dsat,cv}$

$$Q_S = Q_D = -\frac{W_{active} L_{active} C_{ox}}{2} \left(V_{gsteff,cv} - \frac{A_{bulk}'}{2} V_{ds} + \frac{A_{bulk}'^2 V_{ds}^2}{12 \left(V_{gsteff,cv} - \frac{A_{bulk}'}{2} V_{ds} \right)} \right)$$

If $V_{ds} > V_{dsat,cv}$

$$Q_S = Q_D = -\frac{W_{active} L_{active} C_{ox}}{3} V_{gsteff,cv}$$

ii. 40/60 charge partition

If $0<V_{ds} \le V_{dsat,cv}$

$$Q_S = -\frac{W_{active}L_{active}C_{ox}}{2\left(V_{gsteff,cv} - \frac{A_{bulk}'}{2}V_{ds}\right)^2}[V_{gsteff,cv}{}^3 - \frac{4}{3}V_{gsteff,cv}{}^2\left(A_{bulk}'V_{ds}\right) + \frac{2}{3}V_{gsteff,cv}\left(A_{bulk}'V_{ds}\right)^2 - \frac{2}{15}\left(A_{bulk}'V_{ds}\right)^3]$$

$$Q_D = -(Q_G + Q_B + Q_S)$$

If $V_{ds} > V_{dsat,cv}$

$$Q_S = -\frac{2W_{active}L_{active}C_{ox}}{5}V_{gsteff,cv}$$

$$Q_D = -(Q_G + Q_B + Q_S)$$

iii. 0/100 charge partition

If $0<V_{ds} \le V_{dsat,cv}$

$$Q_S = -W_{active}L_{active}C_{ox}\left(\frac{V_{gsteff,cv}}{2} + \frac{A_{bulk}'V_{ds}}{4} - \frac{\left(A_{bulk}'V_{ds}\right)^2}{24\left(V_{gsteff,cv} - \frac{A_{bulk}'}{2}V_{ds}\right)}\right)$$

$$Q_D = -(Q_G + Q_B + Q_S)$$

If $V_{ds} > V_{dsat,cv}$

$$Q_S = -W_{active}L_{active}C_{ox}\frac{2V_{gsteff,cv}}{3}$$

$$Q_D = -(Q_G + Q_B + Q_S)$$

(3) For ***capMod***=2

$$Q_G = -(Q_B + Q_{INV})$$

$$Q_B = Q_{DEP} + Q_{ACC}$$

$$Q_{INV} = Q_S + Q_D$$

$$Q_{DEP} = Q_{DEP0} + \delta Q_{DEP}$$

$$V_{dsat,cv} = \frac{V_{gsteff,cv}}{A_{bulk}'}$$

$$A_{bulk}' = A_{bulk0}\left(1 + \left(\frac{CLC}{L_{active}}\right)^{CLE}\right)$$

$$V_{FBeff} = vfb - 0.5\left\{V_3 + \sqrt{V_3^{\,2} + 4\delta_3 |vfb|}\right\}$$

$$vfb = Vth - \phi s - K1OX\sqrt{\phi s - Vbseff}$$

$$V_3 = vfb - V_{gb} - \delta 3 \qquad \text{where } \delta_3 = 0.02$$

$$Q_{ACC} = -W_{active}L_{active}C_{ox}\left(V_{FBeff} - vfb\right)$$

$$Q_{DEP0} = -W_{active}L_{active}C_{ox}\frac{K1OX^2}{2}\left(-1 + \sqrt{1 + \frac{4\left(V_{gs} - V_{FBeff} - V_{gsteff,cv} - V_{bseff}\right)}{K1OX^2}}\right)$$

$$V_{cveff} = V_{dsat,cv} - 0.5\left\{V_4 + \sqrt{V_4^{\,2} + 4\delta_4 V_{dsat,cv}}\right\}$$

$$V_4 = V_{dsat,cv} - V_{ds} - \delta_4 \qquad \text{where } \delta_4 = 0.02.$$

$$Q_{INV} = -W_{active} L_{active} C_{ox} \left(\left(V_{gsteff,cv} - \frac{A_{bulk}'}{2} V_{cveff} \right) + \frac{A_{bulk}'^2 V_{cveff}^2}{12\left(V_{gsteff,cv} - \frac{A_{bulk}'}{2} V_{cveff} \right)} \right)$$

$$\delta Q_{DEP} = W_{active} L_{active} C_{ox} \left(\frac{1 - A_{bulk}'}{2} V_{cveff} - \frac{(1 - A_{bulk}') A_{bulk}' V_{cveff}^2}{12\left(V_{gsteff,cv} - \frac{A_{bulk}'}{2} V_{cveff} \right)} \right)$$

i. 50/50 charge partition

$$Q_S = Q_D = -\frac{W_{active} L_{active} C_{ox}}{2} \left(V_{gsteff,cv} - \frac{A_{bulk}'}{2} V_{cveff} + \frac{A_{bulk}'^2 V_{cveff}^2}{12\left(V_{gsteff,cv} - \frac{A_{bulk}'}{2} V_{cveff} \right)} \right)$$

ii. 40/60 charge partition

$$Q_S = -\frac{W_{active} L_{active} C_{ox}}{2\left(V_{gsteff,cv} - \frac{A_{bulk}'}{2} V_{cveff} \right)^2} \left(V_{gsteff,cv}^3 - \frac{4}{3} V_{gsteff,cv}^2 \left(A_{bulk}' V_{cveff} \right) + \frac{2}{3} V_{gsteff,cv} \left(A_{bulk}' V_{cveff} \right)^2 - \frac{2}{15} \left(A_{bulk}' V_{cveff} \right)^3 \right)$$

$$Q_D = -\frac{W_{active} L_{active} C_{ox}}{2\left(V_{gsteff,cv} - \frac{A_{bulk}'}{2} V_{cveff} \right)^2} \left(V_{gsteff,cv}^3 - \frac{5}{3} V_{gsteff,cv}^2 \left(A_{bulk}' V_{cveff} \right) + V_{gsteff,cv} \left(A_{bulk}' V_{cveff} \right)^2 - \frac{1}{5} \left(A_{bulk}' V_{cveff} \right)^3 \right)$$

iii. 0/100 charge partition

$$Q_S = -W_{active} L_{active} C_{ox} \left(\frac{V_{gsteff,cv}}{2} + \frac{A_{bulk}' V_{cveff}}{4} - \frac{\left(A_{bulk}' V_{cveff} \right)^2}{24\left(V_{gsteff,cv} - \frac{A_{bulk}'}{2} V_{cveff} \right)} \right)$$

$$(Q_D = -W_{active} L_{active} C_{ox} \left(\frac{V_{gsteff,cv}}{2} - \frac{3A_{bulk}' V_{cveff}}{4} + \frac{\left(A_{bulk}' V_{cveff}\right)^2}{8\left(V_{gsteff,cv} - \frac{A_{bulk}'}{2} V_{cveff}\right)} \right)$$

(4) For ***capMod***=3

$$C_{oxeff} = \frac{C_{ox} C_{cen}}{C_{ox} + C_{cen}}$$

$$C_{cen} = \varepsilon_{si} / X_{DC}$$

$$X_{DC} = \frac{1}{3} L_{debye} \exp[\boldsymbol{ACDE}(\frac{\boldsymbol{N_{CH}}}{2x10^{16}})^{-0.25} \frac{V_{gs} - V_{bs} - v_{fb}}{\boldsymbol{Tox}}] \text{ [cm]}$$

$$L_{debye} = \sqrt{\frac{\varepsilon_{si} v_{t0}}{q N_{CH}}}$$

$$X_{DCeff} = X_{DC\max} - \frac{1}{2}(X_0 + \sqrt{X_0^2 + 4\delta_x X_{DC\max}}) \qquad \text{(In accumulation)}$$

$X_0 = X_{DCmax} - X_{DC} - \delta_x$ where $\delta_x = 10^{-3} \cdot T_{ox}$

$$X_{DCeff} = \frac{1.9x10^{-7}}{1 + [\frac{V_{gsteff,cv} + 4(V_{th} - v_{fb} - 2\phi_B)}{2T_{ox}}]^{0.7}} \text{ [cm] (In strong inversion)}$$

$$Q_{ACC} = W_{active} L_{ative} C_{oxeff} V_{gbacc}$$

$$V_{gbacc} = \frac{1}{2}\left(V_0 + \sqrt{V_0^2 + 4\delta_v |v_{fb}|} \right)$$

where $V_0 = v_{fb} + V_{bs} - V_{gs} - \delta_v$ and $\delta_v = 0.02\text{V}$

$$v_{fb} = V_{th} - 2\phi_B - K_{1OX}\sqrt{2\phi_B - V_{bs}}$$

$$
\begin{aligned}
V_{th} = & V_{th0OX} + K_{1OX}\left(\sqrt{\phi_s - V_{bseff}} - \sqrt{\phi_s}\right) \\
& + K_{1OX}\left(\sqrt{1+\frac{N_{LX}}{L_{eff}}} - 1\right)\sqrt{\phi_s} + \boldsymbol{K_3}\frac{T_{OX}}{W_{eff}' + \boldsymbol{W_0}}\phi_s \\
& - \boldsymbol{D_{VT0}}\left(\exp(-\boldsymbol{D_{VT1}}\frac{L_{eff}}{2l_{t0}}) + 2\exp(-\boldsymbol{D_{VT1}}\frac{L_{eff}}{l_{t0}})\right)(V_{bi} - \phi_s) \\
& - \boldsymbol{D_{VT0w}}\left(\exp(-\boldsymbol{D_{VT1w}}\frac{W_{eff}' L_{eff}}{2l_{tw0}}) + 2\exp(-\boldsymbol{D_{VT1w}}\frac{W_{eff}' L_{eff}}{l_{tw0}})\right)(V_{bi} - \phi_s)
\end{aligned}
$$

$$
Q_{DEP0} = -W_{active}L_{active}C_{oxeff}\frac{K_{1OX}^2}{2}\left[-1 + \sqrt{1 + \frac{4(V_{gs} - v_{fbx} - V_{bseff} - V_{gsteff,cv})}{K_{1OX}^2}}\right]
$$

$$
v_{fbx} = v_{fb} - V_{gbacc}
$$

$$
\begin{aligned}
Q_{INV} = & -W_{active}L_{active}C_{oxeff}\left[V_{gsteff,cv} - \Phi\delta - \frac{1}{2}A_{bulk}'V_{cveff}\right. \\
& \left. + \frac{A_{bulk}'^2 V_{cveff}^2}{12(V_{gsteff,cv} - \Phi\delta - A_{bulk}'V_{cveff}/2)}\right]
\end{aligned}
$$

$$
\delta Q_{DEP} = W_{active}L_{active}C_{oxeff}\left[\frac{1 - A_{bulk}'}{2}V_{cveff} - \frac{(1 - A_{bulk}')A_{bulk}'V_{cveff}^2}{12(V_{gsteff,cv} - \Phi\delta - A_{bulk}'V_{cveff}/2)}\right]
$$

$$
V_{dsat,cv} \quad \frac{V_{gsteff,cv}}{A_{bulk}}
$$

where δ_4=0.02.

i. 50/50 charge partition

ii. 40/60 charge partition

$$Q_S = -\frac{W_{active}L_{active}C_{oxeff}}{2\left(V_{gsteff,cv} - \Phi\delta - \frac{A_{bulk}'}{2}V_{cveff}\right)^2}\{(V_{gsteff,cv} - \Phi\delta)^3 - \frac{4}{3}(V_{gsteff,cv} - \Phi\delta)^2$$

$$(A_{bulk}'V_{cveff}) + \frac{2}{3}(V_{gsteff,cv} - \Phi\delta)(A_{bulk}'V_{cveff})^2 - \frac{2}{15}(A_{bulk}'V_{cveff})^3\}$$

$$Q_D = -\frac{W_{active}L_{active}C_{oxeff}}{2\left(V_{gsteff,cv} - \Phi\delta - \frac{A_{bulk}'}{2}V_{cveff}\right)^2}$$

$$\{(V_{gsteff,cv} - \Phi\delta)^3 - \frac{5}{3}(V_{gsteff,cv} - \Phi\delta)^2(A_{bulk}'V_{cveff})$$

$$+ (V_{gsteff,cv} - \Phi\delta)(A_{bulk}'V_{cveff})^2 - \frac{1}{5}(A_{bulk}'V_{cveff})^3\}$$

iii. 0/100 Charge Partition

$$Q_S = -W_{active}L_{active}C_{oxeff}[\frac{V_{gsteff,cv} - \Phi\delta}{2} + \frac{A_{bulk}'V_{cveff}}{4}$$

$$-\frac{(A_{bulk}'V_{cveff})^2}{24(V_{gsteff,cv} - \Phi\delta - \frac{A_{bulk}'}{2}V_{cveff})}]$$

$$Q_D = -W_{active}L_{active}C_{oxeff}[\frac{V_{gsteff,cv} - \Phi\delta}{2} - \frac{3(A_{bulk}'V_{cveff})}{4}$$

$$+\frac{(A_{bulk}'V_{cveff})^2}{8(V_{gsteff,cv} - \Phi\delta - \frac{A_{bulk}'}{2}V_{cveff})}]$$

B.12 Noise model equations

1. SPICE2 flicker noise model (***noiMod***=1, 4)

$$Sid(f) = \frac{\boldsymbol{KF} Ids^{\boldsymbol{AF}}}{CoxLeff^2 f^{\boldsymbol{EF}}}$$

2. Unified flicker noise model (***noiMod***=2, 3)

1). Strong inversion region (V_{gs}-V_{th}>0.1V):

$$S_{id}(f) = \frac{v_t q^3 I_{ds} \mu_{eff}}{f^{EF} L_{eff}^2 C_{ox} 10^8} [\boldsymbol{N_{OIA}} \log(\frac{N_o + 2x10^{14}}{N_L + 2x10^{14}}) + \boldsymbol{N_{OIB}}(N_o - N_L)$$

$$+0.5\boldsymbol{N_{OIC}}(N_o^2 - N_L^2)] + \frac{q v_t I_{ds}^2 \Delta L_{clm}}{f^{EF} L_{eff}^2 W_{eff}' 10^8} \frac{\boldsymbol{N_{OIA}} + \boldsymbol{N_{OIB}} N_L + \boldsymbol{N_{OIC}} N_L^2}{(N_L + 2x10^{14})^2}$$

$$N_o = \frac{C_{ox}(V_{gs} - V_{th})}{q}$$

$$N_L = \frac{C_{ox}(V_{gs} - V_{th} - V_{ds}')}{q}$$

$$V_{ds}' = MIN(V_{ds}', V_{dsat})$$

$$\Delta L_{clm} = L_{itl} \log\left(\frac{\frac{V_{ds} - V_{dsat}}{L_{itl}} + \boldsymbol{E_M}}{E_{sat}}\right) \quad V_{ds} > V_{dsat}$$

$$\Delta L_{clm} = 0 \qquad V_{ds} \le V_{dsat}$$

$$L_{itl} = \sqrt{3 \boldsymbol{X_J} \boldsymbol{T_{OX}}}$$

$$E_{sat} = 2 \frac{\boldsymbol{v_{SAT}}}{\mu_{eff}}$$

2). Moderate inversion and subthreshold regions (V_{gs}-$V_{th} \le 0.1$V)

$$S_{id}(f) = \frac{S_{limit} S_{wi}}{S_{limit} + S_{wi}}$$

where S_{limit} is the flicker noise, Eq. (7.3.29), calculated at V_{gs}=V_{th}+0.1V.

$$S_{wi} = \frac{\boldsymbol{N_{OIA}} v_t I_{ds}^2}{W_{eff}' L_{eff} f^{EF} 4x10^{36}}$$

3. Modified SPICE2 thermal noise model (***noiMod****=1,3)*

$$S_{id}(f)=\frac{8K_BT}{3}(g_m+g_{ds}+g_{mb})$$

4. BSIM3 thermal noise model (***noiMod***=*2,4)*

$$S_{id}(f)=\frac{4K_BT\mu_{eff}}{L_{eff}^2}Q_{INV}$$

(1) When ***capMod***=0,

a). Linear region ($V_{gs}>V_{th}$, $V_{ds}<V_{dsat,cv}$)

$$Q_{INV}=-W_{active}L_{active}C_{ox}\left(V_{gs}-V_{th}-\frac{A_{bulk}'}{2}V_{ds}+\frac{A_{bulk}'^2V_{ds}^2}{12\left(V_{gs}-V_{th}-\frac{A_{bulk}'}{2}V_{ds}\right)}\right)$$

b). Saturation region ($V_{gs}>V_{th}$, $V_{ds}\geq V_{dsat,cv}$)

$$Q_{INV}=-\frac{2W_{active}L_{active}C_{ox}}{3}\left(V_{gs}-V_{th}\right)$$

(2) When ***capMod***=1,

a). Linear region ($V_{gs}>V_{th}$, $V_{ds}<V_{dsat,cv}$)

$$Q_{INV}=-W_{active}L_{active}C_{ox}\left(V_{gsteff,cv}-\frac{A_{bulk}'}{2}V_{ds}+\frac{A_{bulk}'^2V_{ds}^2}{12\left(V_{gsteff,cv}-\frac{A_{bulk}'}{2}V_{ds}\right)}\right)$$

b). Saturation region ($V_{gs}>V_{th}$, $V_{ds}\geq V_{dsat,cv}$)

$$Q_{INV}=-\frac{2W_{active}L_{active}C_{ox}}{3}V_{gsteff,cv}$$

(3) When ***capMod***=2,

$$Q_{INV} = -W_{active}L_{active}C_{ox}\left(\left(V_{gsteff,cv} - \frac{A_{bulk}'}{2}V_{cveff}\right) + \frac{A_{bulk}'^2 V_{cveff}^2}{12\left(V_{gsteff,cv} - \frac{A_{bulk}'}{2}V_{cveff}\right)}\right)$$

(4) When ***capMod***=3,

$$Q_{INV} = -W_{active}L_{active}C_{ox}[(V_{gsteff,cv} - \phi\delta - \frac{A_{bulk}'}{2}V_{cveff}) + \frac{A_{bulk}'^2 V_{cveff}^2}{12(V_{gsteff,cv} - \phi\delta - \frac{A_{bulk}'}{2}V_{cveff})}]$$

B.13 DC model of the source/drain diodes

$$I_{sbs} = J_s A_S + P_s J_{ssw}$$

$$J_s = \boldsymbol{J_{S0}} \exp[\frac{\frac{E_{g0}}{V_{tm0}} - \frac{E_g}{V_{tm}} + \boldsymbol{X_{TI}} \ln(\frac{T}{\boldsymbol{T_{NOM}}})}{\boldsymbol{N_J}}]$$

$$J_{ssw} = \boldsymbol{J_{S0SW}} \exp[\frac{\frac{E_{g0}}{V_{tm0}} - \frac{E_g}{V_{tm}} + \boldsymbol{X_{TI}} \ln(\frac{T}{\boldsymbol{T_{NOM}}})}{\boldsymbol{N_J}}]$$

$$E_{g0} = 1.16 - \frac{7.02 \times 10^{-4} \boldsymbol{T_{NOM}}^2}{\boldsymbol{T_{NOM}} + 1108.0}$$

$$E_g = 1.16 - \frac{7.02 \times 10^{-4} T^2}{T + 1108.0}$$

If I_{sbs} is not positive,

$$I_{bs} = \boldsymbol{G_{MIN}} V_{bs}$$

If I_{sbs} is larger than zero, the following equations will be used to calculate the S/B junction current I_{bs}, depending on the value of $\boldsymbol{I_{JTH}}$ specified in the model card.

When I_{JTH} is equal to zero,

$$I_{bs} = I_{sbs}[\exp(\frac{V_{bs}}{N_J v_t}) - 1] + G_{MIN} V_{bs}$$

If I_{JTH} is not zero,

$$V_{jsm} = N_J v_t \ln(\frac{I_{JTH}}{I_{sbs}} + 1)$$

If $V_{bs} < V_{jsm}$

$$I_{bs} = I_{sbs}[\exp(\frac{V_{bs}}{N_J v_t}) - 1] + G_{MIN} V_{bs}$$

If $V_{bs} \geq V_{jsm}$

$$I_{bs} = I_{JTH} + \frac{I_{JTH} + I_{sbs}}{N_J v t}(V_{bs} - V_{jsm}) + G_{MIN} V_{bs}$$

B.14 Capacitance model of the source/bulk and drain/bulk diodes

$$C_{apbs} = C_{jbst} + C_{jbsswgt} + C_{jbsswt}$$

$$C_{jbst} = A_s C_{jbs}$$

If $P_s > W_{eff}'$,

$$C_{jbsswgt} = W_{eff}' C_{jbsswg}$$

$$C_{jbsswt} = (P_S - W_{eff}')C_{jbssw}$$

$$C_{apbs} = A_S C_{jbs} + W_{eff}' C_{jbsswg} + (P_S - W_{eff}')C_{jbssw}$$

If $P_s \leq W_{eff}'$,

$$C_{jsbswgt} = P_S C_{jbsswg}$$

$$C_{apbs} = A_S C_{jbs} + P_S C_{jbsswg}$$

$$C_{jbs} = \boldsymbol{C_J}(1 - \frac{V_{bs}}{\boldsymbol{P_B}})^{-\boldsymbol{M_J}} \quad V_{bs}<0$$

$$C_{jbs} = \boldsymbol{C_J}(1 + \boldsymbol{M_J}\frac{V_{bs}}{\boldsymbol{P_B}}) \quad V_{bs}\geq 0$$

$$C_{jbssw} = \boldsymbol{C_{JSW}}(1 - \frac{V_{bs}}{\boldsymbol{P_{BSW}}})^{-\boldsymbol{M_{SWJ}}} \quad V_{bs}<0$$

$$C_{jbssw} = \boldsymbol{C_{JSW}}(1 + \boldsymbol{M_{JSW}}\frac{V_{bs}}{\boldsymbol{P_{BSW}}}) \quad V_{bs}\geq 0$$

$$C_{jbsswg} = \boldsymbol{C_{JSWG}}(1 - \frac{V_{bs}}{\boldsymbol{P_{BSWG}}})^{-\boldsymbol{M_{SWJG}}} \quad V_{bs}<0$$

$$C_{jbsswg} = \boldsymbol{C_{JSWG}}(1 + \boldsymbol{M_{JSWG}}\frac{V_{bs}}{\boldsymbol{P_{BSWG}}}) \quad V_{bs}\geq 0$$

The drain-bulk capacitance model equations are the same but substituting s with d in the subscripts.

B.15 Temperature effects

$$\mu_{eff} = \frac{\mu_0(T/\boldsymbol{T_{NOM}})^{\boldsymbol{ute}}}{1 + (U_a(T) + U_c(T)V_{bseff})(\frac{V_{gsteff} + 2V_{th}}{\boldsymbol{T_{OX}}}) + U_b(T)(\frac{V_{gsteff} + 2V_{th}}{\boldsymbol{T_{OX}}})^2}$$

(***mobMod***=1)

$$\mu_{eff} = \frac{\mu_0(T/\boldsymbol{T_{NOM}})^{\boldsymbol{ute}}}{1 + (U_a(T) + U_c(T)V_{bseff})(\frac{V_{gsteff}}{\boldsymbol{T_{OX}}}) + U_b(T)(\frac{V_{gsteff}}{\boldsymbol{T_{OX}}})^2}$$

(***mobMod***=2)

$$\mu_{eff} = \frac{\mu_0(T/\boldsymbol{T_{NOM}})^{\boldsymbol{ute}}}{1 + [U_a(T)(\frac{V_{gsteff} + 2V_{th}}{\boldsymbol{T_{OX}}}) + U_b(T)(\frac{V_{gsteff} + 2V_{th}}{\boldsymbol{T_{OX}}})^2](1 + U_c(T)V_{bseff})}$$

(***mobMod***=3)

$$U_a(T) = \mathbf{U_A} + \mathbf{U_{A1}}(T / \mathbf{T_{NOM}} - 1)$$

$$U_b(T) = \mathbf{U_B} + \mathbf{U_{B1}}(T / \mathbf{T_{NOM}} - 1)$$

$$U_c(T) = \mathbf{U_C} + \mathbf{U_{C1}}(T / \mathbf{T_{NOM}} - 1)$$

$$V_{th}(T) = V_{th}(\mathbf{T_{NOM}}, L, V_{ds}) + (\mathbf{K_{T1}} + \frac{\mathbf{K_{T1L}}}{L} + \mathbf{K_{T2}}V_{bs})(\frac{T}{\mathbf{T_{NOM}}} - 1)$$

$$\upsilon_{sat}(T) = \mathbf{v_{SAT}} + \mathbf{A_T}(\frac{T}{\mathbf{T_{NOM}}} - 1)$$

$$R_{ds}(T) = \frac{R_{dsw}(T)[1 + \mathbf{P_{RWG}}V_{gsteff} + \mathbf{P_{RWB}}(\sqrt{\phi_s - V_{bseff}} - \sqrt{\phi_s})]}{W_{eff}'^{\mathbf{W_R}}}$$

$$R_{dsw}(T) = \mathbf{R_{DSW}} + \mathbf{P_{RT}}(\frac{T}{\mathbf{T_{NOM}}} - 1)$$

$$J_s = \mathbf{J_{S0}} \exp[\frac{\frac{E_{g0}}{V_{tm0}} - \frac{E_g}{V_{tm}} + \mathbf{X_{TI}} \ln(\frac{T}{\mathbf{T_{NOM}}})}{\mathbf{N_J}}]$$

$$J_{ssw} = \mathbf{J_{S0SW}} \exp[\frac{\frac{E_{g0}}{V_{tm0}} - \frac{E_g}{V_{tm}} + \mathbf{X_{TI}} \ln(\frac{T}{\mathbf{T_{NOM}}})}{\mathbf{N_J}}]$$

$$E_{g0} = 1.16 - \frac{7.02 \times 10^{-4} \mathbf{T_{NOM}}^2}{\mathbf{T_{NOM}} + 1108.0}$$

$$E_g = 1.16 - \frac{7.02 \times 10^{-4} T^2}{T + 1108.0}$$

$$C_j(T) = \mathbf{C_J}[1 + \mathbf{T_{CJ}}(T - \mathbf{T_{NOM}})]$$

$$C_{jsw}(T) = \mathbf{C_{JSW}}[1 + \mathbf{T_{CJSW}}(T - \mathbf{T_{NOM}})]$$

$$C_{jswg}(T) = \mathbf{C_{JSWG}}[1 + \mathbf{T_{CJSWG}}(T - \mathbf{T_{NOM}})]$$

$$P_b(T) = \boldsymbol{P_B} - \boldsymbol{T_{PB}}(T - \boldsymbol{T_{NOM}})$$

$$P_{bsw}(T) = \boldsymbol{P_{BSW}} - \boldsymbol{T_{PBSW}}(T - \boldsymbol{T_{NOM}})$$

$$P_{bswg}(T) = \boldsymbol{P_{BSWG}} - \boldsymbol{T_{PBSWG}}(T - \boldsymbol{T_{NOM}})$$

$$C_{jbs} = C_j(T)\left(1 - \frac{V_{bs}}{P_b(T)}\right)^{-\boldsymbol{M_J}} \quad V_{bs}<0$$

$$C_{jbs} = C_j(T)\left(1 + \boldsymbol{M_J}\frac{V_{bs}}{P_b(T)}\right) \quad V_{bs}\geq 0$$

$$C_{jbssw} = C_{jsw}(T)\left(1 - \frac{V_{bs}}{P_{bsw}(T)}\right)^{-\boldsymbol{M_{JSW}}} \quad V_{bs}<0$$

$$C_{jbssw} = C_{jsw}(T)\left(1 + \boldsymbol{M_{JSW}}\frac{V_{bs}}{P_{bsw}(T)}\right) \quad V_{bs}\geq 0$$

$$C_{jbsswg} = C_{jswg}(T)\left(1 - \frac{V_{bs}}{P_{bswg}(T)}\right)^{-\boldsymbol{M_{JSWG}}} \quad V_{bs}<0$$

$$C_{jbsswg} = C_{jswg}(T)\left(1 + \boldsymbol{M_{JSWG}}\frac{V_{bs}}{P_{bswg}(T)}\right) \quad V_{bs}\geq 0$$

B.16 NQS Model Equations

$$Q_{def}(t) = V_{def}(1 \cdot C_{fact})$$

$$R_{elm} = \frac{L_{eff}^{\,2}}{\boldsymbol{E_{LM}}\mu Q_{ch}} \approx \frac{L_{eff}^{\,2}}{\boldsymbol{E_{LM}}\mu Q_{cheq}}$$

$$\mu = \boldsymbol{\mu_0}(T / \boldsymbol{T_{NOM}})^{\boldsymbol{U_{TE}}}$$

$$\tau_{drift} \approx R_{elm} C_{ox} W_{eff} L_{eff} \approx \frac{C_{ox} W_{eff} L_{eff}^{\,3}}{\boldsymbol{E_{LM}}\mu Q_{cheq}}$$

$$\tau_{diffusion} = \frac{Leff^2}{16\mu v_t}$$

$$I_{D,G,S}(t) = I_{D,G,S}(DC) + \frac{\partial Q_{D,G,S}(t)}{\partial t}$$

$$Q_{def}(t) = Q_{cheq}(t) - Q_{ch}(t)$$

$$\frac{\partial Q_{def}(t)}{dt} = \frac{\partial Q_{cheq}(t)}{dt} - \frac{\partial Q_{def}(t)}{\tau}$$

$$\frac{\partial Q_{D,G,S}(t)}{dt} = D,G,S_{xpart}\frac{\partial Q_{def}(t)}{\tau}$$

$$D_{xpart} = \frac{Q_D}{Q_D + Q_S} = \frac{Q_D}{Q_{cheq}}$$

$$\frac{D_{xpart}}{dV_i} = \frac{S_{xpart}C_{di} - D_{xpart}C_{si}}{Q_{cheq}}$$

where i represents the four terminals (g,s,d,b) and C_{di} and C_{si} are the intrinsic capacitances in the strong inversion. The corresponding value of S_{xpart} can be derived from the fact that $D_{xpart}+S_{xpart}=1$.

In the accumulation and depletion regions, if $\boldsymbol{X_{PART}}<0.5$, $D_{xpart}=0.4$; if $\boldsymbol{X_{PART}}=0.5$, $D_{xpart}=0.5$; if $\boldsymbol{X_{PART}}>0.5$, $D_{xpart}=0$.

$$G_{tau} = \frac{C_{fact}}{\tau}$$

B.17 A note on the poly-gate depletion effect

If the poly-gate depletion effect is included, all V_{gs} in the above DC and AC model equations will be replaced by V_{gs_eff}:

$$V_{gs\_eff} = \boldsymbol{V_{FB}} + \phi_s + \frac{q\varepsilon_{si}\boldsymbol{N_{GATE}T_{OX}^2}}{\varepsilon_{ox}^2}\left(\sqrt{1 + \frac{2\varepsilon_{ox}^2(V_{gs} - \boldsymbol{V_{FB}} - \phi_s)}{q\varepsilon_{si}\boldsymbol{N_{GATE}T_{OX}^2}}} - 1\right)$$

APPENDIX C

Enhancements and Changes in BSIM3v3.1 versus BSIM3v3.0

C.1 Enhancements

Many improvements have been made in the BSIM3v3.1 code, released in Dec. of 1996, relative to the BSIM3v3.0 code released in Oct. of 1995:

(1) Code improvements to avoid any model discontinuity that may be caused by bad values of certain parameters.
(2) Code changes to avoid any algebraic problems such as divide by zero or square root domain.
(3) Bug fixes.
(4) Addition of a new routine to check certain parameters for proper value ranges.
(5) Addition of options for using different noise models.
(6) Modification of the S/B and D/B diode model.
(7) Add capmod=0 for BSIM1-like long channel capacitance model.
(8) Code clean-up

C.2 Detailed changes

The routines B3ld.c, b3temp.c, b3noi.c, b3mpar.c, b3.c, b3mask.c, b3set.c, and bsim3ext.h have been changed. A description of the changes is given in the following:

1. Code improvement to avoid any model discontinuity caused by certain parameters or unusual operation bias conditions.

(1) To avoid any problems caused by 1+dvt2*Vbseff in Vth calculations in b3ld.c

if (T1=1+dvt2*Vbseff)<0.5

T1=(1+3*dvt2*Vbseff)/(3+8*dvt2*Vbseff)

(2) To avoid any problems caused by 1+dvt2w*Vbseff in Vth calculations in b3ld.c

if (T1=1+dvt2w*Vbseff)<0.5

T1=(1+3*dvt2w*Vbseff)/(3+8*dvt2w*Vbseff)

(3)To avoid any problems caused by nfactor*esi/Xdep+theta0*(cdsc+cdscb*Vbseff+Cdscd*Vds) in calculating n in b3ld.c.

if (n=1+nfactor*esi/Xdep+theta0*(cdsc+cdscb*Vbseff+Cdscd*Vds))<-0.5

n=[1+3*nfactor*esi/Xdep+theta0*(cdsc+cdscb*Vbseff+Cdscd*Vds)]/
{3+8*[nfactor*esi/Xdep+theta0*(cdsc+cdscb*Vbseff+Cdscd*Vds)]}

(4) To avoid any problems caused by Abulk0 and Abulk in b3ld.c.

if (Abulk0<0.1)

Abulk0=(0.2-Abulk0)/(3-20*Abulk0)

if (Abulk<0.1)

Abulk=(0.2-Abulk)/(3-20*Abulk)

(5) To avoid any problems caused by dwg and dwb in calculating Weff in b3ld.c.

if (Weff<2.e-8)

Weff=2.e-8(4.e-8-Weff)/(6.e-8-2*Weff)

(6) To avoid any problems caused by Prwg*Vgsteff+Prwb*(sqrt(PHI-VBS)-Sqrt(PHI)) in calculating Rds in b3ld.c.

If {T0=Prwg*Vgsteff+Prwb*(sqrt(PHI-VBS)-Sqrt(PHI))}<-0.9

Rds=Rds0*(0.8+T0)/(17+20*T0)

(7) To avoid problems caused by 1/(1+Keta*Vbseff) in calculating Abulk in b3ld.c.

if (T0=1/(1+Keta*Vbseff)>10

T0=(17+20*Keta*Vbseff)/(0.8+Keta*Vbseff)

(8)To avoid problem caused by the denominator in calculating μeff in b3ld.c.

Denomi=1+T5

if (T5<-0.8)

Denomi=(0.6+T5)/(7+10*T5)

(9) To avoid problems caused by A1*Vgsteff+A2 in calculating lambda in b3ld.c.

if A1>0

Lambda=1-0.5(T1+T2)

```
T1=1-A2-A1*Vgsteff-1.e-4
T2=sqrt(T1*T1+0.004*(1-A2))
else
Lambda=0.5(T1+T2)
T1=A2+A1*Vgsteff-1.e-4
T2=sqrt(T1*T1+0.004*A2)
```

(10) To avoid problems caused by 1/(pdiblcb*Vbseff) in calculating Vadibl in b3ld.c.

```
T7=pdiblcb*Vbseff
T3=1/(1+T7)
if (T7<-0.9)
T3=(17+20*T7)/(0.8+T7)
```

(11) To avoid problems caused by 1+Pvag*Vgsteff/(Esat*Leff) in calculating Va in b3ld.c.

```
T9=Pvag*Vgsteff/(Esat*Leff)
T0=1+T9
if (T9<-0.9)
T0=(0.8+T9)/(17+20T9)
```

(12) To avoid problems caused by eta0+etab*Vbseff in b3ld.c.

```
if (T3=eta0+etab*Vbseff)<1.e-4
T3=(2.e-4-T3)/(3.-2.e-4*T3)
```

(13) To avoid problems caused by Vgs_eff in b3ld.c

```
Vgs_eff=Vgs-Vpoly
if (Vpoly>1.12)
Vgs_eff=Vgs-T5
T5=1.12-0.5*(T7+T6)
T7=1.12-Vpoly-0.05
T6=sqrt(T7*T7+0.224)
```

2. Code change to avoid math problems such as divide by zero or square root domain.

(1) Introducing smoothing functions to avoid any Sqrt Domain errors in calculating T1 in calculation of capacitance when capmod=1 and 2 in b3ld.c.
(2) Code change in calculating the Vascbe in b3ld.c to avoid the problem of divide by zero when pscbe2=0

3. Bug fixing

(1) "ldvt1w and ldvt2w missed their 'w' " at line 519 and 523 in b3mset.c
(2) "undefined vfb parameter when Vth0 is not defined" in b3temp.c
(3) "considering CONSTCtoK twice" at line 53, 69, and 232 in b3noi.c
(4) "here->BSIM3gtg should be here->gtb" in line 1905 in b3ld.c

(5) uninitialized parameters in SizeDepend structure in b3temp.c
(6) "0.5 should be dxpart" in 1901 in b3ld.c.

4. Addition of a new routine to check parameters:

A new routine called as b3check.c has been created to check the parameters before doing the simulation. In this routine, the following parameters are checked: L_{eff}, L_{effCV}, W_{eff}, W_{effCV}, N_{LX}, N_{CH}, V_{BSC}, T_{OX}, D_{VT0}, D_{VT0W}, D_{VT1}, D_{VT1W}, W_0, N_{FACTOR}, C_{DSC}, C_{DSCD}, E_{T0}, B_1, U_0, D_{ELTA}, A_1, A_2, R_{DSW}, vsattemp, P_{CLM}, P_{DIBLC1}, P_{DIBLC2}, C_{LC}. For some parameters such as T_{OX}, X_J, and N_{LX} the simulator outputs "Fatal error" and quits the simulation if the users input any parameters outside the bounds in the code. For some parameters such as E_{T0}, A_1, A_2 etc., the simulator outputs a "Warning" message to let the users know that they are using some unsuitable parameters for their simulation.

(1) Values of parameters outside the following bounds would be treated as "fatal errors":

$L_{eff}<0$
$W_{eff}<0$
$T_{OX}\leq 0$
$N_{CH}\leq 0$
$N_{GATE}<0$
$N_{LX}\leq -L_{eff}$
$W_0=-W_{eff}$
$X_J\leq 0$
$B_1=-W_{eff}$
$U_0\leq 0$
$D_{ELTA}\leq 0$
$N_{SUB}\leq 0$
$D_{VT1}<0$
$D_{VT1W}<0$
$D_{SUB}<0$
$V_{sattemp}\leq 0$
$P_{CLM}<0$
$D_{ROUT}<0$
$C_{LC}<0$

(2) Values of parameters within the following regions would be treated as "warning errors":

$N_{LX}>-L_{eff}$ but <0

N_{CH} <1e15 cm^{-3} or >1E21 cm^{-3}
N_{SUB} <1e14 cm^{-3} or >1E21 cm^{-3}
T_{OX} < 1nm
L_{eff} < 0.05um
L_{effCV} <0.05um
W_{eff} < 0.1um
W_{effCV} < 0.1um
N_{GATE} >0 but <1e18cm^{-3}
D_{VT0} <0
ABS(1e-6/(W_{eff}+W_0)) > 10
ABS(B_0/(W_{eff}+ B_1) > 10
N_{FACTOR} <0
C_{DSC} <0
C_{DSCD} <0
E_{TA0} <0
A_2 <0.01 or A_2 >1
R_{DSW} >0 but <0.001
$V_{sattemp}$ >0 but <1e3
P_{DLBLC1} <0
P_{DLBLC2} <0
C_{GDO} <0
C_{GSO} <0
C_{GBO} <0
P_S and P_D <W_{eff} (when C_{J0} or C_{JSW} is given)

5. Changes in noise routine:

(1) Adding two more options (***noiMod***= 1, 2, 3, 4) for users to use different combinations of thermal and flicker noise models.
noiMod=1: SPICE2 flicker noise model + SPICE2 thermal noise model;
noiMob=2: BSIM3 flicker noise model + BSIM3 thermal noise model;
noiMod=3: SPICE2 thermal noise model + BSIM3 flicker noise model;
noiMod=4: SPICE2 flicker noise model + BSIM3 thermal noise model.
(2) Adding the G_{mb} term in the calculation of SPICE2 thermal noise model equation.

6. Modify the S/B and D/B diode model.

(1) Modify the code for the calculation of source/bulk and drain/bulk diode currents (One new parameter, J_{SSW}, is introduced for the parasitic side junction current).

(2) Modify the code for the calculation of source/bulk and drain/bulk diode parasitic side capacitances (Three new parameters, C_{JSWG}, P_{BSWG}, and M_{JSWG}, are introduced for the sidewall parasitic capacitances at gate side).
(3) Add the code to account for the temperature effect of S/B and D/B diode (two more parameters, X_{TI} and N_J, are introduced).

7. Add ***capMod***=0 for capacitance model.

8. Other code change and clean-ups:

(1) Change the code in b3set.c to calculate C_F according to the equation in the manual as the default value instead of zero when it is not given by the user.
(2) Add an option for users to use high V_{BC} value in the simulation when parameter $K_2>0$
If $K_2<0$
Vbc =0.9*[PHI-(0.5*k1/k2)^2]
else
Vbc=-30 if (Vbm>=-30)
Vbc=Vbm if (Vbm<-30)
(3) Change judgment condition for poly gate depletion effect in b3ld.c.
Poly-gate depletion effect is calculated if $N_{GATE}>N_{CH}$ and $V_{gs}>V_{FB}$+PHI)
(4) Code change for the calculation of V_{gsteff} function in b3ld.c.
(5) One parameter, version, was added for the version control of the code. The default value is 3.1 in the release of BSIM3v3.1.
(6) One flag parameter, paramchk, was added for users to have the option to turn on or off the parameter checking for warning error.
(7)Code clean-ups to improve the calculation efficiency.

APPENDIX D

Enhancements and Changes in BSIM3v3.2 versus BSIM3v3.1

D.1 Enhancements

BSIM3v3.2, released in June of 1998, has the following enhancements and improvements relative to BSIM3v3.1 released in Dec. of 1996:

(1) A new intrinsic capacitance model (the charge thickness model) considering the finite charge layer thickness determined by quantum effects is introduced as ***capMod***=3. It is smooth, continuous and accurate in all operating regions.

(2) Improved modeling of *C-V* characteristics at the transition from weak to strong inversion.

(3) Add the oxide thickness dependence in the threshold voltage (V_{th}) model.

(4) Add the flat-band voltage (V_{FB}) as a new model parameter.

(5) Improved substrate current scalability with channel length.

(6) Restructure the non-quasi-static (NQS) model, adding NQS into the pole-zero analysis and fixing bugs in NQS code.

(7) Add temperature dependence into the diode junction capacitance.

(8) Support a resistance-free diode and current-limiting feature in the DC diode model.

(9) Use the inversion charge from the capacitance models to evaluate the BSIM3 thermal noise.

(10) Elimination of the small negative capacitance of C_{gs} and C_{gd} in the accumulation-depletion regions.

(11) Introduce a separate set of channel-width and length dependence parameters to calculate effective channel length and width for *C-V* models for better fitting of the capacitance data.

(12) Add parameter checking to avoid bad values for certain parameters.

(13) Bug fixes.

D.2 Detailed changes

1. Two model parameters $\boldsymbol{N_{OFF}}$ and $\boldsymbol{V_{OFFCV}}$ are introduced in $V_{gsteff,cv}$ to adjust the *C-V* curve shape when V_{gs} is around V_{th}. $\boldsymbol{N_{OFF}}$ defaults to 1.0 and $\boldsymbol{V_{OFFCV}}$ defaults to 0.0 for backward compatibility; if ($\boldsymbol{N_{OFF}}$<0.1) and ($\boldsymbol{N_{OFF}}$>4.0), or if ($\boldsymbol{V_{OFFCV}}$<-0.5) and ($\boldsymbol{V_{OFFCV}}$>0.5), warning messages will be given. $V_{gsteff,cv}$ has been re-implemented to avoid any potential discontinuities and numerical instabilities.

2. A new parameter $\boldsymbol{T_{OXM}}$ is introduced to represent the Tox dependence in the model parameters $\boldsymbol{K_1}$ and $\boldsymbol{K_2}$. $\boldsymbol{T_{OXM}}$ has a default value of $\boldsymbol{T_{OX}}$. If $\boldsymbol{T_{OX}}$ is equal to or smaller than zero, a fatal error message will be given. The scalability of V_{th} model with respect to $\boldsymbol{T_{OX}}$ is improved.

3. A new parameter $\boldsymbol{V_{FB}}$ has been added for the flat band voltage in the DC model to improve the model accuracy for MOSFETs with different gate materials. $\boldsymbol{V_{FB}}$ defaults to the following for backward compatibility: if vth0 is not given, $\boldsymbol{V_{FB}}$ will be computed from V_{th0}; otherwise $\boldsymbol{V_{FB}}$=-1.0.

4. A new parameter 'alpha1' is added to improve substrate current scalability with the channel length. It defaults to 0.0 for backward compatibility with BSIM3v3.1.

5. The NQS model is re-implemented. A new charge partitioning scheme is used which is physically consistent with that in the quasi-static C-V model and significantly improves the simulation performance. The parameter nqsmod is now an element (instance) parameter, no longer a model parameter in the release of BSIM3v3.2.

6. Temperature dependence in the diode junction capacitance model is added, where both the unit area junction capacitance and built-in potential are now temperature dependent. All new parameters for the temperature effect of junction capacitances are set to zero to be identical to BSIM3v3.1 by default.

7. The DC junction diode model now supports a resistance-free diode model and a current-limiting feature. A current limiting model parameter ijth is introduced which corresponds to two critical voltages: vjsm and vjdm. BSIM3 will calculate vjsm and vjdm depending on the value of I_{JTH}. For the S/B diode, if I_{JTH} is explicitly specified to be zero, BSIM3 will not calculate either vjsm or vjdm; a resistance-free (pure) diode model will be triggered; Otherwise (I_{JTH}>0.0), a current-limiting feature will be used and vjsm and vjdm will be evaluated. I_{JTH} defaults to 0.1. However, users are highly recommended to always explicitly specify I_{JTH}=0.1 in the model cards if they want to use the default value of I_{JTH}. Backward compatibility for the diode I-V model is therefore not kept.

8. The inversion charge equations of the C-V models (***capMod*** 0, 1, 2 or 3) are used to calculate the BSIM3 thermal noise when ***noiMod*** = 2 or 4. The old channel charge equation is removed; backward compatibility is not kept on advice from the Compact Model Council.

9. A zero-bias v_{fb} calculated from V_{th} is used in capmod 1, 2 and 3 when ***version*** = 3.2. If ***version*** < 3.2, the old bias-dependent v_{fb} is kept for ***capMod*** 1 and 2 for backward compatibility; capmod 3 does not support the old bias-dependent v_{fb}.

10. New parameters L_{LC}, L_{WC}, L_{WLC}, W_{LC}, W_{WC}, and W_{WLC} are introduced in BSIM3v3.2. They default to the corresponding DC parameters L_L, L_W, L_{WL}, W_L, W_W and W_{WL}, respectively, for backward compatibility.

11. Parameter checking for some parameters are added.

P_{SCBE2} <= 0.0, the user will be warned of the poor value used.

If (M_{OIN} < 5.0) or (M_{OIN} > 25.0), a warning message will be given.

If (A_{CDE} < 0.4) or (A_{CDE} > 1.6), a warning message will be generated. If (N_{OFF} < 0.1) or (N_{OFF} > 4.0), a warning message will be given.

If (V_{OFFCV} < -0.5) or (V_{OFFCV} > 0.5), a warning message will be given. If (ijth < 0.0), a fatal error occurs.

If (toxm <= 0.0), a fatal error occurs.

The recommended parameter ranges for A_{CDE}, M_{OIN}, N_{OFF}, V_{OFFCV} can be found from above.

12. Summary on backward compatibility

(1) Backward compatibility with BSIM3v3.1

Even when all new model parameters are given their default values, the following could result in inconsistencies between BSIM3v3.2 and BSIM3v3.1:

1) Re-implementation of NQS model;

2) Explicitly specifying ijth=0.1 to use its default value of 0.1;

3) Using Q_{inv} in C-V models for BSIM3 thermal noise evaluation;

4) Zero-bias v_{fb} for ***capMod*** 1 and 2 (through BSIM3version number control); Removing of P_S and P_D clamps;

5) Using L_{active} for A_{bulk}' ;

6) Removing here->BSIM3gbd from "*(here->BSIM3SPdpPtr)"; Removing ckt->CKTgmin from "ceqbs" and "ceqbd" ;

7) Fixing of "vgs = pParam->BSIM3vtho + 0.1" term in b3ld.c file; I_{sub} bug fixing in b3ld.c, b3acid.c and b3pzld.c.

(2) Backward compatibility with BSIM3v3.0

Current users of BSIM3v3.0 may migrate to BSIM3v3.2 directly by passing BSIM3v3.1. The compatibility of BSIM3v3.2 to BSIM3v3.0 is summarized by the above list (Backward compatibility with BSIM3v3.1) plus the summarized information given in Appendix C for the compatibility between BSIM3v3.1 and BSIM3v3.0

Index

R

S

T

U

V

W

X